Quarks Leptons & Gauge Fields

Kerson Huang

Professor of Physics
Massachusetts Institute of Technology

World Scientific Publishing Co Pte Ltd
P.O. Box 128
Farrer Road
Singapore 9128

ISBN 9971-950-03-0

Copyright © 1982 by World Scientific Publishing Co Pte Ltd.
All rights reserved. This book, or parts thereof, may not be
reproduced in any form or by any means, electronic or mechanical,
including photocopying, recording or any information storage
and retrieval system now known or to be invented, without written
permission from the publisher.

Printed and bound by Singapore National Printers (Pte) Ltd.
Typesetted by Polyglot Pte Ltd.

*To
Rosemary
and
the memory of
Gianna Maria Russo*

PREFACE

According to the current view, the basic building blocks of matter are quarks and leptons, which interact with one another through the intermediaries of Yang-Mills gauge fields (gravity being ignored in this context). This means that the forms of the interactions are completely determined by the algebraic structure of certain internal symmetry groups. Thus, the strong interactions are associated with the group $SU(3)$, and is described by a gauge theory called quantum chromodynamics. The electro-weak interactions, as described by the now standard Weinberg-Salam model, is associated with the group $SU(2) \times U(1)$.

This book is a concise introduction to the physical motivation behind these ideas, and precise mathematical formulation thereof. The goal of the book is to explain why and how the mathematical formalism helps us to understand the relevant observed phenomena.

I must mention, and apologize for, the fact that I have not found it possible to discuss "quark confinement" in a sufficiently concise yet intelligible style. The length of exposition usually increases with lack of understanding. Rather than make a poor attempt, I have omitted it entirely. What material included here, I hope, will enable the reader to follow the current literature on the subject.

The audience for which this book is written are graduate students in physics who have some knowledge of the experimental parts of particle physics, and an acquaintance with quantum field theory, including Feynman graphs and the notion of renormalization. This book might serve as a text for a one-semester course beyond quantum field theory. The material, in fact, represents an expanded and updated version of such a course that I offered at M.I.T. in 1978, and at the Institute for High Energy Physics in Peking China, in 1979. The chapter on quarks is based on lectures I gave at the University of Chile in Santiago, Chile, in 1977.

I wish to thank I. Saavedra for the opportunity to lecture in Chile, to thank Chang Wen-yu and S.C.C. Ting for the inducement to give the Peking lectures, from which the present book germinates, and to thank M. Jacob and K.K. Phua for their encouragement in this writing effort. I owe a special debt to my colleagues at M.I.T., especially A. Guth, R. Jackiw, and K. Johnson, from whom I have learned much that is being passed along in this book.

Kerson Huang

Marblehead, Massachusetts, U.S.A.
July 29, 1982

CONTENTS

PREFACE   v

I. INTRODUCTION
1.1 Particles and Interactions   1
1.2 Gauge Theories of Interactions   6
1.3 Notations and Conventions   10

II. QUARKS
2.1 Internal Symmetries   12
     1 Isospin   13
     2 The gauge groups   14
     3 More general internal symmetries: $SU(n)$ 14
     4 Unitary symmetry   15
2.2 Representation of $SU(3)$
     1 The basic representation   17
     2 Young's tableaux   18
     3 Irreducible representations   20
2.3 The Quark Model
     1 Quarks as basic triplets   22
     2 Quarks as building blocks   25
     3 Weight diagrams   25
     4 The composition of hadrons   28
2.4 Color
     1 Independent quark model   28
     2 Color $SU(3)$ group   30
2.5 Electromagnetic and Weak Probes
     1 Electromagnetic interactions   33
     2 Parton model   35
     3 Evidence for color   38
     4 Weak interactions   40
2.6 Charm
     1 The charmed quark   43
     2 The J/ψ and its family   45
     3 Correspondence between quarks and leptons   46

III. MAXWELL FIELD: $U(1)$ GAUGE THEORY
3.1 Global and Local Gauge Invariance   47
3.2 Spontaneous Breaking of Global Gauge Invariance:
     Goldstone Mode   50
3.3 Spontaneous Breaking of Local Gauge Invariance: Higgs Mode   53
3.4 Classical Finite-Energy Solutions   55

		3.5 Magnetic Flux Quantization	56
		3.6 Soliton Solutions: Vortex Lines	58
IV.	**YANG-MILLS FIELDS: NON-ABELIAN GAUGE THEORIES**		
		4.1 Introductory Note	61
		4.2 Lie Groups	
		1 Structure constants	61
		2 Matrix representations	62
		3 Topological properties	64
		4 General remarks	66
		4.3 The Yang-Mills Construction	
		1 Global gauge invariance	67
		2 Local gauge invariance	69
		4.4 Properties of Yang-Mills Fields	
		1 Electric and magnetic fields	72
		2 Dual tensor	73
		3 Path representation of the gauge group	74
		4.5 Canonical Formalism	
		1 Equations of motion	77
		2 Hamiltonian	79
		4.6 Spontaneous Symmetry Breaking	
		1 The Little Group	80
		2 Higgs mechanism	83
V.	**TOPOLOGICAL SOLITONS**		
		5.1 Solitons	86
		5.2 The Instanton	
		1 Topological charge	88
		2 Explicit solution	92
		5.3 The Monopole	
		1 Topological stability	94
		2 Flux quantization	96
		3 Boundary conditions	98
		4 Explicit solution	100
		5 Physical fields	101
		6 Spin from isospin	103
VI.	**WEINBERG-SALAM MODEL**		
		6.1 The Matter Fields	105
		6.2 The Gauge Fields	
		1 Gauging $SU(2) \times U(1)$	108
		2 Determination of constants	111
		3 Interactions	111
		6.3 The General Theory	
		1 Mass terms	113
		2 Cabibbo angle	117
		3 Kobayashi-Maskawa matrix	117
		4 Solitons	119
		6.4 Comments	120

Contents

VII. METHOD OF PATH INTEGRALS
    7.1 Non-Relativistic Quantum Mechanics ... 122
    7.2 Quantum Field Theory ... 127
    7.3 External Sources ... 129
    7.4 Euclidean 4-Space ... 133
    7.5 Calculation of Path Integrals ... 135
    7.6 The Feynman Propagator ... 136
    7.7 Feynman Graphs ... 138
    7.8 Boson Loops and Fermion Loops ... 141
    7.9 Fermion Fields ... 144

VIII. QUANTIZATION OF GAUGE FIELDS
    8.1 Canonical Quantization ... 148
    8.2 Path Integral Method in Hamiltonian Form ... 151
    8.3 Feynman Path Integral: Fadeev-Popov Method ... 152
    8.4 Free Maxwell Field ... 156
        1 Lorentz gauge ... 157
        2 Coulomb gauge ... 159
        3 Temporal and axial gauges ... 161
    8.5 Pure Yang-Mills Fields ... 162
        1 Axial gauge ... 164
        2 Lorentz gauge: Fadeev-Popov ghosts ... 164
    8.6 The θ-World and the Instanton
        1 Discovering the θ-world ... 165
        2 Instanton as tunnelling solution ... 168
        3 The θ-action ... 170

IX. RENORMALIZATION
    9.1 Charge Renormalization ... 173
    9.2 Renormalization in Quantum Electrodynamics
        1 Divergences in Feynman graphs ... 176
        2 Vertex ... 179
        3 Electron propagator ... 180
        4 Photon propagator ... 180
        5 Multiplicative transformations ... 183
        6 Renormalization ... 184
    9.3 Gauge Invariance and the Photon Mass ... 186
    9.4 The Renormalization Group
        1 The group invariant ... 187
        2 Running coupling constant ... 189
        3 Gell-Mann-Low function ... 190
        4 Fixed points ... 192
    9.5 Callan-Symanzik Equation ... 194
    9.6 Example: Massless ϕ^4 Theory ... 197

X. METHOD OF EFFECTIVE POTENTIAL
    10.1 Spontaneous Symmetry Breaking ... 200
    10.2 The Effective Action ... 200
    10.3 The Effective Potential ... 202

	10.4 The Loop Expansion	204
	10.5 One-Loop Effective Potential	207
	10.6 Renormalization	
	1 General scheme	208
	2 Massive case	210
	3 Massless case	210
	10.7 Dimensional Transmutation	211
	10.8 A Non-Relativistic Example	214
	10.9 Application to Weinberg-Salam Model	216
XI.	**THE AXIAL ANOMALY**	
	11.1 Origin of the Axial Anomaly	219
	11.2 The Triangle Graph	220
	11.3 Radiative Corrections	225
	11.4 Anomalous Divergence of the Chiral Current	226
	11.5 Physical Explanation of the Axial Anomaly	228
	11.6 Cancellation of Anomalies	232
	11.7 't Hooft's Principle	236
XII.	**QUANTUM CHROMODYNAMICS**	
	12.1 General Properties	
	1 Lagrangian density	241
	2 Feynman rules	242
	3 Quark-Gluon interactions	244
	4 Gluon Self-interactions	245
	12.2 The Color Gyromagnetic Ratio	249
	12.3 Asymptotic Freedom	
	1 The running coupling constant	251
	2 The vacuum as magnetic medium	254
	3 The Nielsen-Hughes formula	257
	12.4 The Pion as Goldstone Boson	
	1 The low-energy domain	258
	2 Chiral symmetry: an idealized limit	258
	3 PCAC	261
	4 The decay $\pi^0 \to 2\gamma$	262
	5 Extension to pion octet	264
	12.5 The $U(1)$ Puzzle	265
	12.6 θ-Worlds in QCD	
	1 Euclidean action	267
	2 The axial anomaly and the index theorem	268
	3 Chiral limit: Collapse of the θ-worlds	271
	4 Quark mass matrix	272
	5 Strong CP violation	275
CODA		277
INDEX		279

CHAPTER I

INTRODUCTION

1.1 Particles and Interactions
一尺之棰　日取其半　万世不竭
*Take half from a foot-long stick each day;
you will not exhaust it in a million years.*

The thought experiment contemplated in this proposition by an ancient Chinese sophist[1] is an apt allegory for what physicists actually do in the laboratory, in their search for the ultimate constituents of matter.

During the three centuries since the birth of physics in the modern sense, we have done about 60 days' worth of "halving" (down to 10^{-16} cm). At around day 30 (at 10^{-8} cm), we encountered the first granular structure of matter—atoms, which appeared at first to be indivisible. As we know, they turned out to be divisible further into electrons and nuclei; and nuclei could in turn be split into nucleons. Now we are at the stage when constituents of the nucleon—quarks—can be confidently identified. Indications are that the subdividing process will continue. The ancient sophist seems to be right so far.

From an experimental point of view, particles are detectable packets of energy and momentum, be they billiard balls, photons, or lambda hyperons. At each stage of our understanding, we designate certain particles as "fundamental", in the sense that they are the most elementary interacting units in our theories. As our experimental knowledge expands, we have often been forced to revise our views. The necessity for such revisions rests with the stringent requirement we place upon our theories: they must, in principle, be able to predict the quantitative results of all possible experiments.

It is fortunate that, at any given stage, we were able to regard certain particles as provisionally fundamental, without jeopardizing the right to change our mind. The reason is that, according to quantum mechanics, it is a good approximation to ignore those quantum states of a system whose excitation energies lie far above the energy range being studied. For example, a nucleus could be treated phenomenologically as a point mass at energies far below 1 MeV. We have discovered many layers of substructure since the era of atomic physics; but it is a remarkable fact that the dynamical principles learned from that era, as synthesized by relativistic local quantum field theory[2], continues to work up to the present stage.

---
[1] Kungsun Lung (公孙龙), quoted in Chuang Chou, *Chuangtse* (ca. 300 B.C.), chapter 33. (莊子天下篇第三十三).
[2] J. D. Bjorken and S. D. Drell, *Relativistic Quantum Mechanics* (McGraw-Hill, New York, 1964); *Relativistic Quantum Fields* (McGraw-Hill, New York, 1965).

Interactions among experimentally observed particles fall into four types of markedly different strengths: gravitational, weak, electromagnetic, and strong interactions. These are briefly reviewed in Table 1.1.

In the current theoretical view, the weak and electromagnetic interactions are low-energy manifestations of a single unified interaction, and the strong interactions originate in a hidden charge called "color", carried by quarks permanently confined in nucleons and other strongly interacting particles. All these interactions are supposed to be mediated via the exchange of vector mesons with "minimal" coupling, similar to the well-known situation in electrodynamics. We are even in a position to speculate that all the above interactions are really low-energy manifestations of a single "grand unified" interaction, whose simplicity will be directly revealed in experiments only at energies above 10^{17} GeV! Unfortunately, nothing reliable can be said about the microscopic aspects of the gravitational interaction, due to a total lack of experimental information. Important as it may be in an eventual grand synthesis of all the interactions, we will have nothing to say about gravity in this book.

Basic to the theoretical classification of particles is the assumption that physical laws are invariant under Poincaré transformations, i.e., Lorentz transformations and space-time translations. A particle, be it "fundamental" or composite, is defined as a state of a quantum field that transforms under elements of the Poincaré group according to a definite irreducible representation. This implies that a particle has definite mass and spin, and that to each particle is associated an antiparticle of the same mass and spin[3]. The assumption

Table 1.1 THE FOUR TYPES OF INTERACTIONS

Interaction	Gravitational	Weak	Electromagnetic	Strong
Manifestation	Celestial mechanics	β-radio-activity	Everyday world	Nuclear binding
Quantum view	graviton between mass 1 and mass 2: $\sqrt{\gamma}m_1 \cdots \sqrt{\gamma}m_2$	$p, \bar{\nu}, e$ via G from n	photon between charge 1 and charge 2: $q_1 \cdots q_2$	(?) $g_{\text{eff}} \cdots g_{\text{eff}}$ nucleon
Static potential	$-\dfrac{\gamma m_1 m_2}{r}$ r = distance between sources	—	$\dfrac{q_1 q_2}{4\pi r}$	$-g_{\text{eff}} \dfrac{e^{-\mu r}}{4\pi r}$ $\dfrac{\hbar}{\mu c} \sim 10^{-13}$ cm
Coupling strength	$\dfrac{\gamma m_p^2}{\hbar c} = 5.76 \times 10^{-36}$ m_p = proton mass	$Gm_p^2 = 1.01 \times 10^{-5}$	$\dfrac{e^2}{4\pi\hbar c} = \dfrac{1}{137.036}$ e = electron charge	$\dfrac{g^2}{4\pi\hbar c} \cong 10$

---

[3] E. P. Wigner, *Ann. Math.* **40**, 149 (1934).

Introduction

of microcausality in local quantum field theory further implies a connection between spin and statistics: particles with integer spin are bosons, and those with half-integer spin are fermions[4]. The interactions among particles are required to be invariant under the Poincaré group; this imposes non-trivial conditions on possible local quantum field theories.[5]

In addition to Poincaré invariance, which is a space-time symmetry, there are also internal symmetries having to do with space-time-independent transformations of particle states. The invariance of interactions under internal symmetry groups gives rise to further quantum numbers that label particle states, such as electric charge, baryon number, isospin, etc.

A partial list of known particles, classified according to mass, spin, internal quantum numbers, and the types of interactions they have, is shown in Fig. 1.1.

"Hadrons" denote bosons and fermions having strong interactions, and "leptons" denote fermions without strong interactions[a]. Among the hadrons, "mesons" are bosons with baryon number 0, and "baryons" are fermions with baryon number different from 0[b]. Of all these particles (apart from the photon not shown in Fig. 1.1), only electrons and nucleons are relevant to our everyday experience. One might go a little further and include neutrinos as important catalysts for the generation of solar power, and μ mesons are free gifts from heaven[c]. Everything else is created primarily in high-energy accelerators.

Two striking features should be mentioned. First, all the leptons appear to be point-like particles, the latest experimental upperbound on their "radii" being 10^{-16} cm.[6] This is particularly remarkable for the τ, which is about twice as heavy as the proton. Secondly, there is a wild proliferation of hadrons. As noted by Hagedorn[7], a plot of the density of hadronic states against mass suggests an exponential growth, as shown in Fig. 1.2.

If this trend continues to asymptotically large masses, there would exist an "ultimate temperature" of about 160 MeV (2×10^{13} K), beyond which no system could be heated[8]. If the growth were faster than exponential, the partition function of statistical mechanics would not exist. Thus, the density of hadronic states seems to be growing at the maximum rate consistent with thermodynamics.

Even if we had not detected experimentally a finite radius for the proton (which we have, at about 10^{-13} cm)[9], the sheer number of the hadrons would make it absurd to suppose that they are all "fundamental". A key to the inner

---

[a] All observed bosons so far have strong interactions except the photon. Historically, leptons were so named because they were light; but this is no longer true with the discovery of the τ.

[b] The reason that all baryons are fermions, while all mesons are bosons, comes from baryon conservation in relativistic field theory, i.e., fermion fields must occur bilinearly in the Lagrangian, but bosons can occur linearly.

[c] "Who ordered these?" asked I. I. Rabi.

[4] R. F. Streater and A. S. Wightman, *PCT, Spin and Statistics, and All That* (W. A. Benjamin, New York, 1964).
[5] N. N. Bogolubov, G. G. Logunov, and I. T. Todorov, *Introduction to Axiomatic Quantum Field Theory* (W. A. Benjamin, Reading, Mass., 1975).
[6] D. P. Barber *et al.*, *Phys. Rev. Lett.* **43**, 1915 (1979).
[7] R. Hagedorn, *N. Cim.* **56A**, 1027 (1968).
[8] K. Huang and S. Weinberg, *Phys. Rev. Lett.* **25**, 895 (1970).
[9] R. Hofstadter and R. W. McAllister, *Phys. Rev.* **98**, 217 (1955).

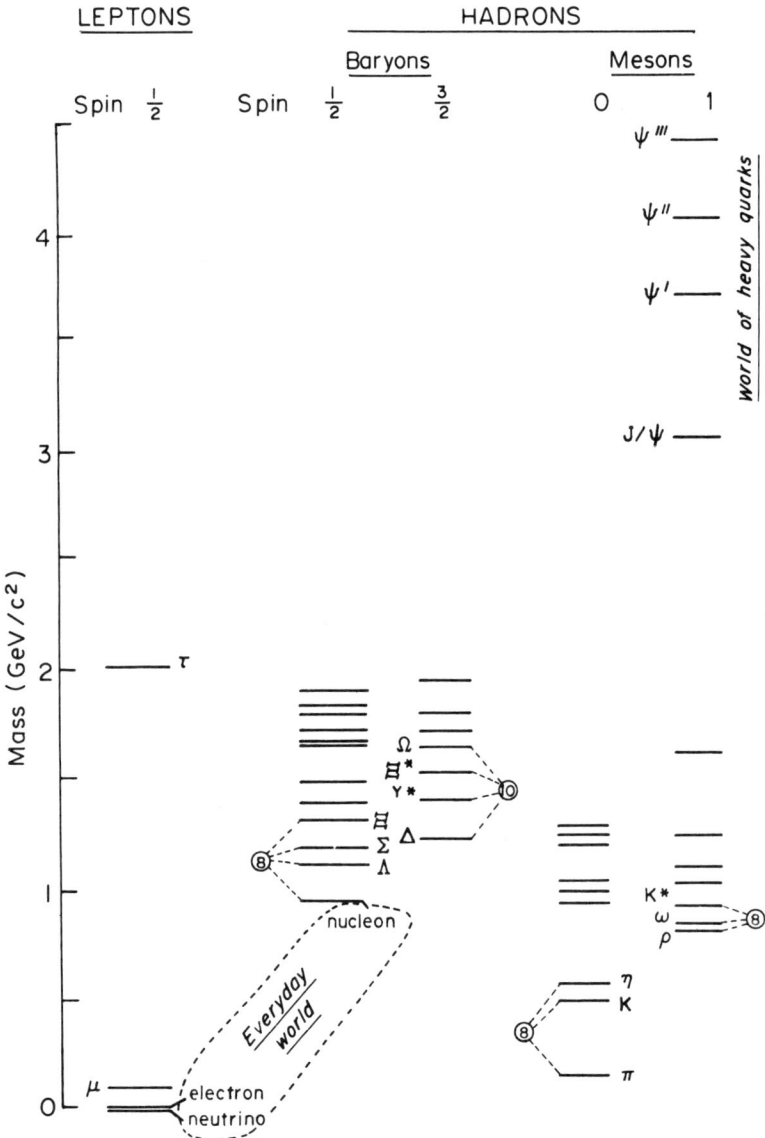

Fig. 1.1 Particle mass spectrum

Introduction

structure of hadrons are the multiplet structures (e.g., **8** and **10** in Fig. 1.1) identifiable with irreducible representations of an internal symmetry group $SU(3)$. This is the first lead to the notion of quarks as hadronic constituents, namely, they form a fundamental representation of $SU(3)$. A more detailed discussion of the evidence for quarks and their interactions will be given in Chapter II.

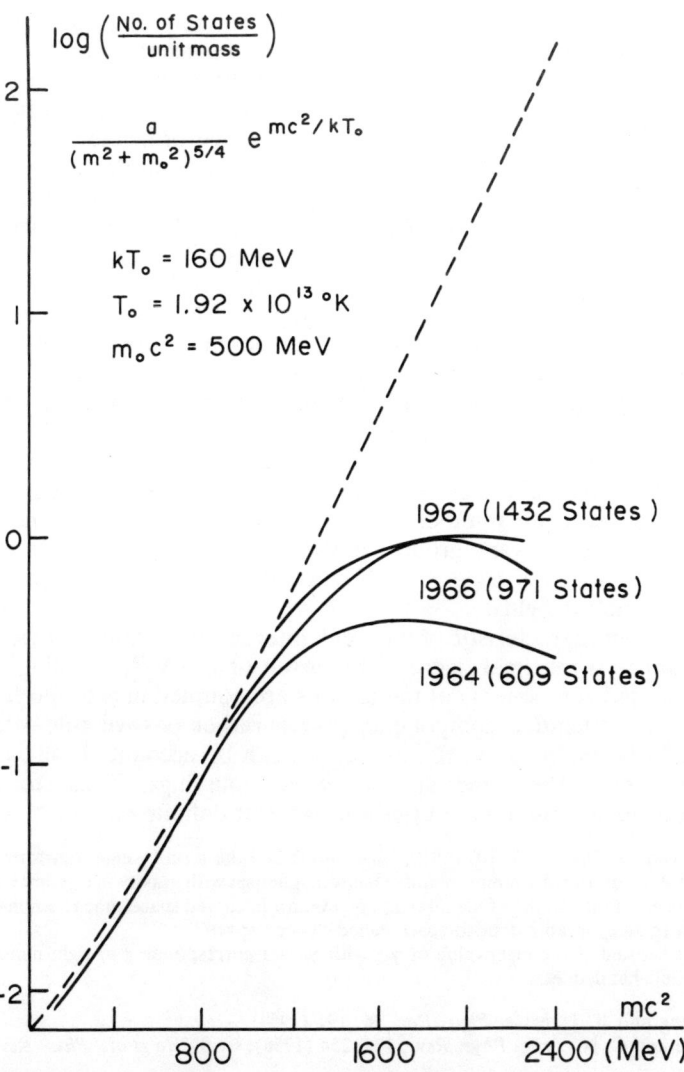

Fig. 1.2 Number of hadronic states as function of mass

1.2 Gauge Theories of Interactions

In the current view, all the interactions are derived from a "gauge principle" similar to that in electromagnetism. We recall that the coupling of the electromagnetic field A^μ to a charged matter field ψ can be derived through the following prescription: replace $\partial^\mu \psi$ in the matter Lagrangian by the covariant derivative $(\partial^\mu + ieA^\mu)\psi$, where e is the electric charge of ψ. Before we "turn on" the coupling (i.e., for $e = 0$), the matter Lagrangian must be invariant under constant phase changes of ψ, called "global gauge transformations". What the prescription does is to enlarge this symmetry to a "local gauge invariance", i.e., invariance under arbitrary space-time dependent phase changes of ψ (correlated with corresponding gauge transformations of A^μ)[d]. The original global gauge invariance implies the existence of a conserved matter current j^μ, and the prescription leads to an interaction of the form $ej^\mu A_\mu$, in conformity with Maxwell's theory. Under the usual assumptions of canonical field theory, the prescription is unique, and is called the "gauge principle".

We may restate the gauge principle as follows. Consider a matter system originally invariant under a global $U(1)$ group of gauge transformations. We "gauge" this symmetry, i.e., enlarge it to a local $U(1)$ gauge invariance. This means that an independent $U(1)$ gauge group shall be associated with each space-time point. To do this it is necessary to introduce a vector gauge field, to which the matter field current becomes coupled. The coupling constant is the electric charge, the generator of $U(1)$. The original global symmetry can be gauged only if it is an exact symmetry.

We shall use a generalized gauge principle formulated by Yang and Mills[10], which applies to a multicomponent matter field. Instead of $U(1)$, the gauge group is now a larger group of transformations that mix the different components of the matter field. There will now be more than one gauge field—the Yang-Mills fields. Their number is equal to the number of generators of the gauge group. The relevant group for the weak, electromagnetic and strong interactions is $SU(2) \times U(1) \times SU(3)$. To define this group, we must first describe the matter fields.

A well-known characteristic of the weak interactions is that they violate parity conservation to a maximal degree[11] by virtue of the V-A coupling[12]. That is, only left-handed components of the leptons are coupled in the charge-changing sector; the right-handed components play a rather passive role—to provide mass. Similarly, hadronic weak interactions can be accounted for by assuming that quarks have the same kind of weak couplings. Thus, to the weak interactions, the elementary entities are states of definite chirality[e], which have

---

[d] H. Weyl, *Ann. d. Physik*, **59**, 101 (1919), first introduced the term "gauge transformation" in an interesting but unsuccessful attempt to unify electromagnetism with gravity in a geometric theory, by extending the non-integrability of the direction of a vector in curved space-time to a non-integrability of its length (gauge) in an extended space called "gauge space".

[e] Chirality is defined as the eigenvalue of γ_5, with $\gamma_5 = 1$ corresponding to right-handedness, and $\gamma_5 = -1$ to left-handedness.

[10] C. N. Yang and R. L. Mills, *Phys. Rev.* **96**, 191 (1954).
[11] T. D. Lee and C. N. Yang, *Phys. Rev.*, **104**, 254 (1956); C. S. Wu et al., *Phys. Rev.*, **105**, 1413 (1957).
[12] R. P. Feynman and M. Gell-Mann, *Phys. Rev.*, **109**, 193 (1958).

Introduction

zero mass. (An eigenstate of finite mass is a superposition of left and right-handed states with equal weight). Glashow[13] first proposed a unified gauge theory of electroweak interactions based on a gauge group $SU(2) \times U(1)$, which mixes different massless chiral states. However, the fact that physical particles have finite masses seems to violate this symmetry. The seeming impasse was overcome by Weinberg[14] and Salam[15] by appealing to the notion of "spontaneous symmetry breaking". In the now-standard Weinberg-Salam model, "Higgs fields" are introduced to implement this idea, though they may be phenomenological parameters to be replaced by something more basic in a future theory. It is fair to say that at present we have no deep understanding of where masses come from.

The symmetries to be gauged refer to transformations among massless quarks and leptons of definite chirality. They come in at least six "flavors" (the sixth one being not yet experimentally confirmed). The lepton flavors are (e, ν), (μ, ν'), (τ, ν''), where the ν's denote massless left-handed neutrinos. The quark flavors bear a one-to-one correspondence to the above: (u, d), (s, c), (t, b). The parentheses group the particles into three families, which are indistinguishable copies as far as the weak interactions are concerned[f]. In addition, each quark flavor comes in three (and only three) "colors", while leptons have no color. Thus, the elementary particles are

$$\text{quarks: } q_{fn} \begin{cases} (f = 1, \ldots, 6) & \text{(flavor index)} \\ (n = 1, 2, 3) & \text{(color index)} \end{cases}$$
$$\text{leptons: } l_f \quad (f = 1, \ldots, 6) \quad \text{(flavor index)}$$

It is understood that, for example, q_{fn} denotes collectively $(q_R)_{fn}$ and $(q_L)_{fn}$, the right and left-handed components respectively, each regarded as an independent particle.

We list the quarks and leptons more explicitly in Table 1.2, and postulate the following internal symmetries:

(a) **Color $SU(3)$:** With respect to the color index, the three quarks of each flavor form a triplet representation of a "color group" $SU(3)$. The leptons are color singlets[g].

(b) **Weak isospin $SU(2)$:** In each family, the left-handed components of the upper and lower particles (e.g., ν_L and e_L) form a doublet representation of a "weak isospin group" $SU(2)$. All right-handed particles are $SU(2)$ singlets.

(c) **Weak hypercharge $U(1)$:** There is a $U(1)$ symmetry, called "weak hypercharge", associated with simultaneous phase changes of each particle. The relative phases are fixed by definite "weak hypercharge" assignments.

The gauge group is then $SU(2) \times U(1) \times SU(3)$, a direct product of the three mutually commuting groups defined above. Gauging this group necessitates the

---

[f] Rabi's question on p. 3 can be generalized, but remains unanswered.
[g] This means that the theory is invariant under the group in question, and that the particles transform under the group according to the representations specified.

[13] S. L. Glashow, *Nucl. Phys.* **22**, 579 (1961).
[14] S. Weinberg, *Phys. Rev. Lett.*, **19**, 1264 (1964).
[15] A. Salam, in *Elementary Particle Theory*, ed. N. Svarthholm (Almquist and Wiksell, Stockholm, 1968).

introduction of 12 vector gauge fields, one for each group generator, as listed in Table 1.3. The resulting interactions are described schematically by the Feynman vertices shown in Fig. 1.3.

The gauge fields generally have self-interactions because, unlike the photon, they generally carry "charge" by virtue of the non-Abelian nature of the group. It is to be noted that there are other exact symmetries of the theory, such as baryon number and lepton number, which are not gauged.

The theory so far has a serious defect, namely, all particles are massless. One cannot remedy this by simply including conventional mass terms in the Lagrangian, because such terms violate the $SU(2) \times U(1)$ symmetry, which we assume to be exact. Conventional vector boson mass terms also lead to non-renormalizable theories. A way out is to regard the masses as arising from "spontaneous breaking" of the $SU(2) \times U(1)$ symmetry, through couplings to scalar "Higgs fields". This will be fully explained in later chapters. It suffices to mention here that, by this method, one can obtain a renormalizable theory in which all particles can acquire arbitrary finite masses. The photon and the neutrinos can be arranged to remain massless in a natural way.

Since mass and chirality do not commute, physical particles are not necessarily members of $SU(2) \times U(1)$ multiplets. This leads to a mixing of flavors across families. For the same reason, the weak hypercharge $U(1)$ is not necessarily the electromagnetic $U(1)$. These points will be taken up in detail in Chapter VI.

Table 1.2 INTERNAL SYMMETRIES OF QUARKS AND LEPTONS

Family	Flavor f	Quarks q_{fn} Color: $n = 1, 2, 3$	Leptons l_f	
I	1	u_1 u_2 u_3	ν	$\updownarrow SU(2)$
	2	d_1 d_2 d_3	e	
II	3	c_1 c_2 c_3	ν'	$\updownarrow SU(2)$
	4	s_1 s_2 s_3	μ	
III	5	t_1 t_2 t_3	ν''	$\updownarrow SU(2)$
	6	b_1 b_2 b_3	τ	

$\longleftrightarrow$
$SU(3)$

Table 1.3 THE GAUGE FIELDS

Gauge group	Number of Generators	Gauge Fields
Color $SU(3)$	8	G_a^μ $(a = 1, \ldots, 8)$ (gluons)
Weak isospin $SU(2)$	3	W_i^μ $(i = 1, 2, 3)$
Weak hypercharge $U(1)$	1	W_0^μ

Introduction

On the other hand, color multiplets are mass eigenstates, because right and left-handed quarks can have the same color. Experimental evidence indicates that color $SU(3)$ does not suffer spontaneous breakdown.

The electroweak interactions that result from gauging $SU(2) \times U(1)$ reproduce all known phenomena and predict new ones. Chief among these is the existence of "neutral currents", which has been verified experimentally. However, the gauge vector bosons have not yet been discovered, possibly because they have high masses, about 80 GeV/c² according to theory.

Due to the structure of color $SU(3)$, the quark-gluon coupling tends to vanish at large momenta (or small distances)—a phenomenon known as "asymptotic freedom". Thus, one should be able to detect quasi-free quarks inside a hadron by using probes that impart large momentum transfers to quarks. This has indeed been successfully demonstrated experimentally.

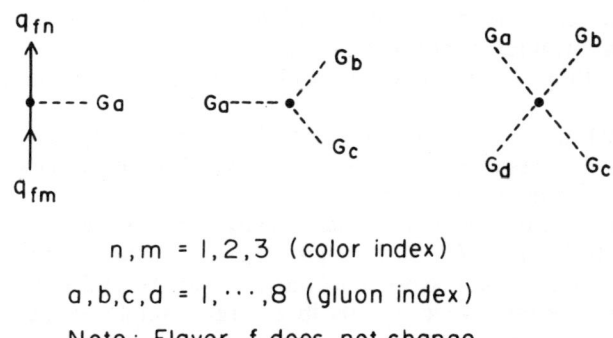

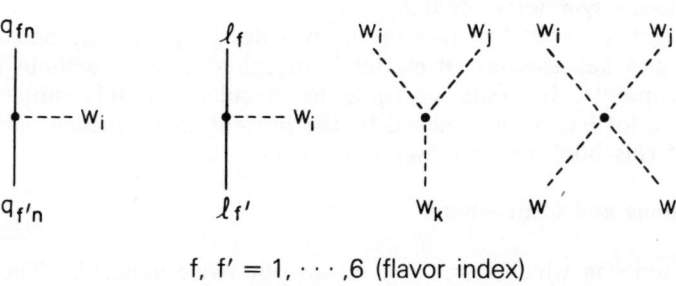

Fig. 1.3 Interaction vertices in gauge theory of strong and electroweak interactions

By the same token, one would expect the quark-gluon coupling to grow indefinitely as one approaches the zero-momentum (or infinite-distance) limit. There are theoretical indications that this is true, and that it leads to "quark confinement". That is, quarks (and gluons) do not exist in isolation as physical states, but only as components of bound states (hadrons) that are color singlets. Phenomenological models incorporating the idea of quark confinement have been very successful in explaining the properties of hadrons; but a proof of this idea from first principles is still lacking.

The idea of deriving all the interactions from a gauge principle has intrinsic aesthetic appeal, and the gauge theory based on $SU(2) \times U(1) \times SU(3)$ has led to many predictions subsequently verified by experiments. However, many questions remain unanswered, both from the experimental and the theoretical points of view.

Experimentally speaking, the most important open question concerns the existence of the heavy electroweak gauge bosons (and, to a lesser degree, the Higgs boson). Until these are actually observed, the electroweak gauge theory has not passed the acid test.

Theoretically, there are many more puzzling questions:
— Why do left-handed and right-handed particles behave differently under internal symmetry transformations?
— Why is there a one-to-one correspondence between quark and lepton flavors?
— Why are there so many families? "Who ordered these?"
— Where do masses *really* come from, or, what *really* happens in spontaneous symmetry breaking?
— What is the principle that tells us to gauge certain exact symmetries [i.e., $SU(2) \times U(1) \times SU(3)$], but not others, like baryon number?

Present attempts to address some of these questions tend to the direction of imbedding the present gauge groups in a larger simple gauge group, while keeping all the basic theoretical concepts intact. Known as "grand unified theories"[16], they provide answers to some questions. For example, the ungauged symmetries such as baryon number are explicitly broken. But most questions are left unanswered, while others become even more puzzling, such as the question of spontaneous symmetry breaking.

One cannot help but feel that the answer lies in a radically new direction, perhaps a new substructure of matter, unimaginable to us without new hints from experiments. It would be futile to speculate on this subject. In the meantime, a lot has been achieved by the present gauge theory, and it is the purpose of this book to introduce that to the reader.

1.3 Notations and Conventions

We use units in which $\hbar = c = 1$ unless otherwise indicated. The diagonal metric tensor in Minkowski space-time has the diagonal elements

$$g^{00} = -g^{11} = -g^{22} = -g^{33} = 1.$$

---

[16] For a review, see P. Langacker, *Phys. Reports*, **72**, 185 (1981).

Introduction

The contravariant space-time 4-vector x^μ has components designated by
$$x^\mu = (x^0, x^1, x^2, x^3) \equiv (x^0, \mathbf{x}),$$
with the corresponding covariant vector given by
$$x_\mu = (x^0, -\mathbf{x}).$$

Some frequently used differential operators are
$$\partial_\mu \equiv \frac{\partial}{\partial x^\mu} = \left(\frac{\partial}{\partial x^0}, \nabla\right),$$

$$\partial^\mu \equiv \frac{\partial}{\partial x_\mu} = \left(\frac{\partial}{\partial x^0}, -\nabla\right),$$

$$\Box^2 \equiv \partial_\mu \partial^\mu = \left(\frac{\partial}{\partial x^0}\right)^2 - \nabla^2,$$

$$A \overset{\leftrightarrow}{\partial}_\mu B \equiv (\partial_\mu A) B - A(\partial_\mu B).$$

The Dirac matrices γ^μ are chosen so that γ^0 is hermitian, while $\gamma^k (k = 1, 2, 3)$ are anti-hermitian:
$$(\gamma^0)^\dagger = \gamma^0, \quad (\gamma^0)^2 = 1,$$
$$(\gamma^k)^\dagger = -\gamma^k, \quad (\gamma^k)^2 = -1 \quad (k = 1, 2, 3).$$

We define γ_5 as the hermitian matrix
$$\gamma_5 = -i\gamma^0 \gamma^1 \gamma^2 \gamma^3, \quad (\gamma_5)^2 = 1.$$

In addition, we use the notation:
$$\alpha^k = \gamma^0 \gamma^k \quad (k = 1, 2, 3), \quad (\alpha^k)^\dagger = \alpha^k,$$
$$\sigma^k = \varepsilon^{klm} \gamma^l \gamma^m / 2i, \quad (\sigma^k)^\dagger = \sigma^k,$$

from which follows
$$\boldsymbol{\alpha} = \gamma_5 \boldsymbol{\sigma}.$$

A standard representation is the following:
$$\gamma^k = \begin{pmatrix} 0 & \underline{\sigma}^k \\ -\underline{\sigma}^k & 0 \end{pmatrix}, \quad \gamma^0 \equiv \beta = \begin{pmatrix} 1 & 0 \\ 0 & -1 \end{pmatrix},$$

$$\alpha^k = \begin{pmatrix} 0 & \underline{\sigma}^k \\ \underline{\sigma}^k & 0 \end{pmatrix}, \quad \sigma^k = \begin{pmatrix} \underline{\sigma}^k & 0 \\ 0 & \underline{\sigma}^k \end{pmatrix},$$

$$\gamma_5 = \begin{pmatrix} 0 & 1 \\ 1 & 0 \end{pmatrix},$$

where underlined symbols denote 2×2 matrices:
$$\underline{\sigma}^1 = \begin{pmatrix} 0 & 1 \\ 1 & 0 \end{pmatrix}, \quad \underline{\sigma}^2 = \begin{pmatrix} 0 & -i \\ i & 0 \end{pmatrix}, \quad \underline{\sigma}^3 = \begin{pmatrix} 1 & 0 \\ 0 & -1 \end{pmatrix}.$$

CHAPTER II
QUARKS

2.1 Internal Symmetries

The hadrons we know all fall into multiplets that seem to reflect underlying internal symmetries. To express this fact in a simple and concrete way, it was hypothesized that hadrons are composed of more elementary constituents with certain basic symmetries, called quarks.

The actual existence of quarks has been indirectly confirmed, in experiments that probe hadronic structure by means of electromagnetic and weak interactions, and with the discovery of heavy quark "atoms". The available evidence is consistent with the picture that hadrons participate in these interactions not as elementary entities, but through quarks. On the other hand, we have not completely understood why quarks have not been seen individually, although there are plausible explanations.

Internal symmetry refers to the fact that particles occur in families, called multiplets, that have degenerate or nearly degenerate masses. Each multiplet is looked upon as the realization of an irreducible representation of some internal symmetry group. One tries to identify such groups by the patterns of multiplets observed experimentally. If the masses in a multiplet are not exactly the same, one says that the associated symmetry is only an approximate one. Among hadrons, we have long recognized the internal symmetries I, S, B, and Q, with varying degrees of exactness, as indicated in Table 2.1

We define the hypercharge Y by

$$Y \equiv B + S, \tag{2.1}$$

and state the empirical rule

$$Q = I_3 + \tfrac{1}{2}Y. \tag{2.2}$$

Table 2.1 INTERNAL SYMMETRIES OF HADRONS

Symbol	Quantum number	Symmetry group	Interactions conserving it	Interactions violating it
I	Isospin	$SU(2)$	strong	em, weak
S	Strangeness	$U(1)$	strong, em	weak
B	Baryon number	$U(1)$	all	none
Q	Charge	$U(1)$	all	none

Quarks

1 Isospin

Let us review the familiar case of isospin. Experimental evidence suggests that the nucleons and the π-mesons may be grouped into the following multiplets:

$$N = \begin{pmatrix} p \\ n \end{pmatrix}, \qquad \pi = \begin{pmatrix} \pi_1 \\ \pi_2 \\ \pi_3 \end{pmatrix}, \tag{2.3}$$

where π_1, π_2, π_3 are related to the observed pions by $\pi^{\pm} = 2^{-1/2}(\pi_1 \pm \pi_2)$, and $\pi^0 = \pi_3$. Members of each multiplet have nearly equal masses, and the small mass differences can be thought of as electromagnetic corrections.

Apart from electromagnetic corrections, systems of nucleons and pions are invariant under matrix transformations representing the isospin $SU(2)$ group:

$$\begin{aligned} N &\to N + \delta N, & \delta N &= -i\omega_\alpha I_\alpha N \\ \pi &\to \pi + \delta \pi, & \delta \pi &= -i\omega_\alpha I_\alpha \pi \end{aligned} \tag{2.4}$$

where ω_α are arbitrary infinitesimal real numbers, and the components of isospin I_α ($\alpha = 1, 2, 3$) are the generators of $SU(2)$, obeying the commutation relation

$$[I_\alpha, I_\beta] = i\varepsilon_{\alpha\beta\gamma} I_\gamma. \tag{2.5}$$

The nucleon doublet forms a basis for a 2-dimensional representation[a], in which $2I_\alpha$ is represented by the 2×2 Pauli matrix τ_α:

2: $I_\alpha = \frac{1}{2}\tau_\alpha, \quad \tau_1 = \begin{pmatrix} 0 & 1 \\ 1 & 0 \end{pmatrix}, \quad \tau_2 = \begin{pmatrix} 0 & -i \\ i & 0 \end{pmatrix}, \quad \tau_3 = \begin{pmatrix} 1 & 0 \\ 0 & -1 \end{pmatrix}.$

$$I_\alpha N = \frac{1}{2} \tau_\alpha N. \tag{2.6}$$

$$\delta N_i = -\frac{i}{2}[\omega_1(\tau_1)_{ij} + \omega_2(\tau_2)_{ij} + \omega_3(\tau_3)_{ij}] N_j.$$

The π-meson triplet forms a basis for a 3-dimensional irreducible representation:

3: $(I_\alpha)_{\beta\gamma} = -i\varepsilon_{\alpha\beta\gamma}.$
$(I_\alpha \pi)_\beta = (I_\alpha)_{\beta\gamma} \pi_\gamma = i\varepsilon_{\alpha\beta\gamma} \pi_\gamma.$
$\delta \pi_\beta = i\varepsilon_{\beta\alpha\gamma} \omega_\alpha \pi_\gamma,$
or, $\quad \delta \pi = i\omega \times \pi. \tag{2.7}$

The representation **3** is special, in that its dimensionality equals the number of generators, and that the matrix representation for I_α is obtainable directly from the structure constants $\varepsilon_{\alpha\beta\gamma}$ in (2.5). It is called the *adjoint representation*. Other irreducible representations of $SU(2)$ are familiar from the theory of angular momentum. Their possible dimensionalities are $2I + 1$ ($I = 0, 1/2, 1, 3/2, \ldots$).

---

[a] We use **n** to denote either an n-dimensional irreducible representation of a group, or the n-dimensional vector space on which the representation is realized.

The adjoint representation can also be represented in an alternative form as follows: if x_i ($i = 1, 2$) transforms as **2**, i.e.,

2: $\quad \delta x_i = -\dfrac{i}{2}\omega_\alpha(\tau_\alpha)_{ij}x_j,$ (2.8)

then y_α ($\alpha = 1, 2, 3$), which transforms as **3**, may be represented as

$$y_\alpha = (x^\dagger I_\alpha x) = \tfrac{1}{2}x_i^*(\tau_\alpha)_{ij}x_j,$$
$$\text{or,} \quad y = \tfrac{1}{2}(x^\dagger \tau x). \tag{2.9}$$

The equivalence between (2.9) and (2.7) can be shown by the following calculation:

$$\begin{aligned}\delta y_\alpha &= (\delta x^\dagger I_\alpha x) + (x^\dagger I_\alpha \delta x) \\ &= i\omega_\beta (x^\dagger [I_\beta, I_\alpha] x) \\ &= -\omega_\beta \varepsilon_{\beta\alpha\gamma}(x^\dagger I_\gamma x) \\ &= -i\omega_\beta(-i\varepsilon_{\beta\alpha\gamma})y_\gamma.\end{aligned} \tag{2.10}$$

2 The Gauge Groups

The quantum numbers B, Q, S label one-dimensional representations of mutually commuting groups isomorphic to $U(1)$, the unitary group of dimension one. The group operation is multiplication of the particle state by a phase factor:

$B: \quad \psi \to e^{-i\alpha B}\psi, \quad B = \text{baryon number of } \psi,$
$Q: \quad \psi \to e^{-i\beta Q}\psi, \quad Q = \text{charge of } \psi,$
$S: \quad \psi \to e^{-i\gamma S}\psi, \quad S = \text{strangeness of } \psi,$ (2.11)

where α, β, γ are arbitrary real numbers.

3 More General Internal Symmetries: $SU(n)$

More generally, particles may fall into multiplets forming representations of $SU(n)$, the group isomorphic to that of all $n \times n$ special unitarity complex matrices U (det $U = 1$, $U^\dagger U = 1$). The condition det $U = 1$ singles out a connected subgroup of the group of matrices. The requirement $U^\dagger U = 1$ insures that the norms of particle states are preserved under the group transformations.

A general $n \times n$ complex matrix has $2n^2$ arbitrary real parameters. The requirement $U^\dagger U = 1$ imposes n^2 conditions, and det $U = 1$ imposes one condition. Hence, there remains $n^2 - 1$ arbitrary parameters. Correspondingly, $SU(n)$ has $n^2 - 1$ generators L_α obeying

$$[L_\alpha, L_\beta] = if_{\alpha\beta\gamma}L_\gamma. \tag{2.12}$$

An arbitrary infinitesimal element of the group is given by

$$U = 1 - i\omega_\alpha L_\alpha, \tag{2.13}$$

where ω_α are arbitrary infinitesimal real numbers. The generators L_α can be taken to be hermitian. The structure constants $f_{\alpha\beta\gamma}$ can be taken to be real and

Quarks

completely antisymmetric with respect to α, β, γ. The smallest non-trivial irreducible representation (the fundamental representation) is of dimensionality n by definition. There always exists the adjoint representation, whose dimensionality equals the number of generators, with

$$(L_\alpha)_{\beta\gamma} = -if_{\alpha\beta\gamma}.$$

The possible dimensionalities of other irreducible representations depend on n.

4 Unitary Symmetry

The approximate symmetries corresponding to isospin and hypercharge conservation can be enlarged into the so-called "unitary symmetry", associated with the group $SU(3)$. This symmetry, however, is approximate even with respect to the strong interactions. The motivation for the enlargement comes from the observation that hadrons can be grouped into larger multiplets containing isospin multiplets. For example, one can recognize an octet of spin 1/2 baryons (the N-octet), a decaplet of spin 3/2 baryons (the Δ^- decaplet), an octet of spin 0 mesons (the π-octet) and an octet of spin 1 mesons (the ρ-octet). These are indicated in Fig. 1.1 and displayed in Fig. 2.1 in the form of Y-I_3 plots.

Note that baryons and antibaryons form their own separate multiplets, whereas mesons and antimesons are in the same multiplet. The identification of $SU(3)$ as the relevant symmetry group is based on the fact that **8** and **10** are possible irreducible representations of that group, of which **8** is the adjoint representation.

The violation of unitary symmetry by the strong interactions is reflected in the large mass splittings within the multiplets. Gell-Mann and Okubo have proposed a mass formula[1], based on the assumption that the interaction that violates unitary symmetry transforms like Y under $SU(3)$:

$$M(m, n) = a + bY - c[2I(I + 1) - \tfrac{1}{2}Y^2 + \tfrac{4}{3}(n - m)Y] \\ - \tfrac{1}{3}m(m + 2) - \tfrac{1}{3}n(n + 2) + \tfrac{1}{9}(m - n)^2, \quad (2.14)$$

where a, b, c, are empirical constants, and (m, n) labels the irreducible representation: **8** corresponds to $(1, 1)$, and **10** corresponds to $(0, 3)$ (see Sec. 2.2). This formula accounts reasonably well for the observed mass splittings, and is historically important for establishing the case for unitary symmetry.

It is natural to surmise that the basic representation **3** might also be realized, and this leads to the quark hypothesis. The main motivation, of course, is to decrease the number of fundamental particles; but the quark hypothesis also gives a natural explanation for the difference between baryon and meson multiplets with respect to the inclusion of antiparticles, if quarks are assumed to be baryons (see Sec. 2.3). The mechanism for the violation of unitary symmetry will find a simple and concrete origin in the mass differences among different kinds of quarks.

---

[1] See M. Gell-Mann and Y. Ne'eman, *The Eightfold Way* (W. A. Benjamin, New York, 1964).

16 *Quarks, Leptons and Gauge Fields*

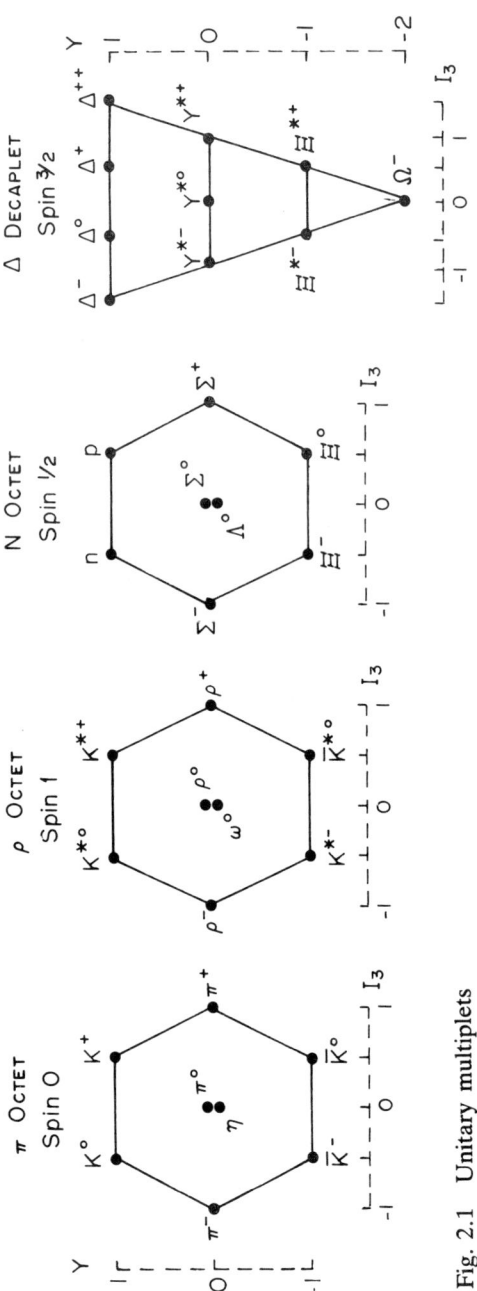

Fig. 2.1 Unitary multiplets

2.2 Representation of $SU(3)$[2]

1 The basic representation

The group $SU(3)$ has 8 generators L_α ($\alpha = 1, \ldots, 8$), satisfying

$$[L_\alpha, L_\beta] = if_{\alpha\beta\gamma}L_\gamma, \qquad (2.15)$$

where $f_{\alpha\beta\gamma}$ are real constants completely antisymmetric in α, β, γ. The basic representation is **3**, in which the generators are written in the form

$$L_\alpha = \tfrac{1}{2}\lambda_\alpha \quad (\alpha = 1, \ldots, 8), \qquad (2.16)$$

where λ_α are 3×3 hermitian matrices. They act on basis vectors, of the form

$$x = \begin{pmatrix} x_1 \\ x_2 \\ x_3 \end{pmatrix} \qquad (2.17)$$

An infinitesimal element of the group is represented by the transformation

$$x' = Sx,$$
$$S = 1 - \frac{i}{2}\omega_\alpha\lambda_\alpha, \qquad (2.18)$$

where ω_α ($\alpha = 1, \ldots, 8$) are arbitrary infinitesimal real numbers. A set of matrices satisfying (2.15) is given in Table 2.2. By construction, the first three are respectively $2I_1$, $2I_2$, $2I_3$. The last one, which is diagonal and commutes with isospin, is identified with $\sqrt{3}Y$. These are called "Gell-Mann matrices", and are generalizations of the Pauli matrices.

The structure constants $f_{\alpha\beta\gamma}$ can be calculated by explicit commutation of matrices λ_α, and the results are listed in Table 2.3. They hold, of course, for any

Table 2.2   GELL-MANN MATRICES

$$\lambda_1 = \begin{pmatrix} 0 & 1 & 0 \\ 1 & 0 & 0 \\ 0 & 0 & 0 \end{pmatrix} \quad \lambda_4 = \begin{pmatrix} 0 & 0 & 1 \\ 0 & 0 & 0 \\ 1 & 0 & 0 \end{pmatrix} \quad \lambda_7 = \begin{pmatrix} 0 & 0 & 0 \\ 0 & 0 & -i \\ 0 & i & 0 \end{pmatrix}$$

$$\lambda_2 = \begin{pmatrix} 0 & -i & 0 \\ i & 0 & 0 \\ 0 & 0 & 0 \end{pmatrix} \quad \lambda_5 = \begin{pmatrix} 0 & 0 & -i \\ 0 & 0 & 0 \\ i & 0 & 0 \end{pmatrix} \quad \lambda_8 = \frac{1}{\sqrt{3}}\begin{pmatrix} 1 & 0 & 0 \\ 0 & 1 & 0 \\ 0 & 0 & -2 \end{pmatrix}$$

$$\lambda_3 = \begin{pmatrix} 1 & 0 & 0 \\ 0 & -1 & 0 \\ 0 & 0 & 0 \end{pmatrix} \quad \lambda_6 = \begin{pmatrix} 0 & 0 & 0 \\ 0 & 0 & 1 \\ 0 & 1 & 0 \end{pmatrix}$$

---

[2] For a general reference, see W. Miller, *Symmetry Groups and Their Applications* (Academic Press, New York, 1972).

representation of the group. However, other properties given in Table 2.3 are valid only for the fundamental representation.

We note that, in the construction given in Table 2.2, λ_4 and λ_5 form two members of another set of Pauli-like matrices (called U-spin). The same is true for λ_6 and λ_7 (called V-spin). We construct the missing member in each set by taking appropriate linear combinations of λ_3 and λ_8. Thus, the original set of generators may be replaced by two other equivalent sets. These are listed in Table 2.4.

2 Young's tableaux

Taking the direct product of **3** with itself any number of times, we immediately obtain representations of higher dimensions 3×3, $3 \times 3 \times 3$, etc. These representations, however, are reducible.

To decompose these reducible representations into irreducible ones, we decompose the corresponding product spaces into invariant irreducible subspaces[b]. Then, the representations induced by the fundamental representation in these subspaces are irreducible representations.

Let $x_{i_1 \ldots i_n}$ be a tensor that transforms like the product $x_{i_1} \ldots x_{i_n}$. It can be decomposed into tensors of different symmetry classes with respect to a permutation of the indices $i_1, \ldots, i_n$. By definition, a tensor belonging to a definite *symmetry class* is obtained from $x_{i_1 \ldots i_n}$ through the following construction:

Table 2.3 PROPERTIES OF λ_α

$[\lambda_\alpha, \lambda_\beta] = 2if_{\alpha\beta\gamma}\lambda_\gamma$		$\{\lambda_\alpha, \lambda_\beta\} = \frac{4}{3}\delta_{\alpha\beta} 1 + 2d_{\alpha\beta\gamma}\lambda_\gamma$	
$\alpha\beta\gamma$	$f_{\alpha\beta\gamma}$ (antisymmetric)	$\alpha\beta\gamma$	$d_{\alpha\beta\gamma}$ (symmetric)
123	1	118	$1/\sqrt{3}$
147	1/2	146	1/2
156	−1/2	157	1/2
246	1/2	228	$1/\sqrt{3}$
257	1/2	247	−1/2
345	1/2	256	1/2
367	−1/2	338	$1/\sqrt{3}$
458	$\sqrt{3}/2$	344	1/2
678	$\sqrt{3}/2$	355	1/2
Tr $\lambda_\alpha = 0$		366	−1/2
Tr $\lambda_\alpha\lambda_\beta = 2\delta_{\alpha\beta}$		377	−1/2
Tr $\lambda_\alpha[\lambda_\beta, \lambda_\gamma] = 4if_{\alpha\beta\gamma}$		448	$-1/(2\sqrt{3})$
Tr $\lambda_\alpha\{\lambda_\beta, \lambda_\gamma\} = 4id_{\alpha\beta\gamma}$		558	$-1/(2\sqrt{3})$
		668	$-1/(2\sqrt{3})$
		778	$-1/(2\sqrt{3})$
		888	$-1/\sqrt{3}$

[b] "Invariant" means the space goes into itself under the group transformations. "Irreducible" means it does not contain a smaller invariant subspace.

Quarks

(a) First, pick n_1 of the indices, and symmetrize among them. Display this operation symbolically in the picture below:

☐☐···☐ n_1 boxes (Fill in the chosen indices)

(b) Next, pick $n_2 \leq n_1$ of the remaining indices, and symmetrize among them:

☐☐···☐ n_2 boxes (Fill in the chosen indices)

(c) Repeat the procedure until all indices have been used. Stack up the rows to form the following tableau, called a *Young's tableau*:

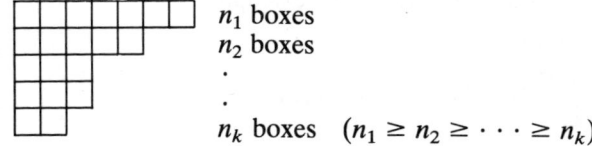

n_1 boxes
n_2 boxes
.
.
n_k boxes $(n_1 \geq n_2 \geq \cdots \geq n_k)$

The set of integers $\{n_1, \ldots, n_k\}$ is any possible partition of n:

$$n = n_1 + \cdots + n_k, \quad (n_1 \geq n_2 \geq \cdots \geq n_k). \tag{2.19}$$

Table 2.4 EQUIVALENT SETS OF $SU(3)$ GENERATORS

(1) $\underbrace{\lambda_1, \lambda_2, \lambda_3}_{2(I\text{ spin})}, \lambda_4, \lambda_5, \lambda_6, \lambda_7, \lambda_8\ (=\sqrt{3}Y)$

(2) $\lambda_1, \lambda_2, \underbrace{\rho_3, \lambda_4, \lambda_5}_{2(U\text{-spin})}, \lambda_6, \lambda_7, \rho'\ (=\sqrt{3}Q)$

(3) $\lambda_1, \lambda_2, \lambda_4, \lambda_5, \underbrace{\varepsilon, \lambda_6, \lambda_7}_{2(V\text{-spin})}, \varepsilon'$

where

$$\rho = \begin{pmatrix} 1 & 0 & 0 \\ 0 & 0 & 0 \\ 0 & 0 & -1 \end{pmatrix} = \frac{1}{2}(\lambda_3 + \sqrt{3}\lambda_8) = I_3 + \frac{3}{2}Y$$

$$\rho' = \frac{1}{\sqrt{3}}\begin{pmatrix} 2 & 0 & 0 \\ 0 & -1 & 0 \\ 0 & 0 & -1 \end{pmatrix} = \frac{1}{2}(\sqrt{3}\lambda_3 + \lambda_8) = \sqrt{3}\left(I_3 + \frac{1}{2}Y\right) = \sqrt{3}Q$$

$$\varepsilon = \begin{pmatrix} 0 & 0 & 0 \\ 0 & 1 & 0 \\ 0 & 0 & -1 \end{pmatrix} = \frac{1}{2}(\sqrt{3}\lambda_8 - \lambda_3) = \frac{3}{2}Y - I_3$$

$$\varepsilon' = \frac{1}{\sqrt{3}}\begin{pmatrix} -1 & 0 & 0 \\ 0 & 2 & 0 \\ 0 & 0 & -1 \end{pmatrix} = \frac{1}{2}(\lambda_8 - \sqrt{3}\lambda_3) = \sqrt{3}\left(\frac{1}{2}Y - I_3\right)$$

(d) Finally, antisymmetrize the indices in each column of the tableau separately and independently.

The geometrical pattern of a Young's tableau characterizes a symmetry class. The number of classes is therefore equal to the number of possible partitions of n. Tensors belonging to a given symmetry class can differ from one another only in the choice of indices in the various boxes of the Young's tableau. It is a matter of convention that the rows are permuted before the columns. In this convention, the indices appearing in each column of a Young's tableau are antisymmetric among themselves; but the indices appearing in each row are symmetric under a permutation if and only if the indices being permuted are not antisymmetrized with indices of a different row. For example:

These are not independent tensors:

$$\begin{array}{|c|c|c|}\hline 1 & 3 & 4 \\\hline 2 \\\hline\end{array} = - \begin{array}{|c|c|c|}\hline 2 & 3 & 4 \\\hline 1 \\\hline\end{array}, \quad \text{(where 1 means } i_1, \text{ etc.).}$$

These are not independent tensors:

$$\begin{array}{|c|c|c|}\hline 1 & 3 & 4 \\\hline 2 \\\hline\end{array} = \begin{array}{|c|c|c|}\hline 1 & 4 & 3 \\\hline 2 \\\hline\end{array}$$

These are independent tensors:

$$\begin{array}{|c|c|c|}\hline 1 & 3 & 4 \\\hline 2 \\\hline\end{array} \neq \begin{array}{|c|c|c|}\hline 3 & 1 & 4 \\\hline 2 \\\hline\end{array}$$

The central theorem, which we give without proof, is the following:

(a) Tensors in a given symmetry class form an invariant irreducible space. Hence the group representation induced in this space by the fundamental representation is irreducible.

(b) The irreducible representations generated through all symmetry classes are exhaustive.

3 Irreducible Representations

Since a tensor index takes on only three values $i = 1, 2, 3$, a column in a Young's tableau can have no more than three boxes. The tableau

represents a tensor of rank 0, and corresponds to the 1-dimensional trivial representation (i.e., it is invariant under the group). Such a tableau may be omitted, if it occurs as part of a larger tableau:

$$1 = \square = \square = \square = \cdots \tag{2.20}$$

Consequently, the most general Young's tableau for $SU(3)$ has at most two rows:

$$\begin{array}{|c|c|c|c|c|c|c|c|}\hline k_1 & k_2 & \cdots & k_m & i_1 & i_2 & \cdots & i_n \\\hline l_1 & l_2 & \cdots & l_m \\\hline\end{array} \equiv \begin{pmatrix} k_1 & \cdot & \cdot & k_m \\ l_1 & \cdot & \cdot & l_m \end{pmatrix} i_1 \ldots i_n \tag{2.21}$$

Quarks

An irreducible representation is therefore completely specified by two integers (m, n).

It is convenient to replace the antisymmetric pair (k_λ, l_λ) by another index j_λ, in the following manner:

$$x^{j_1 \ldots j_m}_{i_1 \ldots i_n} \equiv \begin{pmatrix} k_1 & \ldots & k_m \\ l_1 & \ldots & l_m \end{pmatrix} i_1 \ldots i_n \, \varepsilon^{j_1 k_1 l_1} \ldots \varepsilon^{j_m k_m l_m}. \qquad (2.22)$$

This tensor is symmetric in $\{j_1 \ldots j_m\}$ and in $\{i_1 \ldots i_n\}$. It is straightforward to show that

$$\sum_{j=1}^{3} x^{j \, j_2 \ldots j_m}_{j \, i_2 \ldots i_n} = 0. \qquad (2.23)$$

Thus, components of the tensor $x^{j_1 \ldots j_m}_{i_1 \ldots i_n}$ are not all independent. The number of independent components is the dimension $D(m, n)$ of the irreducible representation (m, n).

We now calculate $D(m, n)$. Suppose, among $\{j_1 \ldots j_m\}$, 1 occurs σ_1 times, 2 occurs σ_2 times, and 3 occurs σ_3 times. Then the number of possible sets $\{j_1 \ldots j_m\}$ is equal to

$$P_m = \text{No. of ordered sets } \{\sigma_1, \sigma_2, \sigma_3 | \sigma_1 + \sigma_2 + \sigma_3 = m\}. \qquad (2.24)$$

Any possible set $\{\sigma_1, \sigma_2, \sigma_3\}$, with $\sigma_2 > 0$, can be chosen by choosing *two* of the dots in the picture below, $\left[\text{in } \binom{m+1}{2} \text{ ways} \right]$:

$$\begin{array}{c} \sigma_1 \downarrow \quad \sigma_2 \downarrow \quad \sigma_3 \\ \bullet\,|\,1\,|\,2\,|\,3\,|\,4\,|\,\ldots\,|\,m\,|\,\bullet \end{array}$$

Any possible set $\{\sigma_1, \sigma_2, \sigma_3\}$ can be chosen by choosing only *one* dot, $\left[\text{in } \binom{m+1}{1} \text{ ways} \right]$. Hence,

$$P_m = \binom{m+1}{2} + \binom{m+1}{1} = \frac{1}{2}(m+1)(m+2). \qquad (2.25)$$

The total number of ways of choosing $\{j_1 \ldots j_m\}$ and $\{i_1 \ldots i_n\}$ independently is $P_m P_n$. The requirement (2.23) imposes $P_{m-1} P_{n-1}$ conditions. Therefore $D(m, n) = P_m P_n - P_{m-1} P_{n-1}$, or

$$D(m, n) = \tfrac{1}{2}(n+1)(m+1)(n+m+2). \qquad (2.26)$$

The representations (m, n) and (n, m) have the same dimensionality, and are said to be conjugate to each other:

$$\overline{(m, n)} \equiv (n, m). \qquad (2.27)$$

The representation (n, n) is self-conjugate, with dimensionality $D(n, n) = (n + 1)^3$.

The matrices representing L_α in the irreducible representation (m, n) can be worked out from the transformation law,

$$x'^{j_1' \ldots j_m'}_{i_1' \ldots i_n'} = (S^*_{j_1 j_1'} \ldots S^*_{j_m j_m'})(S_{i_1 i_1'} \ldots S_{i_n i_n'}) x^{j_1 \ldots j_m}_{i_1 \ldots i_n}, \qquad (2.28)$$

where S_{ij} is given in (2.18). This states that *an upper index transforms just like a lower index, except that λ_α is replaced by $-\lambda_\alpha^*$.*

The antisymmetric tensor ε^{ijk} can be used to raise or lower indices:

$$x^i = \varepsilon^{ijk} A_{jk}, \quad (A_{jk} = -A_{kj}), \qquad (2.29)$$

where A_{jk} transforms like $x_j x_k$.

In Table 2.5 we list some irreducible representations, in increasing order of dimensionality. Table 2.6 illustrates the rules for decomposing product representations into sums of irreducible representations.

2.3 The Quark Model[3]

1 Quarks as basic triplets

We think of the vector components x_i for the fundamental representation **3** as particle states called quarks. Those for the conjugate representation $\bar{\mathbf{3}}$ are called antiquarks[c]. Then all higher representations can be regarded as composite states of quarks and/or antiquarks.

Actually, from the point of view of representing $SU(3)$, all we need is **3**, because $\bar{\mathbf{3}}$ can be generated by $\mathbf{3} \times \mathbf{3} = \bar{\mathbf{3}} + \mathbf{6}$, as shown in Table 2.6. However, we want to be able to distinguish between particle and antiparticle. To do this, we assign quarks and antiquarks to **3** and $\bar{\mathbf{3}}$ respectively, and assign them equal and opposite baryon numbers B, which is the generator of a $U(1)$ group that commutes with $SU(3)$. We assign to quarks $B = 1/3$, and to antiquarks $B = -1/3$. Then $\bar{\mathbf{3}}$ occurring in $\mathbf{3} \times \mathbf{3}$ is distinct from an antiquark, for it has $B = 2/3$. [Of course, under $SU(3)$ alone, it does transform like an antiquark].

If we take quarks seriously, we must eventually face the question of their interactions. Whether we regard quarks as real objects or mere mathematical constructions, the quantum numbers of their bound states are determined purely group-theoretically.

In accordance with current convention, we name the quarks u, d, s; which stand respectively for "up", "down" and "strange". These are said to be the different "flavors" of a quark. Similarly, the antiquarks are named $\bar{u}, \bar{d}, \bar{s}$. More explicitly, we write

$$x = \begin{pmatrix} x_1 \\ x_2 \\ x_3 \end{pmatrix} \equiv \begin{pmatrix} u \\ d \\ s \end{pmatrix}, \quad \bar{x} = \begin{pmatrix} x^1 \\ x^2 \\ x^3 \end{pmatrix} \equiv \begin{pmatrix} \bar{u} \\ \bar{d} \\ \bar{s} \end{pmatrix}. \qquad (2.30)$$

---

[c] Note that **3** and $\bar{\mathbf{3}}$ are distinct, unlike the situation in $SU(2)$, where **2** and $\bar{\mathbf{2}}$ are equivalent.

[3] For a general reference, see J. J. J. Kokkedee, *The Quark Model* (W. A. Benjamin, New York, 1969).

Quarks

Thus, a u-quark corresponds to $\begin{pmatrix} u \\ 0 \\ 0 \end{pmatrix}$, a d-quark corresponds to $\begin{pmatrix} 0 \\ d \\ 0 \end{pmatrix}$, etc. We define isospin and hypercharge Y by

$$I_\alpha = \tfrac{1}{2}\lambda_\alpha \quad (\alpha = 1, 2, 3),$$
$$Y = \lambda_8/\sqrt{3}. \qquad (2.31)$$

Table 2.5   IRREDUCIBLE REPRESENTATIONS OF $SU(3)$

(m, n)	$D(m, n)$	Tableau	Tensor
	1		1
(0, 1)	3		$x_i\,(i = 1, 2, 3)$
(1, 0)	$\bar{3}$		x^i
(0, 2)	6		x_{ij}
(2, 0)	$\bar{6}$		x^{ij}
(1, 1)	$8 = \bar{8}$		$x^i_j\left(\sum_{i=1}^{3} x^i_i = 0\right)$ Adjoint rep.
(0, 3)	10		x_{ijk}
(3, 0)	$\overline{10}$		x^{ijk}
(1, 2)	15		$x^i_{jk}\left(\sum_{i=1}^{3} x^i_{ik} = 0\right)$
(2, 1)	$\overline{15}$		$x^{ij}_k\left(\sum_{i=1}^{3} x^{ij}_i = 0\right)$
(0, 4)	15'		x_{ijkl}
(4, 0)	$\overline{15'}$		x^{ijkl}
(1, 3)	24		$x^i_{jkl}\left(\sum_{i=1}^{3} x^i_{ikl} = 0\right)$
(3, 1)	$\overline{24}$		$x^{ijk}_l\left(\sum_{i=1}^{3} x^{ijk}_i = 0\right)$
(2, 2)	$27 = \overline{27}$		$x^{ij}_{kl}\left(\sum_{i=1}^{3} x^{ij}_{il} = 0\right)$

Two of these matrices, namely I_3 and Y, are diagonal in the quark basis:

$$I_3 = \begin{pmatrix} 1/2 & 0 & 0 \\ 0 & -1/2 & 0 \\ 0 & 0 & 0 \end{pmatrix}, \quad Y = \begin{pmatrix} 1/3 & 0 & 0 \\ 0 & 1/3 & 0 \\ 0 & 0 & -2/3 \end{pmatrix}. \quad (2.32)$$

The square of isospin, as well as strangeness S and charge Q, are all diagonal:

$$I(I+1) \equiv I_1^2 + I_2^2 + I_3^2 = \begin{pmatrix} 3/4 & 0 & 0 \\ 0 & 3/4 & 0 \\ 0 & 0 & 3/4 \end{pmatrix},$$

$$S \equiv Y - B = \begin{pmatrix} 0 & 0 & 0 \\ 0 & 0 & 0 \\ 0 & 0 & -1 \end{pmatrix}, \quad (2.33)$$

$$Q \equiv I_3 + \tfrac{1}{2}Y = \begin{pmatrix} 2/3 & 0 & 0 \\ 0 & -1/3 & 0 \\ 0 & 0 & -1/3 \end{pmatrix}.$$

By (2.28), the I_3 and Y for antiquarks are the negatives of those for quarks.

Table 2.6 DECOMPOSITION OF PRODUCT REPRESENTATIONS

(a) $3 \times 3 = \square \times \square = \boxminus + \square\square = \bar{3} + 6$

(b) $\bar{3} \times 3 = \boxminus \times \square = \boxminus + \boxminus\boxminus = 1 + 8$

(c) $6 \times 3 = \square\square \times \square = \boxminus\boxminus + \square\square\square = 8 + 10$

(d) $3 \times 3 \times 3 = (\bar{3} + 6) \times 3 = 1 + 8 + 8 + 10$

(e)* $8 \times 8 = \begin{array}{|c|c|}\hline 1 & 2 \\\hline 3 \\\cline{1-1}\end{array} \times \begin{array}{|c|c|}\hline 1' & 2' \\\hline 3' \\\cline{1-1}\end{array} = 1 + 8 + 8 + 10 + \overline{10} + 27$

$\begin{array}{|c|c|c|c|}\hline 1 & 1' & 2 & 2' \\\hline 3 & 3' \\\cline{1-2}\end{array}$ $\begin{array}{|c|c|c|c|}\hline 1 & 1' & 2 & 2' \\\hline 3 \\\cline{1-1} 3' \\\cline{1-1}\end{array}$ $\begin{array}{|c|c|c|}\hline 1 & 3' & 2 \\\hline 3 & 1' & 2' \\\hline\end{array}$ (1 means i_1 etc)

27   10 $\overline{10}$

$\begin{array}{|c|c|c|}\hline 1' & 2' & 3' \\\hline 1 & 2 \\\cline{1-2} 3' \\\cline{1-1}\end{array}$ $\begin{array}{|c|c|c|}\hline 1' & 2' & 3' \\\hline 1 & 2 \\\cline{1-2} 3 \\\cline{1-1}\end{array}$ $\begin{array}{|c|c|}\hline 1 & 1' \\\hline 2 & 2' \\\hline 3 & 3' \\\hline\end{array}$

8   8   1

* Arrange the six labelled blocks in all possible combinations, but preserve antisymmetry in (1, 3) or (1', 3'), if both labels remain in the final tableau.

Quarks

Consequently, they also have opposite signs for S and Q. However, $I(I + 1)$ is the same for quarks and antiquarks. These quantum numbers are tabulated in Table 2.7.

2 Quarks as building blocks

The irreducible representation (m, n) is realized in the space of the tensors $x^{j_1 \cdots j_m}_{i_1 \cdots i_n}$, whose independent components may be regarded as a multiplet of $D(m, n)$ particles, composed of quarks and/or antiquarks. We require each multiplet to have a definite baryon number that is the same for all particles in the multiplet. The quark and antiquark content of a multiplet is then uniquely determined.

In the tensor $x^{j_1 \cdots j_m}_{i_1 \cdots i_n}$, each lower index can be associated either with a quark, or an antisymmetric pair of antiquarks, according to (2.29). The same rule applies to an upper index, with quark and antiquark interchanged. The different choices are distinguished by baryon number. These possibilities are summarized in Table 2.8.

A straightforward application of (2.28) leads to the following theorem:

(a) If λ is a generator that is diagonal in the fundamental representation 3, then it is diagonal in any irreducible representation.

(b) If the eigenvalues of λ in the fundamental representation are $c_i(i = 1, 2, 3)$, then in the irreducible representation (m, n), where λ acts on $x^{j_1 \cdots j_m}_{i_1 \cdots i_n}$, the eigenvalues of λ are

$$(c_{i_1} + \cdots + c_{i_n}) - (c_{j_1} + \cdots + c_{j_m}).$$

In the quark model, the theorem merely states that in forming a composite state of quarks and antiquarks, their I_3 and Y are additive. It follows that charge and strangeness are also additive.

3 Weight diagrams

A convenient way to display the structure of a multiplet is to show all its components on a I_3-Y plot, called the *weight diagram* of the irreducible

Table 2.7 QUANTUM NUMBERS OF QUARKS

	Quarks			Antiquarks		
	u	d	s	$\bar{u}$	$\bar{d}$	$\bar{s}$
I	1/2	1/2	0	1/2	1/2	0
I_3	1/2	−1/2	0	−1/2	1/2	0
Y	1/3	1/3	−2/3	−1/3	−1/3	2/3
Q	2/3	−1/3	−1/3	−2/3	1/3	1/3
B	1/3	1/3	1/3	−1/3	−1/3	−1/3
S	0	0	−1	0	0	1

representation. The weight diagrams for the quark **3** and the antiquark **$\bar{3}$** are shown in Fig. 2.2.

The weight diagram for (m, n) is constructed by identifying $D(m, n)$ lattice sites (not necessarily distinct), on the Y-I_3 plot, as follows. In Fig. 2.3, treat the vectors $\mathbf{x}_1$, $\mathbf{x}_2$, $\mathbf{x}_3$ as lattice displacement vectors. Then $x_{i_1 \cdots i_n}^{j_1 \cdots j_m}$ is located at $(\mathbf{x}_{i_1} + \cdots + \mathbf{x}_{i_n}) - (\mathbf{x}_{j_1} + \cdots + \mathbf{x}_{j_m})$. The possible lattice sites form a diamond lattice.

Since different components may occupy the same site, each occupied site is characterized by a degeneracy. To obtain the correct degeneracy, it is important

Table 2.8 CORRESPONDENCE BETWEEN $SU(3)$ INDEX AND QUARK CONTENT

Each lower index i:

$$\text{Either} \quad i = \begin{pmatrix} 1 \\ 2 \\ 3 \end{pmatrix} \leftrightarrow \begin{pmatrix} u \\ d \\ s \end{pmatrix} \qquad B = 1/3$$

$$\text{or}^* \quad i = \begin{bmatrix} (23) \\ (31) \\ (12) \end{bmatrix} \leftrightarrow \begin{bmatrix} (\bar{d}\bar{s}) \\ (\bar{s}\bar{u}) \\ (\bar{u}\bar{d}) \end{bmatrix} \qquad B = -2/3$$

Each upper index j:

$$\text{Either} \quad j = \begin{pmatrix} 1 \\ 2 \\ 3 \end{pmatrix} \leftrightarrow \begin{pmatrix} \bar{u} \\ \bar{d} \\ \bar{s} \end{pmatrix} \qquad B = -1/3$$

$$\text{or}^* \quad j = \begin{bmatrix} (23) \\ (31) \\ (12) \end{bmatrix} \leftrightarrow \begin{bmatrix} (ds) \\ (su) \\ (ud) \end{bmatrix} \qquad B = 2/3$$

*See Eq. 2.34 for definition of (ds) etc.

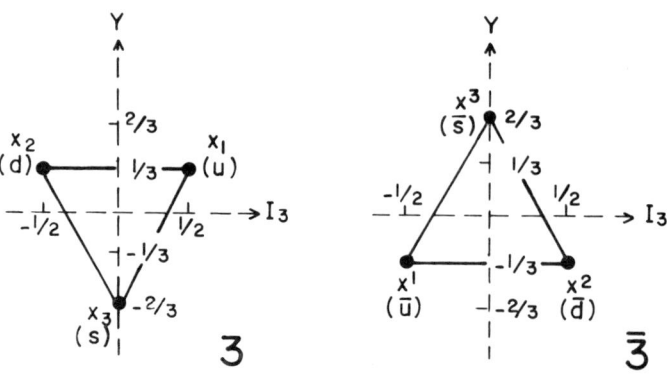

Fig. 2.2 Weight diagrams for **3** and **$\bar{3}$**

Quarks

to observe that a permutation of the upper or the lower indices among themselves does not lead to a distinct component. Furthermore, the condition (2.23) must be used to eliminate redundant components. Instead of giving general rules, we shall just work out some physically relevant examples.

The weight diagram, being a property of $SU(3)$ alone, knows nothing about baryon number. That is, multiplets of different baryon number have the same weight diagram, as long as they correspond to the same irreducible representation of $SU(3)$. The baryon number merely determines the quark content of the members of the multiplet.

Since no hadron of fractional baryon number has ever been observed, we shall consider as physical only multiplets with integer B, and assume that for some reason not understandable within this model, those with fractional B do not occur in nature. By this criterion, a multiplet is admissible only if it corresponds to a Young's tableau whose total number of boxes is divisible by 3 (since one box corresponds to one quark). From Table 2.5, we see that these are **8**, $\overline{\mathbf{10}}$, **10**, **27**, . . . ; and this is consistent with experimental facts.

The weight diagram of **8**, $\overline{\mathbf{10}}$ and **10** are given in Fig. 2.4. For **10** and $\overline{\mathbf{10}}$ the lattice sites have no degeneracies. For **8**, the central site has a two-fold degeneracy (it is occupied by x_1^1, x_2^2, x_3^3; but only two are independent because $x_1^1 + x_2^2 + x_3^3 = 0$). The degenerate states are chosen to be $2^{-1/2}(x_1^1 + x_2^2)$ and $2^{-1/2}(x_1^1 - x_2^2)$. By noting that under $SU(3)$, they respectively transform like $\bar{u}u + \bar{d}d$ and $\bar{u}u - \bar{d}d$, it is clear that $2^{-1/2}(x_1^1 - x_2^2)$ is an isosinglet, while $2^{-1/2}(x_1^1 + x_2^2)$ forms an isotriplet with x_1^2 and x_2^1.

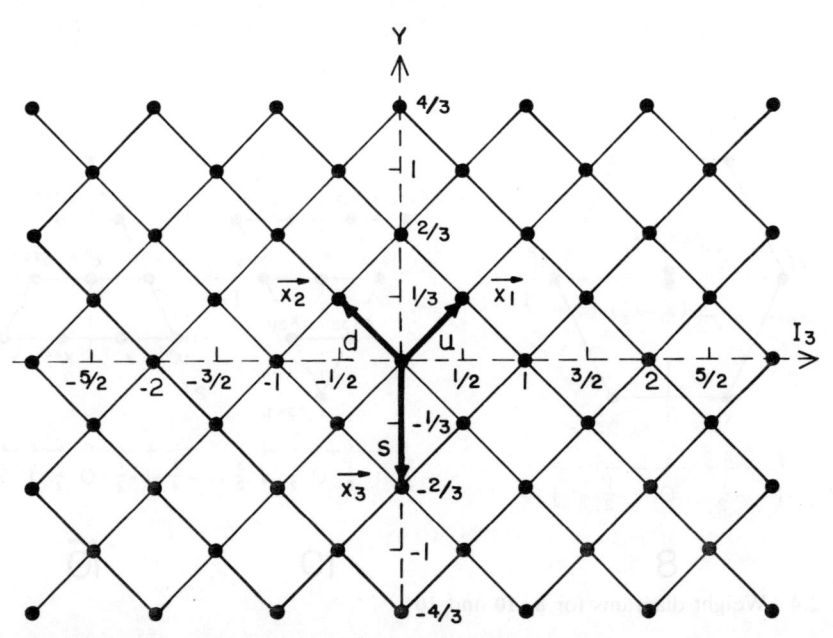

Fig. 2.3 Lattice on which weight diagrams are constructed

4 The composition of hadrons

We now examine the possible quark contents of the multiplets **8**, **$\overline{10}$**, and **10**. According to Table 2.8, **8** may have either $B = 0$ or $B = \pm 1$.

For $B = 0$, each member of the multiplet is composed of a quark and an antiquark. The quark contents are indicated in Fig. 2.5(a). We see that particles and antiparticles are included in the multiplet. Thus, the π-octet and ρ-octet can both be accounted for.

For $B = 1$, the quark contents of **8** are indicated in Fig. 2.5(b), where parentheses mean, for example,

$$(ud) \equiv u(1)\,d(2) - u(2)\,d(1), \tag{2.34}$$

where the labels 1 and 2 identify the two different quarks. That is, $\{u(1), d(1), s(1)\}$ and $\{u(2), d(2), s(2)\}$ are two independent vectors on which $SU(3)$ acts. The case $B = -1$ corresponds to the antiparticle multiplet. Thus we can account for the N-octet and the separate $\overline{N}$-octet.

The **10** with $B = 1$ is shown in Fig. 2.5(c). This accounts for the Δ-decaplet. Correspondingly, $\overline{10}$ with $B = -1$ accounts for the separate $\overline{\Delta}$-decaplet.

In principle, there could be **10**'s with $B = 0$ ($qq\bar{q}\bar{q}$), $B = -1$ ($q\bar{q}\bar{q}\bar{q}\bar{q}$), and $B = -2$ ($\bar{q}\bar{q}\bar{q}\bar{q}\bar{q}\bar{q}$), where q stands for a quark. These states are called "exotic", and do not seem to occur among hadronic states of low masses.

2.4 Color

1 Independent quark model[4]

The quark model so far is purely algebraic. It is just a way of representing $SU(3)$, with the baryon number trivially tagged on. In order to say something

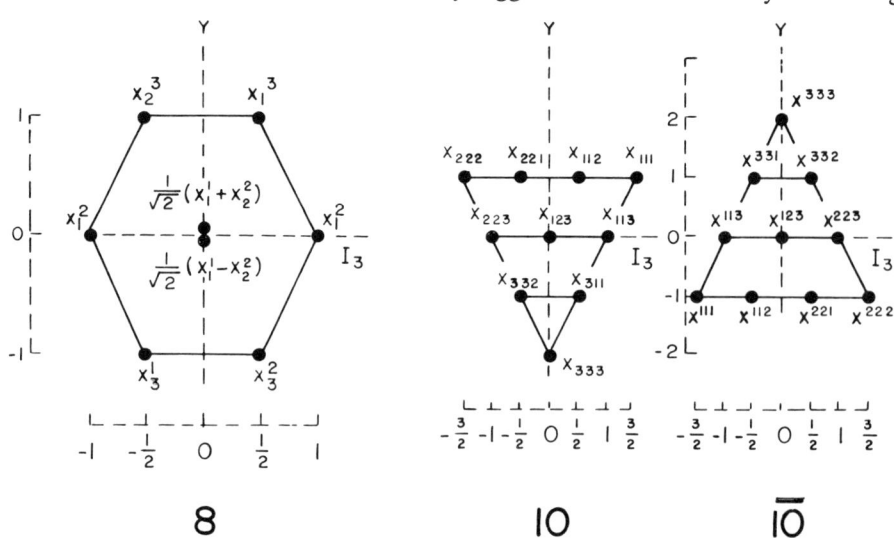

Fig. 2.4 Weight diagrams for **8**, **10** and $\overline{10}$

[4] For a general reference, see J. J. J. Kokkedee, *op. cit.*

Quarks

about the masses and the interactions of hadrons, we have to add dynamical content to the model, i.e., to construct a theory in space-time. At this point, it makes a difference whether or not we consider quarks to be real objects.

An extension of the model, still without dynamical content, is the $SU(6)$ quark model. One associates each quark with spin angular momentum, which generates the group $[SU(2)]_{\text{spin}}$ and embeds $[SU(3)]_{\text{flavor}} \times [SU(2)]_{\text{spin}}$ in a larger symmetry group $SU(6)$. The enlarged basis then consists of a sextet of quarks: $\{u(\uparrow), d(\uparrow), s(\uparrow), u(\downarrow), d(\downarrow), s(\downarrow)\}$. Among the irreducible representations of $SU(6)$, one finds a **35** with $B = 0$ (from $6 \times 6 = 1 + 35$). This can be identified with the union of the spin 0 π-octet and the spin 1 nonet consisting of the ρ-octet and the σ-singlet. One also finds a **56** with $B = 1$, identifiable with the union of the spin 1/2 N-octet and the spin 3/2 Δ-decaplet:

$$6 \times 6 \times 6 = \mathbf{56} + \mathbf{70} + \mathbf{70} + \mathbf{20}, \qquad (2.35)$$

where corresponding Young's tableaux are indicated. As we can see, the **56** is completely symmetric under a permutation of the spin-flavors of the quarks.

The $SU(6)$ symmetry, however, distinguishes between spin and orbital angular momentum. This distinction can be made only in nonrelativistic models, and is inconsistent with Lorentz invariance. A straightforward relativistic extension of $SU(6)$ would be an embedding of $SU(6)$ in a larger group that contains the Poincaré group. Since internal symmetry would then be joined with space-time symmetries, such an extension might have dynamical content. However, all efforts along such directions have failed. In fact, there are theorems stating that a non-trivial imbedding is impossible. Nevertheless, the fact that $SU(6)$ correctly predicts the existence of the meson **35** and the baryon **56** may not be accidental. Perhaps we can view it as a sort of approximation to the non-relativistic limit of a more correct relativistic theory.

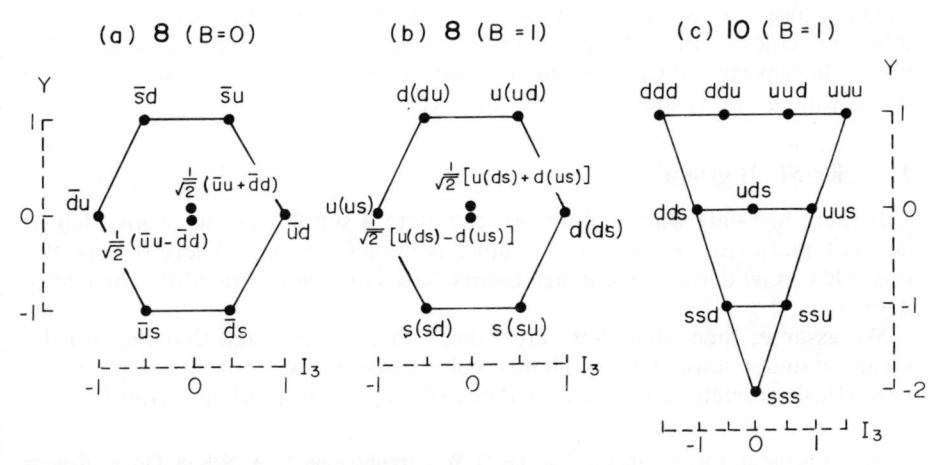

Fig. 2.5 Quark contents of **8** and **10**

From a phenomenological point of view, we can give the quark model some dynamical content by assuming that quarks are real particles moving in some effective potential, much like nucleons in the nuclear shell model. We call this an *independent quark model*, in which quarks occupy single-particle orbitals, which are described by spatial wave functions and have definite energies. Masses of hadrons can then be calculated in terms of the energies of occupied orbitals and the quark masses. The observed violation of unitary symmetry can be attributed to the fact that the s-quark has different mass from that of the u and d-quarks, and one can then understand the Gell-Mann-Okubo mass formula in a more concrete way.

Since quarks have spin 1/2, it is natural to assume that they are fermions and therefore obey the Pauli exclusion principle. That is, the total wave function of a hadron should be antisymmetric under the simultaneous interchange of flavor, spin, and orbital between two quarks. If we take the suggestion from (2.35) that the wave function of a baryon is symmetric under interchange of spin-flavor, then it must be antisymmetric under interchange of orbitals. Thus, three quarks in a baryon must be in different orbitals, only one of which can be the lowest S state (unless there are unsuspected degeneracies). On the other hand, magnetic moment calculations from the model agree with experiments only if there is no orbital contribution, suggesting that the three quarks are in the S state. This presents some sort of paradox.

The paradox is reinforced by considering the charge radii of hadrons. In a meson, the quark and antiquark can both be in the S state, and will be so, because that corresponds to the lowest energy. Since the charge radii of mesons are observed to be nearly the same as those for baryons, the three quarks in a baryon should all be in the S state.

One can always get out of the paradox by saying that $SU(6)$ is completely irrelevant, or that the independent quark model is totally wrong; but that would not lead us anywhere. Instead, one might resolve the paradox within the $SU(6)$ independent quark model by postulating that there are degeneracies which have so far been overlooked. The quantum number labelling this new degeneracy is called "color". If we assume that, for each flavor and spin, quarks come in different "colors", then all three quarks can be in the same state, as long as they have different colors. Clearly, this requires at least three colors. We now show the advantage of assuming exactly three colors.

2 Color $SU(3)$ group[5]

If there are only three colors, we can state a simple rule to insure that no isolated quark can be seen: *only "colorless" states can exist*. There is hope that this rule can be derived in a gauge theory based on color—quantum chromodynamics.

We assume, then, that there are exactly three colors, and that the world is invariant under color change. That is, color space is a representation space for a new $SU(3)$ symmetry group, denoted by $[SU(3)]_{\text{color}}$. A quark now transforms as

---

[5] A review of the idea of color may be found in O. W. Greenberg and C. A. Nelson, *Physics Reports*, 32C, 71 (1977).

Quarks

q_{in}, with

   flavor index: $i = 1, 2, 3$ (or u, d, s), (2.36)
   color index: $n = 1, 2, 3$ (or red, yellow, green).

These indices transform respectively under $[SU(3)]_{\text{flavor}}$ and $[SU(3)]_{\text{color}}$. The antiquark $\bar{q}_{in}$ transforms like q^{in}.

The empirical observation that there are no particles of fractional baryon number is guaranteed by the rule that *any physical state must be a color singlet*. This immediately implies that the number of quarks making up a baryon must be divisible by 3, as each quark corresponds to a square in a Young's tableau for $[SU(3)]_{\text{color}}$. (Note that this requirement does not depend on the number of flavors).

Since a color singlet corresponds to the Young's tableau in (2.20), three quarks in a baryon must be completely anti-symmetric in their color indices. Therefore, by the Pauli principle, they must be completely symmetric with respect to all other indices. Similarly, six quarks in a baryon (e.g., the deuteron) must consist of two color singlets. For a meson composed of quark and antiquark, its state must transform under $[SU(3)]_{\text{color}}$ as

$$\sum_{n=1}^{3} q^{in} q_{jn} = \sum_{n=1}^{3} \bar{q}_{in} q_{jn}. \tag{2.37}$$

As an illustration, let us write down the wave function for a proton in a non-relativistic independent quark model. We adopt a suitable notation for calculating matrix elements with respect to the wave function. The coordinates of a quark are collectively denoted by

$$z = \{n, i, s, \mathbf{r}\}, \quad \text{(coordinates)} \tag{2.38}$$

which are respectively the color, flavor, spin and position coordinates. The coordinates of the three quarks are denoted respectively by z_1, z_2, z_3, with corresponding subscripts on n, i, s, and $\mathbf{r}$. The single-quark quantum numbers are denoted collectively by

$$\lambda = \{N, k, \sigma, l\}, \quad \text{(quantum numbers)} \tag{2.39}$$

which are respectively the color, flavor, spin and spatial quantum numbers. A single-quark wave function is written as

$$\psi_\lambda(z) = C_N(n) F_k(i) X_\sigma(s) R_l(\mathbf{r}). \tag{2.40}$$

Under internal symmetry operations, n and i transform like $SU(3)$ lower indices (i.e., like those on q_{in}), and s transforms like an $SU(2)$ index. The factors in (2.40) are chosen to be

$$\begin{aligned} C_N(n) &= \delta_{Nn}, \\ F_k(i) &= \delta_{ki}, \\ X_\sigma(s) &= \delta_{\sigma s}, \\ R_l(\mathbf{r}) &= R(r). \quad \text{(S-wave orbital)} \end{aligned} \tag{2.41}$$

It is convenient to write out all the components of C_N, F_k, χ_σ in the form of column vectors, and name these vectors

$$C_1 = r = \begin{pmatrix} 1 \\ 0 \\ 0 \end{pmatrix}, \quad F_1 = u = \begin{pmatrix} 1 \\ 0 \\ 0 \end{pmatrix}, \quad \chi_1 = \uparrow = \begin{pmatrix} 1 \\ 0 \end{pmatrix},$$

$$C_2 = y = \begin{pmatrix} 0 \\ 1 \\ 0 \end{pmatrix}, \quad F_2 = d = \begin{pmatrix} 0 \\ 1 \\ 0 \end{pmatrix}, \quad \chi_2 = \downarrow = \begin{pmatrix} 0 \\ 1 \end{pmatrix}.$$

$$C_3 = g = \begin{pmatrix} 0 \\ 0 \\ 1 \end{pmatrix}, \quad F_3 = s = \begin{pmatrix} 0 \\ 0 \\ 1 \end{pmatrix}. \tag{2.42}$$

We can combine flavor and spin by writing

$$F_k \chi_\sigma = u_\uparrow, u_\downarrow, d_\uparrow, d_\downarrow, s_\uparrow, s_\downarrow. \tag{2.43}$$

Furthermore, as a shorthand, write

$$\begin{aligned} r(1) &\equiv r(n_1), \quad \text{etc.} \\ d_\uparrow(1) &\equiv d(i_1)[\uparrow(s_1)], \quad \text{etc.} \\ R(1) &\equiv R(r_1), \quad \text{etc.} \end{aligned} \tag{2.44}$$

The Pauli principle requires the total proton wave function $\Psi(1, 2, 3)$ to be completely antisymmetric under a permutation of 1, 2, 3. Therefore, $\Psi(1, 2, 3)$ is obtained by antisymmetrizing a linear combination of terms of the form $\psi_\lambda(1)\psi_\lambda(2)\psi_\lambda(3)$, the linear combination being dictated by unitary symmetry and spin. Requiring Ψ to be a color singlet, we write

$$\Psi(1, 2, 3) = \left[\sum_P \delta_P P_{123} r(1) y(2) g(3) \right] R(1) R(2) R(3) \Phi(1, 2, 3), \tag{2.45}$$

where P_{123} is a permutation of 1, 2, 3, and δ_P is its signature. The flavor-spin wave function $\Phi(1, 2, 3)$ must be completely symmetric in 1, 2, 3, and is chosen to be

$$\Phi(1, 2, 3) = \sum_P P_{123} u_\uparrow(1)[u_\uparrow(2) d_\downarrow(3) - d_\uparrow(2) u_\downarrow(3)]. \tag{2.46}$$

The choice is unique: it must be antisymmetric in one of the u-d pairs under $[SU(3)]_{\text{flavor}}$ in order that the proton be a member of **8**. The three spins must combine so as to give spin 1/2. The overall symmetrization then automatically makes the proton a member of **56** with respect to $SU(6)$. This, of course, is a consequence of the Pauli principle, and the assumption that the quarks are in the same spatial orbital. Writing out all the terms in (2.46), we obtain

$$\begin{aligned} \Phi(1, 2, 3) = &\; 2u_\uparrow u_\uparrow d_\downarrow + 2u_\uparrow d_\downarrow u_\uparrow + 2d_\downarrow u_\uparrow u_\uparrow \\ &- u_\uparrow d_\downarrow u_\downarrow - u_\downarrow u_\downarrow d_\uparrow - u_\downarrow d_\uparrow u_\uparrow \\ &- d_\uparrow u_\uparrow u_\downarrow - u_\downarrow u_\uparrow d_\uparrow - d_\uparrow u_\downarrow u_\uparrow, \end{aligned} \tag{2.47}$$

Quarks

where it is understood that the coordinates in each term stand in the same order: 1, 2, 3.

As another example, the π^+ wave function in the same model is given by

$$\pi^+(1, 2) = [\bar{r}(1)r(2) + \bar{y}(1)y(2) + \bar{g}(1)g(2)] \\ \cdot [\bar{d}_\uparrow(1)u_\downarrow(2) - \bar{d}_\downarrow(1)u_\uparrow(2)]. \quad (2.48)$$

2.5 Electromagnetic and Weak Probes

If we assume that hadrons are made of quarks, then the electromagnetic and weak interactions of hadrons should be derived from those of quarks. That is, the basic electromagnetic and weak currents should be quark currents. Thus, electrons and neutrinos might "see" the quarks inside a hadron, through their electromagnetic and weak interactions respectively. In fact, experimental results from electron and neutrino scattering from nucleons are consistent with such an interpretation, and provide indirect evidence for quarks being dynamical objects. However, since we do not really know how to do dynamical calculations involving quarks, the interpretations of these experiments are necessarily based on intuitive models, and are plausible rather than conclusive.

1 Electromagnetic interactions

The quark picture assumes that the electromagnetic interactions of hadrons arise from those of the quarks. Thus the electromagnetic interaction Lagrangian density is written

$$\mathcal{L}_{em} = -eA^\mu(x)[J_\mu(x) + j_\mu(x)], \quad (2.49)$$

where $A^\mu(x)$ is the Maxwell field, e the magnitude of the electronic charge ($e^2/4\pi\hbar c = 1/137$), and $j_\mu(x)$ is the usual Dirac current of electrons and muons. The quark electromagnetic current is given by

$$J_\mu(x) = \bar{q}_{in}(x)Q_{ij}\gamma_\mu q_{jn}(x) \\ = \tfrac{2}{3}\bar{u}_n\gamma_\mu u_n - \tfrac{1}{3}\bar{d}_n\gamma_\mu d_n - \tfrac{1}{3}\bar{s}_n\gamma_\mu s_n, \quad (2.50)$$

where the color index n is summed from 1 to 3, and Q_{ij} is the quark charge matrix given in (2.33). Since photons couple directly to the quarks, one might be able to "X-ray" a hadron to see them. Similar techniques employing neutrons as probes have been used to study the momentum distribution of atoms in liquid helium at very low temperatures[6].

Consider the inclusive electron-proton scattering

$$e + p \rightarrow e + X, \quad (2.51)$$

where X is anything. The matrix element for this process is represented by the

---

[6] K. Huang, in *Selected Topics in Physics, Astronomy and Biophysics*, A. DeLaredo and N. Jurisic, eds., (Reidell Publishing Co., Dordrecht, Holland, 1973) pp. 175–213.

Feynman graph in Fig. 2.6. The energy loss of the electron in the laboratory frame is denoted by ν. The 4-momentum transfer carried by a virtual photon is denoted by q_μ. The squared mass is negative: $q^2 < 0$. A Lorentz invariant expression for ν is

$$\nu = P \cdot q/M, \tag{2.52}$$

where P_μ is the proton 4-momentum, and M the proton mass. The interesting kinematic region for our purpose is the so-called deep-inelastic limit:

$$\begin{aligned} &\nu \to \infty, \\ &-q^2 \to \infty, \\ &x \equiv -q^2/2M\nu \quad \text{(fixed)}, \\ &(0 \le x \le 1). \end{aligned} \tag{2.53}$$

The matrix element, as defined in Fig. 2.6, is

$$\mathcal{M} = [\bar{u}(\mathbf{k}')(-ie\gamma^\mu)u(\mathbf{k})]\frac{(-ie)}{q^2}\langle X|J_\mu|P\rangle, \tag{2.54}$$

where $J_\mu \equiv J_\mu(0)$, and the electron and proton states are covariantly normalized to E/m particles per unit volume. The laboratory differential cross section is

$$\frac{d\sigma}{dE'd\Omega} = \frac{4\alpha^2 M^2}{q^4}\frac{E'}{E} I^{\alpha\beta} W_{\alpha\beta}, \tag{2.55}$$

where $\alpha = e^2/4\pi\hbar c$, and $d\Omega$ is the solid angle in which the final electron

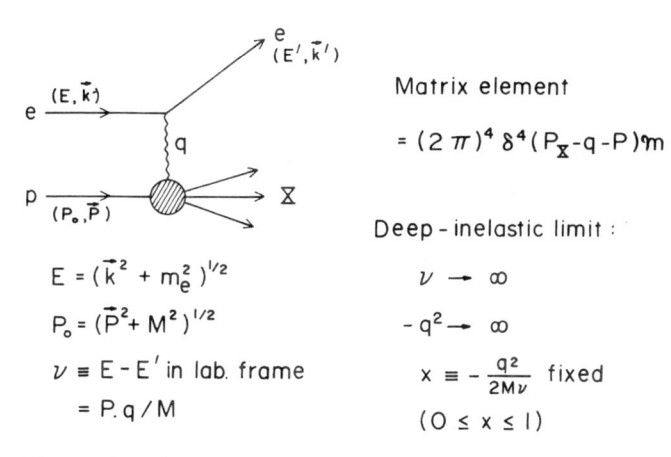

Fig. 2.6 Kinematics of electron-proton deep inelastic scattering

Quarks

emerges. The tensor $I^{\alpha\beta}$ comes from electron spin averages:

$$I^{\alpha\beta} \equiv \frac{1}{2}\text{Tr}\left[\gamma^\alpha \frac{\not{k}' + m_e}{2m_e} \gamma^\beta \frac{\not{k} + m_e}{2m_e}\right]$$

$$= \frac{1}{2m_e^2}[k'^\alpha k^\beta + k^\alpha k'^\beta + g^{\alpha\beta}(m_e^2 - k'\cdot k)]. \quad (2.56)$$

The hadronic structure is entirely contained in the tensor

$$W_{\alpha\beta} \equiv (2\pi)^3 \sum_X \langle P|J_\alpha|X\rangle\langle X|J_\beta|P\rangle \delta^4(P_X - P - q). \quad (2.57)$$

Lorentz invariance and gauge invariance imply that $W_{\alpha\beta}$ can be expressed in terms of two Lorentz invariant form factors W_1 and W_2:

$$W_{\alpha\beta} = -W_1\left(g_{\alpha\beta} - \frac{q_\alpha q_\beta}{q^2}\right) + \frac{W_2}{M^2}\left(P_\alpha - q_\alpha \frac{q\cdot P}{q^2}\right)\left(P_\beta - q_\beta \frac{q\cdot P}{q^2}\right). \quad (2.58)$$

If $W_{\alpha\beta}$ is given, we can extract W_1 and W_2 through the following formulas:

$$W_1 = \frac{1}{2}\left[C_2 - \left(1 - \frac{\nu^2}{q^2}\right)C_1\right]\left(1 - \frac{\nu^2}{q^2}\right)^{-1}, \quad C_1 \equiv W^\alpha{}_\alpha,$$
$$W_2 = \frac{1}{2}\left[3C_2 - \left(1 - \frac{\nu^2}{q^2}\right)C_1\right]\left(1 - \frac{\nu^2}{q^2}\right)^{-1}, \quad C_2 \equiv \frac{P^\alpha P^\beta}{M^2} W_{\alpha\beta}. \quad (2.59)$$

In the deep-inelastic limit,

$$W_1 \approx -\frac{1}{2}C_1 + \frac{Mx}{\nu}C_2,$$
$$W_2 \approx \frac{Mx}{\nu}\left(-C_1 + \frac{6Mx}{\nu}C_2\right). \quad (2.60)$$

After a certain amount of algebra, we can express the laboratory differential cross section (2.55) in the form

$$\frac{d\sigma}{dq^2\, d\nu} = \frac{4\pi\alpha^2}{q^4}\frac{E'}{E}\left(W_2 \cos^2\frac{\theta}{2} + 2W_1 \sin^2\frac{\theta}{2}\right), \quad (2.61)$$

where θ is the laboratory scattering angle, and where the electron mass has been neglected. The main experimental results[7] show that $W_1 \neq 0$, and that W_2 has the scaling property such that in the deep-inelastic limit, νW_2 is a function of x only. We shall argue that these results indicate that the proton is composed of spin 1/2 objects that are point-like with respect to photons.

2 Parton model

We have no way of calculating $W_{\alpha\beta}$ from first principles, because we have no theory for the states $|P\rangle$ and $|X\rangle$ in terms of the quark fields that appear in the

---

[7] J. I. Friedman and H. Kendall, *Ann. Rev. Nucl. Sci.*, **22**, 203 (1972).

current J_α. However, Feynman has suggested a simple intuitive model for this, the "parton" model, which we now describe.[8,9]

Suppose the proton is made up of bound objects that appear point-like to the photon. Then the proton can exist in a transitory virtual state consisting of these free objects. The lifetime of the virtual state is inversely proportional to the difference between the virtual state energy and the proton energy. In a Lorentz frame in which the proton is moving arbitrarily fast, the relativistic time-dilation can make this lifetime arbitrarily long. We call such a frame an ∞-*momentum frame*, and think of it as the limit of a sequence of Lorentz frames. In such a frame, a photon falling on the proton would see a collection of free point charges, which Feynman calls "partons". (Since they are defined in an ∞-momentum frame, partons do not necessarily have meaning in the proton rest frame). It is now imagined that the photon is absorbed by one of the partons. The absorption process lasts for a time of the order of the photon energy, which we can control by selecting ν in the experiment. In the laboratory, we can in principle make ν as large as we please, so that the characteristic time ν^{-1} is as small as we please. However, this time also dilates when we pass to an ∞-momentum frame, and it is not obvious whether a parton can live long enough as a free particle to absorb the photon. More detailed elementary considerations, which we shall not go into, show that the parton lifetime is indeed much longer than the absorption time, when the kinematic conditions (2.53) for deep-inelastic scattering is fulfilled, with $x > 0$.

Accepting the foregoing picture, we can calculate $W_{\alpha\beta}$ as follows. The matrix element $\langle X|J_\beta|P\rangle$ describes the absorption of a photon by a parton. We assume that the absorption takes such a relatively short time that both $|P\rangle$ and $|X\rangle$ can be described as free parton states. In the product $\langle P|J_\alpha|X\rangle\langle X|J_\beta|P\rangle$, the photon is absorbed, then re-emitted. We assume that both processes involve the same parton. (If two different partons were involved, their required momentum correlation would mean that one of them had very high momentum before interacting with the photon, and we deem this very unlikely). Accordingly, each parton contributes to $W_{\alpha\beta}$ singly and additively, as indicated schematically in Fig. 2.7.

In an ∞-momentum frame, all masses can be neglected. Hence we take the partons to be massless Dirac particles, with wave functions covariantly normalized to $2p_0$ particles per unit volume. Then the one parton contribution to $W_{\alpha\beta}$ is

$$\bar{W}_{\alpha\beta} \equiv (2\pi)^3 \frac{1}{2} \sum_{\text{spins}} \sum_{\mathbf{p}'} \langle \mathbf{p}|J_\alpha|\mathbf{p}'\rangle\langle \mathbf{p}'|J_\beta|\mathbf{p}\rangle \delta^4(p' - p - q)$$

$$= Q^2 \int \frac{d^3p'}{2p_0'} \delta^4(p' - p - q) \frac{1}{2}\text{Tr}(\gamma_\alpha \not{p}' \gamma_\beta \not{p})$$

$$= Q^2 \delta\left(p \cdot q + \frac{1}{2}q^2\right)(2p_\alpha p_\beta + q_\alpha p_\beta + q_\beta p_\alpha - g_{\alpha\beta} q \cdot p), \qquad (2.62)$$

---

[8] R. P. Feynman, *Phys. Rev. Lett.* **23**, 1415 (1969).
[9] J. D. Bjorken and E. A. Paschos, *Phys. Rev.* **158**, 1975 (1969).

Quarks

where p^α is the parton 4-momentum, and Q its charge in units of e. Now put

$$p^\alpha = yP^\alpha \quad (0 \leq y \leq 1). \tag{2.63}$$

That is, assume that all transverse momenta are negligible, and that no partons move oppositely to the proton. Then, using (2.60), we obtain the one-parton contributions to W_1 and W_2:

$$\begin{aligned} \nu\tilde{W}_2 &= 2Q^2 M x^2 \, \delta(y - x), \\ \tilde{W}_1 &= \nu\tilde{W}_2/2Mx. \end{aligned} \tag{2.64}$$

Suppose the proton state contains $f_i(y)\,dy$ parton states of the type i in the interval dy. Then

$$W_{1,2} = \sum_i \int_0^1 dy\, f_i(y)\, \tilde{W}_{1,2}. \tag{2.65}$$

The question is, how is $f_i(y)$ normalized? In our convention, a parton state has $2p_0$ partons per unit volume, while a proton state has P_0/M protons per unit volume. Therefore, in one proton, the number of partons of type i, in the interval dy, is $f_i(y)$ multiplied by $2p_0/(P_0/M) = 2My$:

$$n_i(y)\,dy = 2My f_i(y)\,dy. \tag{2.66}$$

By (2.63), $n_i(y)$ must satisfy the condition

$$\sum_i \int_0^1 dy\, y n_i(y) = 1. \tag{2.67}$$

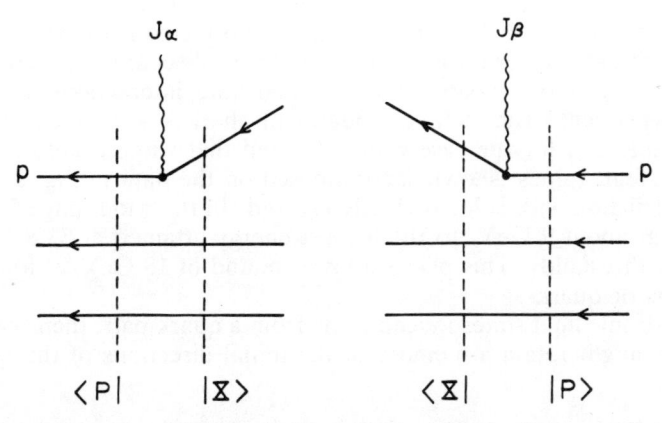

Fig. 2.7 The same parton absorbs and reemits the photon. Other partons are merely spectators.

In terms of this normalized parton distribution, (2.64) becomes

$$\nu W_2 = \sum_i Q_i^2 \, xn_i(x),$$

$$W_1 = \nu W_2/2Mx. \qquad (2.68)$$

We see that νW_2 has the desired scaling property. If we had used spin 0 partons, we would have gotten $W_1 = 0$ (though W_2 would be the same). The model of spin 1/2 partons is therefore consistent with experiments, and we identify the partons with quarks or antiquarks. Experiments indicate that $n_i(x) \xrightarrow[x \to 0]{} x^{-1}$. Hence, the total number of partons is infinite. Thus, as seen by a high-energy photon, the proton is made of three quarks plus an infinite "sea" of quark-antiquark pairs.

3 Evidence for color

The most direct experimental evidence that quarks have color comes from measurements of the total cross section for the annihilation of electron-positron pairs in colliding beam experiments. The results are expressed in terms of the cross section ratio

$$R \equiv \frac{e^+e^- \to \text{Hadrons}}{e^+e^- \to \mu^+\mu^-}, \qquad (2.69)$$

and the data up to c.m. energy 35.8 GeV is shown in Fig. 2.8.[10,11] Note that heavy lepton final states such as $\tau^+\tau^-$ are included, because they decay into hadrons. The final states e^+e^- and $\mu^+\mu^-$ are excluded.

The gross features of the data can be understood as follows. The reactions occur predominantly through annihilation of the initial state into a single virtual photon, which then materializes into the final states, through production of lepton pairs and quark pairs (for these are the only particles coupled directly to the photon). The total cross section should be proportional to the sum of squared charges of all the leptons and quarks that can be energetically pair-produced, barring resonances and final-state interactions. Possible resonances are independently known, and can be subtracted if we wish. As long as the various thresholds for these productions are well-separated, there should be a plateau between thresholds, in which final-state interactions are relatively energy-independent. The sums of squares of charges at various plateaus are given in Table 2.9, for the case with color and that without color.

These plateau values, shown superimposed on the data in Fig. 2.8, indicate that the prediction with color is clearly favored. In the wide range from the $\bar{b}b$ threshold, at about 9 GeV, to the highest energy attained at 35.8 GeV, there are no new thresholds. This places a lower bound of 18 GeV/c² for masses of new leptons or quarks.

If the hadronic final states indeed come from a quark pair, then their angular distribution might retain a memory of the initial directions of the quark pair.

---

[10] Particle Data Group, *Rev. Mod. Phys.* **52**, 556 (1980).
[11] Mark-J Collaboration, *The First Year of Mark-J*, MIT Lab for Nucl. Sci. Report 107 (April 1979).

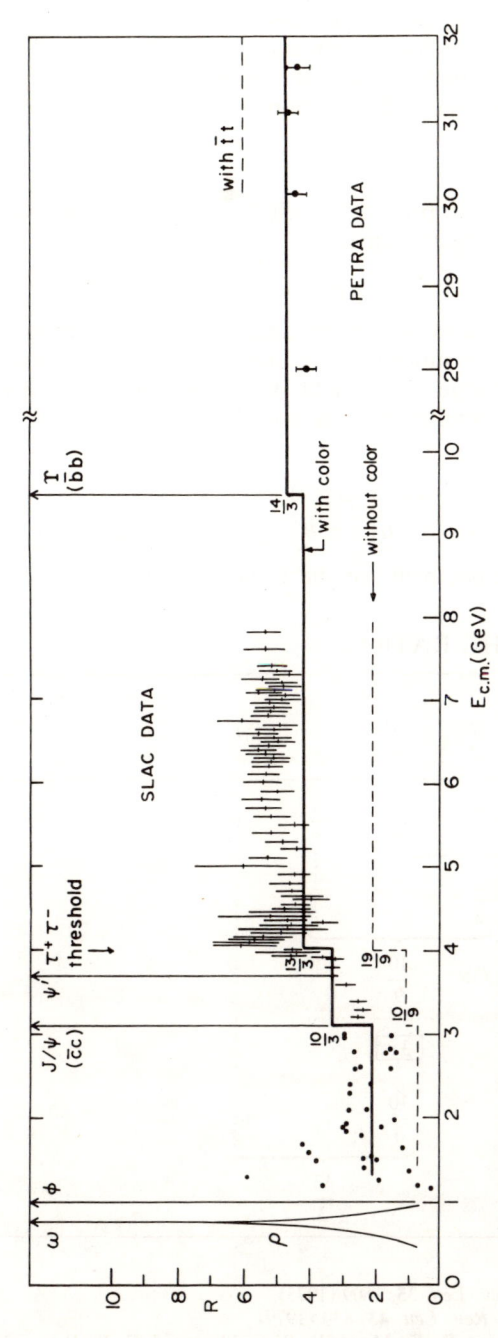

Fig. 2.8 The ratio $R = (e^+e^- \to \text{Hadrons})/(e^+e^- \to \mu^+\mu^-)$. The SLAC and PETRA data have different systematic errors, which are not indicated. Error bars for points below 3 GeV are large, and omitted for clarity. Points above 28 GeV are samplings of data.

This has been found to be the case, in the phenomena of "quark jets"[12]. If, as we believe, color symmetry is a gauge symmetry that gives rise to the strong interactions, then the gauge bosons (gluons) can also be emitted by the quarks, resulting in three or more hadronic jets of a distinctive character (gluons have no electromagnetic interactions). This phenomena has also been observed.[13]

4 Weak interactions[14]

A concise summary of our present knowledge of the weak interactions is given by the interaction Lagrangian density

$$\mathcal{L}_{wk} = gW_\alpha(x)[J^\alpha_{wk}(x) + j^\alpha_{wk}(x)] + \text{h.c.}, \qquad (2.70)$$

where $J^\alpha_{wk}(x)$ is the quark weak current, and $j^\alpha_{wk}(x)$ is the lepton weak current. From now on we drop the subscript "wk" for brevity. These weak currents are coupled to a massive charge vector field $W_\alpha(x)$. Although the W-mesons have not been detected experimentally, it is useful to assume that they exist.

The weak currents are given by

$$\begin{aligned} j^\alpha &= \bar{e}\gamma^\alpha(1 - \gamma_5)\nu + \bar{\mu}\gamma^\alpha(1 - \gamma_5)\nu', \\ J^\alpha &= [\bar{d}\gamma^\alpha(1 - \gamma_5)u]\cos\theta + [\bar{s}\gamma^\alpha(1 - \gamma_5)u]\sin\theta, \end{aligned} \qquad (2.71)$$

where e and ν stand respectively for the Dirac field operators for the electron

Table 2.9 THE RATIO R

$e^+ \searrow Q \nearrow$ $e^- \nearrow \searrow$	u $\bar{u}$	d $\bar{d}$	s $\bar{s}$	c $\bar{c}$	τ^- τ^+	b $\bar{b}$	(?) ... ...
Q^2	$\dfrac{4}{9}$	$\dfrac{1}{9}$	$\dfrac{1}{9}$	$\dfrac{4}{9}$	1	$\dfrac{1}{9}$	

$$\underbrace{}_{\tfrac{2}{3}}$$

WITHOUT COLOR $\dfrac{10}{9}$

....

$\boxed{3\times}$

$\dfrac{10}{3}$

WITH COLOR $\dfrac{13}{3}$

---

[12] G. Hansen et al., Phys. Rev. Lett. **35**, 1609 (1975).
[13] O. P. Barber et al., Phys. Rev. Lett. **43**, 830 (1979).
[14] For a general reference, see R. E. Marshak, Riazuddin, and C. P. Ryan, *Theory of Weak Interactions in Particle Physics* (Wiley-Interscience, New York, 1969).

Quarks

and the electron-neutrino; and μ and ν' respectively for those of the negative muon and the muon-neutrino. The angle θ, called the Cabibbo angle, has an experimental value of $\theta \cong 1/4$.

The matrix $1 - \gamma_5$ acting on a Dirac spinor picks out the component with left-handed chirality. For a massless Dirac particle, or a massive Dirac particle of sufficiently high momentum, this means negative helicity (i.e., spin direction is opposite to momentum). We note that for any Dirac spinor ψ,

$$\bar{\psi}\gamma^\alpha(1-\gamma_5)\psi = \bar{\psi}(1+\gamma_5)\gamma^\alpha\psi = [(1-\gamma_5)\psi]^\dagger \gamma_0\gamma^\alpha\psi. \qquad (2.72)$$

Thus, only left-handed spinors enter into the weak currents. These currents give rise to 4 basic Feynman vertices, as shown in Fig. 2.9. The decay of the μ^- meson corresponds to the Feynman graph of Fig. 2.10.

Since in low-energy processes the W-meson carries small momentum, its propagator may be taken to be m_W^{-2}. We can then obtain the following relation between the coupling constant g and the experimental Fermi coupling

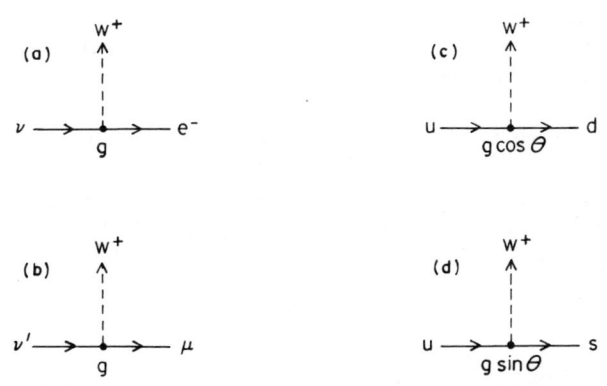

Fig. 2.9 Weak vertices

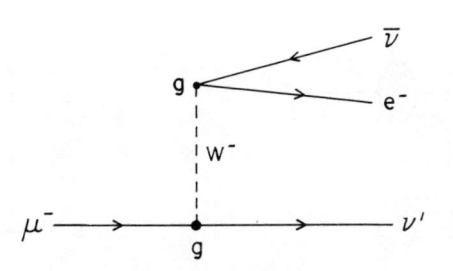

Fig. 2.10 μ^- decay

constant G:

$$g^2/m_W^2 = G/\sqrt{2},$$
$$G = 1.01 \times 10^{-5} m_p^{-2}, \qquad (2.73)$$

where m_p is the proton mass.

Weak decays of hadrons occur through the quark vertices (c) and (d) of Fig. 2.9. Their structure immediately gives, to lowest order in $m_p^2 G$ ($\approx 10^{-5}$), the following selection-rules for hadronic decays which were established experimentally:

(a) The change in the hadron's strangeness does not exceed 1 in magnitude, i.e., $\Delta S = 0, \pm 1$.

(b) When there are leptons in the final state, and when $\Delta S = \pm 1$, the change in the hadron's charge is equal to the change in its strangeness, i.e., $\Delta Q = \Delta S = \pm 1$.

(c) When the hadron's strangeness does change, the change in the hadron's isospin is $\pm 1/2$, i.e., $|\Delta I| = 1/2$. (This rule, however, is violated by electromagnetic interactions, and in reality holds only to order $\alpha \sim 1\%$).

Some Feynman graphs for hadronic decays are shown in Fig. 2.11 in which the shaded blobs denote strong interactions.

We recall that the quarks (u, d) form an isodoublet, while s is an isosinglet. The vector part of the quark current can be written in the form

$$V^\alpha \equiv g \cos \theta (\bar{u} \ \bar{d}) \gamma^\alpha \begin{pmatrix} 0 & 1 \\ 0 & 0 \end{pmatrix} \begin{pmatrix} u \\ d \end{pmatrix} = \sqrt{2} g \cos \theta (\bar{q} \gamma^\alpha I_+ q), \qquad (2.74)$$

where $I_+ = (I_1 + iI_2)/\sqrt{2}$ is the isospin raising operator. This is a succinct statement of the "conserved vector-current hypothesis" (CVC) of Feynman and Gell-Mann (in the form subsequently amended by the introduction of the

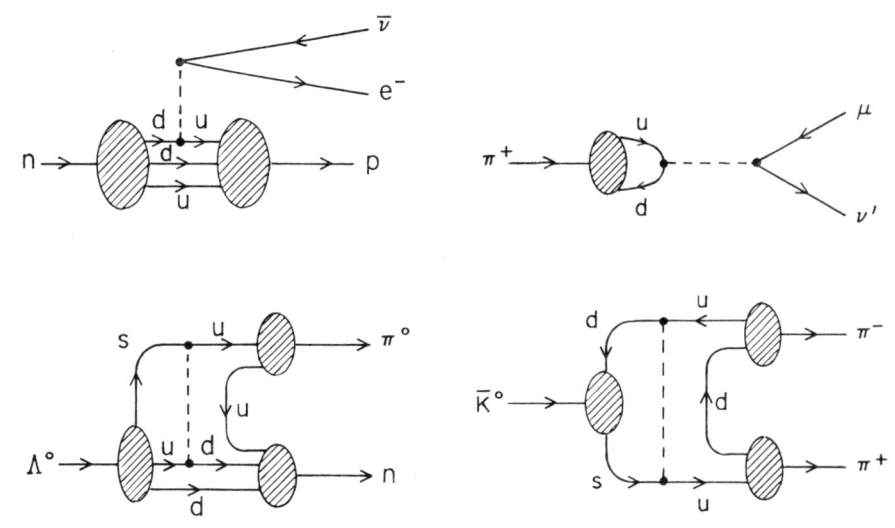

Fig. 2.11 Hadronic decays

Cabibbo angle, as required by experiments). Because strong interactions conserve isospin, the coupling constant g here is unaffected by strong renormalization effects, when V^α is sandwiched between hadronic states. On the other hand, the coupling constant in front of the axial vector part of the current is altered by such effects (experimentally, by a factor 1.25 in the nucleon state).

From the foregoing, we see that the quark hypothesis brings order into the structure of weak interactions. It neatly summarizes the selection rules, and disentangles weak effects from strong.

By analogy with the case of electromagnetic interactions, we can probe hadronic structure through deep-inelastic neutrino scattering from nucleons, and again try out the parton idea. The analysis in this case is more complicated than the electromagnetic case, because we no longer have gauge invariance, and because neutrino and antineutrino have definite and distinct helicities. We shall forego a discussion of this case. Available data are consistent with the predictions of a simple parton picture.[15]

2.6 Charm

1 The Charmed Quark

Glashow, Iliopoulos and Maiani[16] (GIM) proposed that in addition to u, d, s, there should be another quark flavour which they named "charm" (c). The motivation is to rid the old theory of certain undesirable higher order weak effects, i.e., violations of the ΔS and $\Delta S = \Delta Q$ rules, and the occurrence of unobserved decays, such as $\bar{K}^0 \to \mu^+ + \mu^-$, through effective neutral currents. Their conjecture appears to have been independently and brilliantly confirmed, by the subsequent discovery of new families of hadrons beginning with the J/ψ. With their scheme, there emerges a remarkable one-to-one correspondence between quarks and leptons.

Experimentally, the decay mode $\bar{K}^0 \to \mu^+ + \mu^-$ is not observed; its branching ratio being less than 10^{-6}. In the three-flavoured quark theory, this decay mode occurs at an order of g^4, through the Feynman graph of Fig. 2.12. The matrix element is divergent.

Although the theory can be made renormalizable by elaborating it with the addition of other fields (e.g., the Weinberg-Salam model), for qualitative purposes we keep the present model, but introduce a cutoff Λ. Taking the W-propagator to be

$$\Delta^{\alpha\beta}(k) = \frac{1}{k^2 - m_W^2}\left(g^{\alpha\beta} + \frac{k^\alpha k^\beta}{m_W^2}\right), \tag{2.75}$$

and assuming the integration over momentum k' in Fig. 2.12 converges by virtue of strong interactions, we see through simple power-counting that the degree of divergence is $g^4 \Lambda^2 \sim G^2 \Lambda^2$. To keep the branching ratio within the experimental

---

[15] G. B. West, *Phys. Reports* **18C**, 263 (1975).
[16] S. L. Glashow, J. Iliopoulos, and L. Maiani, *Phys. Rev.* **D2**, 1285 (1970).

upper bound, a cutoff $\Lambda \lesssim 3\text{GeV}$ is necessary, and this value seems unreasonably small. In any event, it seems more satisfactory, from a theoretical point of view, to suppress the decay through a dynamical mechanism, rather than through a cutoff, which merely relegates the explanation to things left out in the model.

The GIM proposal is to introduce a new quark c, with weak couplings chosen so as to cancel the Feynman graph of Fig. 2.12, in the limit of completely degenerate quark masses. The required couplings are indicated in Fig. 2.13.

In the real world, where unitary symmetry is inexact, the s quark should have a different mass from that of the u and d quarks. Presumably, the c quark will have a still different mass, and the c-u mass difference will render the cancellation in Fig. 2.13 incomplete. To calculate the true rate of $\bar{K}^0 \to \mu^+ + \mu^-$, different ingredients would be involved. To give quark masses real meaning, one would have to have a model of quark binding. To calculate higher order weak processes in a meaningful way, one has to work with a renormalizable theory of weak interactions. As a rough guess, however, one might expect that the c-u mass difference, defined in some effective way, takes the place of the cutoff Λ. The experimental bound cited earlier should then put this mass difference at $\lesssim 3 \text{ GeV}/c^2$.

The three flavors s, d, u were introduced to realize $[SU(3)]_{\text{flavor}}$. To add one more flavor means enlarging the group to $[SU(4)]_{\text{flavor}}$, which will be even more badly broken than $SU(3)$. Just as in going from $SU(2)$ to $SU(3)$ we added a new

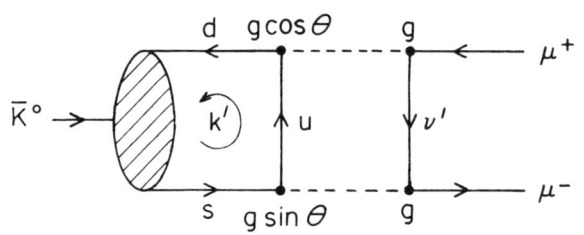

Fig. 2.12 Lowest-order matrix element for $\bar{K}^0 \to \mu^+ + \mu^-$, an unobserved decay mode

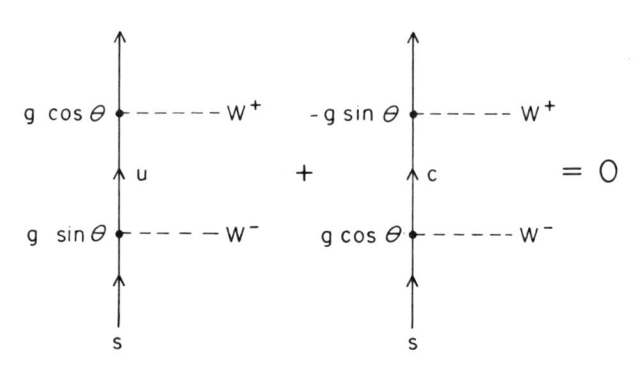

Fig. 2.13 Introducing the charmed quark c

Quarks

hadronic quantum number hypercharge, we now add another hadronic quantum number charm (C). Like Y, it is assumed to be an additive quantum number conserved by the strong interactions. The charmed quark, by definition, is assigned $C = 1$, while u, d, s are assigned $C = 0$. The charmed quark is also supposed to come in the *same* three colors as the other quarks.

Other quantum numbers of the charmed quark c are assigned as follows, motivated by the structure of the vertices involving c in Fig. 2.13:

Spin: $s = 1/2$ To conserve angular momentum.
Isospin: $I = 0$ As there is only one c.
Charge: $Q = 2/3$ To conserve charge.
Baryon number: $B = 1/3$ To conserve baryon number.
Strangeness: $S = -1$ To preserve the ΔS and $\Delta S = \Delta Q$ rules.
Hypercharge: $Y \equiv B + S = -2/3$.

According to these assignments, we have the generalized relation

$$Q = I_3 + \tfrac{1}{2}Y + C. \tag{2.76}$$

2 The J/ψ and its family

In 1974, Ting and Richter, with their respective collaborating teams, independently discovered the J/ψ.[17] This particle is remarkable in that its mass (3.1 GeV/c^2) is more than three times that of the proton, and yet its lifetime is a thousand times longer than the hadrons we had been familiar with (full width = 67 KeV). In Ting's words, "It's like stumbling upon a village inhabited by people who live to be ten thousand."[18] Since then, other particles of the same family have been found: ψ (3684), ψ (3950), ψ (4150), ψ (4400). All are spin 1 mesons with $I = 0$, negative parity and G-parity.

It is now established that these particles are states of "charmonium", i.e., bound states of $\bar{c}c$. The J/ψ is the charmonium ground state. One can account for the J/ψ family by adopting a non-relativistic independent quark model, with a linear potential between c and $\bar{c}$, and taking the mass of c to be roughly half the J mass, i.e., 1.5 GeV/c^2 (thus making the non-relativistic assumption self-consistent). The higher ψ states are supposed to be radial excitations. According to such a model, there should also be orbital excitations, and these have been identified experimentally. Charmed mesons and baryons, i.e., bound states like ($c\bar{d}$) and (cud) have also been found. The unusual narrowness of J/ψ has not been fully understood. It is thought to have a dynamical origin.

A new family of vector mesons, the Υ family, starting at a mass of 9 GeV/c^2, has since been discovered, giving evidence of a new quark flavor b.[19]

There is thus a rich new heavy-quark spectroscopy, which forms a separate topic we cannot go into here[20]. For our purpose, the clarification of this

---

[17] Nobel lectures by Ting and Richter: S. C. C. Ting, *Rev. Mod. Phys.* **49**, 236 (1977); B. Richter, *Ibid*, **49**, 251 (1977).
[18] S. C. C. Ting, (private communications).
[19] S. W. Herb et al., *Phys. Rev. Lett.* **39**, 252 (1977).
[20] T. Appelquist, R. M. Barnett and K. D. Lane, *Ann Rev. Nucl. Part. Sci.* **28**, 387 (1978).

spectroscopy through the quark model constitutes strong evidence for the reality of quarks.

3 Correspondence between quarks and leptons

With the inclusion of the charmed quark, the quark weak current in (2.71) is amended to read

$$J^\alpha = (\bar{d}\cos\theta + \bar{s}\sin\theta)\gamma^\alpha(1 - \gamma_5)u \\ + (\bar{s}\cos\theta - \bar{d}\sin\theta)\gamma^\alpha(1 - \gamma_5)c. \tag{2.77}$$

Apparently, the quark pair (d, s), which was defined with respect to quantum numbers conserved by the strong interactions, partakes in the weak interactions through a linear recombination (d_θ, s_θ), which is just a rotation through the Cabibbo angle:

$$\begin{pmatrix} d_\theta \\ s_\theta \end{pmatrix} = \begin{pmatrix} \cos\theta & \sin\theta \\ -\sin\theta & \cos\theta \end{pmatrix} \begin{pmatrix} d \\ s \end{pmatrix}. \tag{2.78}$$

In terms of these, the quark current and the lepton current have strikingly similar forms:

$$J^\alpha = \bar{d}_\theta \gamma^\alpha(1 - \gamma_5)u + \bar{s}_\theta \gamma^\alpha(1 - \gamma_5)c, \\ j^\alpha = \bar{e}\gamma^\alpha(1 - \gamma_5)\nu + \bar{\mu}\gamma^\alpha(1 - \gamma_5)\nu'. \tag{2.79}$$

With the discovery of τ and b, the similarity persists, provided that one postulates a τ-neutrino, ν'', and another quark flavor t. (These will be discussed in Chap. VI). We have no deeper understanding of this symmetry at present. It cannot help but suggest that the leptonic and quark flavors are manifestations of a deeper unity, and that these particles that seem elementary to us at this stage might yet be reducible to something simpler.

CHAPTER III
MAXWELL FIELD: $U(1)$ GAUGE THEORY

3.1 Global and Local Gauge Invariance

The Maxwell field, or electromagnetic field, is coupled solely to conserved currents of matter fields. This property may be derived from a "gauge principle", namely, the electromagnetic coupling arises from an extension of "global gauge invariance" to "local gauge invariance".

To illustrate the gauge principle, we consider a complex scalar field $\phi(x)$ whose classical Lagrangian density[a] in the absence of electromagnetic coupling has the form

$$\mathcal{L}_0(\phi(x), \partial^\mu \phi(x)) = \partial_\mu \phi^* \partial^\mu \phi - V(\phi^* \phi), \tag{3.1}$$

which is obviously invariant under a constant phase change of $\phi(x)$:

$$\phi(x) \to U\phi(x),$$
$$U = e^{-i\alpha}, \tag{3.2}$$

where α is an arbitrary real constant. This transformation is called a "global gauge transformation", and the theory is said to have global gauge invariance under the group $U(1)$. By Noether's theorem[1], there is a conserved current:

$$j^\mu = \text{const. } \phi^* \overleftrightarrow{\partial}^\mu \phi,$$
$$\partial_\mu j^\mu = 0. \tag{3.3}$$

Now we consider local gauge transformations:

$$\phi(x) \to U(x)\phi(x),$$
$$U(x) = e^{-i\alpha(x)}, \tag{3.4}$$

where $\alpha(x)$ is an arbitrary real function. That is, the gauge transformations at different points of space-time are independent of one another. We note that $\partial^\mu \phi(x)$ transforms in the same manner as $\phi(x)$ under a global gauge transformation; but it acquires an extra term under a local gauge transformation:

$$\partial^\mu \phi(x) \to U(x)\partial^\mu \phi(x) + \phi(x)\partial^\mu U(x). \tag{3.5}$$

Therefore, $\mathcal{L}_0(\phi, \partial^\mu \phi)$ is not invariant under a local gauge transformation.

---

[a] The theory has to be quantized. For expedience we work with the classical theory, and simply read off some simple properties in the quantized version whenever possible.

[1] See R. Jackiw in S. B. Treiman, R. Jackiw, and D. J. Gross, *Lectures in Current Algebra and Its Applications* (Princeton University Press, Princeton, 1972) pp. 101ff.

To make the theory locally gauge invariant, all we have to do is to replace $\partial^\mu \phi(x)$ by a suitable generalization that transforms in the same manner as $\phi(x)$. We define a vector field $A^\mu(x)$, called a "gauge field", which transforms under the local gauge transformation (3.4) according to the rule

$$A^\mu(x) \to A^\mu(x) + \frac{1}{e}\partial^\mu \alpha(x), \tag{3.6}$$

where e is a given real number that fixes the scale of $A^\mu(x)$ relative to $\phi(x)$. Then, the "covariant derivative" defined by

$$D^\mu \phi(x) \equiv [\partial^\mu + ieA^\mu(x)]\phi(x) \tag{3.7}$$

will transform in the same manner as $\phi(x)$:

$$\begin{aligned} D^\mu \phi(x) &\to U(x) D^\mu \phi(x), \\ [D^\mu \phi(x)]^* &\to U^*(x)[D^\mu \phi(x)]^*. \end{aligned} \tag{3.8}$$

By replacing $\partial^\mu \phi$ by $D^\mu \phi$, we obtain a new Lagrangian $\mathcal{L}(\phi, D^\mu \phi)$, which is obviously invariant under local gauge transformations. However, it contains the gauge field $A^\mu(x)$ as an external field. To define a closed dynamical system in the canonical sense, it is necessary to add a term involving $\partial^\nu A^\mu$ quadratically. The only gauge Lorentz scalar of this type is proportional to $F^{\mu\nu} F_{\mu\nu}$, where

$$F^{\mu\nu}(x) \equiv \partial^\mu A^\nu(x) - \partial^\nu A^\mu(x) \tag{3.9}$$

is called the field strength tensor. Thus we arrive at a Lagrangian density for a closed dynamical system invariant under local gauge transformations:

$$\mathcal{L} = -\tfrac{1}{4} F^{\mu\nu} F_{\mu\nu} + \mathcal{L}_0(\phi, D^\mu \phi). \tag{3.10}$$

The factor $-1/4$ in the first term is purely conventional. The global $U(1)$ symmetry, which is now enlarged into a local symmetry, is said to have been "gauged". Within the framework of canonical field theory, the procedure is unique. The classical equations of motion following from (3.10) are just Maxwell's equations for scalar electrodynamics:

$$\begin{aligned} \partial_\mu F^{\mu\nu} &= j^\nu, \\ D_\mu D^\mu \phi &= -\partial V/\partial \phi^*, \\ (D_\mu D^\mu \phi)^* &= -\partial V/\partial \phi, \end{aligned} \tag{3.11}$$

where

$$\begin{aligned} j^\nu &= ie[\phi^* D^\mu \phi - (D^\mu \phi)^* \phi] = ie\phi^* \overleftrightarrow{\partial}^\mu \phi - 2e^2 \phi^* \phi A^\mu, \\ \partial_\nu j^\nu &= 0, \end{aligned} \tag{3.12}$$

where the last statement is required by the equations of motion and the antisymmetry of $F^{\mu\nu}$. Thus we have "derived" electromagnetism.

Although the symmetry has been enlarged, no additional conserved currents arise. An application of Noether's theorem will yield the obvious statement that the most general conserved current is $j^\mu(x)$ plus a term of the form $\partial_\nu [F^{\mu\nu}(x) f(x)]$, where $f(x)$ is an arbitrary function. The gauge principle merely determines the form of coupling between the matter field and the gauge field.

Maxwell Field: U(1) Gauge Theory

Summarizing the local gauge transformation (3.4) and (3.6) in the form

$$\phi(x) \to e^{-ie\omega(x)}\phi(x),$$
$$A^\mu(x) \to A^\mu(x) + \partial^\mu\omega(x), \quad (3.13)$$

where $\omega(x)$ is an arbitrary real function, we see that the charge e (up to a constant factor) acts as the generator of the gauge group $U(1)$. In the original system with only global gauge invariance, the charge of a particle must be identified as e everywhere in space-time. However, in the enlarged system, the charge may be taken to be $\pm e$ independently at every space-time point, because the sign of $\omega(x)$ is free to change. Whatever the convention we choose, the physical content of the theory will be the same, for the gauge field maintains the correct bookkeeping.

In the most general case, we may begin with a set of matter fields $\{\phi_1(x), \ldots \phi_n(x)\}$ (which may be boson or fermion fields), that furnishes a representation (generally reducible) of the group $U(1)$. The generator of $U(1)$ is now represented by an $n \times n$ diagonal matrix, whose eigenvalues $Q_1, \ldots, Q_n$ are the respective charges of the matter fields. A global gauge transformation is an element of $U(1)$ represented by the transformation

$$\begin{pmatrix} \phi_1 \\ \phi_2 \\ \vdots \\ \phi_n \end{pmatrix} \to \begin{pmatrix} e^{-iQ_1\omega} & 0 & 0 & \cdots \\ 0 & e^{-iQ_2\omega} & 0 & \cdots \\ \cdots & \cdots & \cdots & \cdots \\ \cdots & & & e^{-iQ_n\omega} \end{pmatrix} \begin{pmatrix} \phi_1 \\ \phi_2 \\ \vdots \\ \phi_n \end{pmatrix}, \quad (3.14)$$

where ω is an arbitrary real constant. More compactly, we rewrite (3.14) in matrix form as

$$\phi(x) \to U\phi(x),$$
$$U = e^{-iQ\omega}. \quad (3.15)$$

The Lagrangian density $\mathcal{L}_0(\phi, \partial^\mu\phi)$ is assumed to be invariant under (3.15). To enlarge the symmetry to a local one, we replace $\partial^\mu\phi$ by

$$D^\mu\phi(x) = [\partial^\mu + iQA^\mu(x)]\phi(x). \quad (3.16)$$

The Lagrangian density of the enlarged system has the same form as (3.10).

As a representation of $U(1)$, (3.15) places no constraint on the values of the charges $Q_1, \ldots, Q_n$. It leaves unexplained the experimental fact that all observed charges in nature are multiples of a common unit (charge quantization). A possible scenario is that the electromagnetic $U(1)$ group is a subgroup of a *compact* symmetry group G. Then the matrix U in (3.15) will have to be a periodic function of ω, and charge will be quantized[2].

The matter Lagrangian density $\mathcal{L}_0$ may have other $U(1)$ symmetries that are not gauged, for example baryon number and lepton number. These will remain as global symmetries of $\mathcal{L}$, as long as they commute with the electromagnetic $U(1)$.

---

[2] C. N. Yang, *Phys. Rev.* D **1**, 2360 (1970).

It may seem that the theory predicts that the photon is massless, by the following argument. The equation of motion for the gauge field reads $\Box^2 A^\mu = j^\mu$ in Lorentz gauge ($\partial_\mu A^\mu = 0$). Suppose $\phi = 0$ in the lowest state[b] of the system, as gauge invariance would naively require (for ϕ is not gauge invariant, and naively one might expect the lowest state to be unique). Then the low-lying modes of the gauge field satisfy $\Box^2 A^\mu = 0$, which leads to a plane-wave solution of wave vector k^μ, with $k^2 = 0$. These modes correspond to massless photons in quantum theory. Now, we can take j^μ into account by perturbation expansions in powers of e, resulting in a scheme that describes the interactions in terms of emissions and absorptions of massless photons. Quantum electrodynamics is such a scheme, and its predictions have been tested experimentally to extremely high precision. Therefore, it might seem that the masslessness of the photon is a consequence of the theory, in particular the gauge invariance of the theory.

However, the argument above breaks down if perturbation theory is invalid. Such is obviously the case when $\phi \neq 0$ in the lowest state[c], a situation known as "spontaneous symmetry breaking". In that case the photon does develop a mass, as we shall discuss in more detail in Sec. 3.3.

Even if $\phi = 0$ in the lowest state, the validity of perturbation theory cannot be proven mathematically; it remains a logical possibility that the photon may have mass[3]. In fact, the photon does acquire mass in the Schwinger model[4]—quantum electrodynamics in one spatial dimension. It can almost be proved[5] that in quantum electrodynamics the photon is massless if the electron charge is smaller than a critical value, but acquires mass otherwise. The critical value is unknown, although one would like to believe that it is greater than the physical electronic charge. Thus, the masslessness of the photon in conventional quantum electrodynamics is not due to any known principle in the theory, but due to the assumption that conventional perturbation theory is valid.

3.2 Spontaneous Breaking of Global Gauge Invariance: Goldstone Mode

A symmetry of a system is said to be "spontaneously broken" if the lowest state of the system is not invariant under the operations of that symmetry. Far from being an esoteric situation, this is a common occurrence in the macroscopic world. For example, rotational symmetry is spontaneously broken in a ferromagnet, and translational symmetry is spontaneously broken in an infinite crystalline solid.

Here, we discuss the spontaneous breaking of global gauge invariance in a relativistic field theory. As a simple example, we turn again to the complex

---

[b] Here we use the term "state" loosely, to denote either a classical solution or a state of the quantized theory.
[c] In the quantized theory, the statement becomes $\langle \phi \rangle \neq 0$, where $\langle \phi \rangle$ is the vacuum expectation value of ϕ.

[3] J. Schwinger, *Phys. Rev.* **125**, 397 (1962).
[4] J. Schwinger, *Phys. Rev.* **128**, 2425 (1962).
[5] K. Wilson, *Phys. Rev.* **D10**, 2445 (1974); A. Guth, *Phys. Rev.* **D21**, 2291, (1980).

Maxwell Field: U(1) Gauge Theory

scalar field defined by (3.1), with no electromagnetic coupling. The classical Hamiltonian of the system is

$$H = \int d^3x[\pi^*\pi + \nabla\phi^*\cdot\nabla\phi + V(\phi^*\phi)], \tag{3.17}$$

where $\pi = \partial\phi/\partial t$. A solution to the equations of motion with the lowest possible energy corresponds to a constant $\phi(x) = \phi_0$, such that $V(\phi_0^*\phi_0)$ is at its minimum possible value. This clearly minimizes H, and is a solution of the field equations, because for time-independent ϕ, H is proportional to the action. If $\phi_0 \neq 0$, the solution is clearly not invariant under a change of phase, hence global gauge invariance is spontaneously broken. The lowest state is then infinitely degenerate, corresponding to the fact that the phase of ϕ_0 is arbitrary.

To be definite let us choose

$$V(\phi^*\phi) = \mu^2\phi^*\phi + \lambda(\phi^*\phi)^2 + \text{const.} \tag{3.18}$$

The reason for choosing a quartic polynomial form is that the corresponding quantum field theory is renormalizable, in a perturbation scheme based on expansions in powers of λ. The equation of motion now reads

$$(\Box^2 + \mu^2)\phi = -2\lambda\phi^*\phi. \tag{3.19}$$

The free field case corresponds to $\lambda = 0$, $\mu^2 > 0$. In this case the classical modes above the lowest state are plane waves of wave vectors k^μ, with $k^2 = \mu^2$, and correspond to single-particle states of mass μ in the quantum theory.

If $\lambda < 0$, the theory does not exist because the Hamiltonian has no lower bound. If $\lambda < 0$ and $\mu^2 > 0$, then V has the form shown in Fig. 3.1(a). It may appear that the system can have a metastable state, with the field contained (in field space) within the local minimum of V. This would be true in particle quantum mechanics, but not in quantum field theory. In the latter case the decay rate of such an initial state in infinite space is infinite because it has finite value per unit volume.

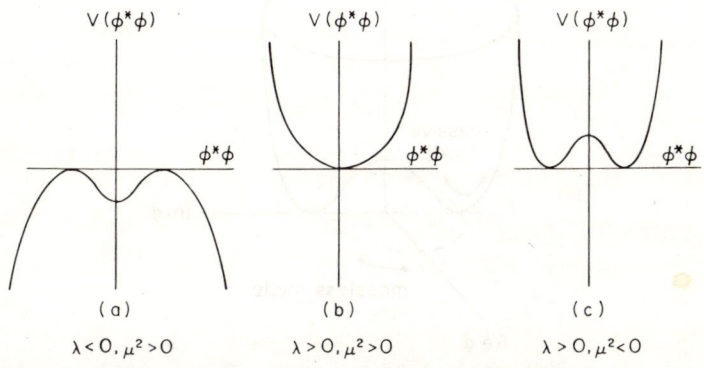

Fig. 3.1 The potential term for various choices of parameters

If $\lambda > 0$, we can have either $\mu^2 > 0$ or $\mu^2 < 0$, with the corresponding forms of V shown in Figs. 3.1(b) and 3.1(c). In the former case, the lowest solution is $\phi = 0$, and there is no spontaneous symmetry breaking. In the latter case, $\phi \neq 0$, which is what we want to study.

For convenience we rewrite (3.18) in the form

$$V(\phi^*\phi) = \lambda(\phi^*\phi - \phi_0^2)^2, \quad (\phi_0 \neq 0). \tag{3.20}$$

The lowest state corresponds to $\phi^*\phi = \phi_0^2$, or

$$\phi(x) = \phi_0 e^{i\alpha_0}, \tag{3.21}$$

where α_0 is an arbitrary real constant.

The low-lying states of the quantum theory can be deduced from the low-lying classical modes by inspection. Let us put

$$\phi(x) = [\phi_0 + \eta(x)] e^{i\alpha(x)}, \tag{3.22}$$

so that the complex fields $\phi^*(x)$, $\phi(x)$ are now replaced by the real fields $\eta(x)$ and $\alpha(x)$, in terms of which the Lagrangian density reads

$$\mathscr{L}_0 = \partial^\mu \eta \, \partial_\mu \eta - \lambda(2\phi_0 - \eta)^2 \eta^2 + (\phi_0 + \eta)^2 \, \partial^\mu \alpha \, \partial_\mu \alpha. \tag{3.23}$$

Assuming η to be small, and dropping terms higher than second order ones, we obtain

$$\mathscr{L}_0 \simeq [\partial^\mu \eta \, \partial_\mu \eta - 4\lambda \phi_0^2 \eta^2] + \phi_0^2 \, \partial^\mu \alpha \, \partial_\mu \alpha + O(\eta^3). \tag{3.24}$$

The terms in the brackets describe free scalar particles of mass $2\phi_0\sqrt{\lambda}$. The next term describes free massless scalar particles. The neglected terms describe the interactions among the particles. These classical modes are illustrated in Fig. 3.2, where V is shown as a function of Re ϕ and Im ϕ. We can see the direct connection between $\phi_0 \neq 0$ and the existence of a massless mode.

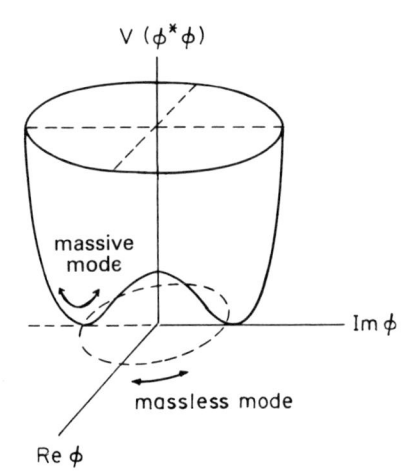

Fig. 3.2 Classical modes in the scalar field with symmetry-breaking potential

In the quantum theory, each value of α_0 in (3.21) gives a possible vacuum state. The transition amplitude between vacua with different values of α_0 vanishes for infinite spatial volume.

The statement that "spontaneous breaking of a continuous global symmetry implies the existence of a massless spin zero particle" is known as *Goldstone's theorem*[6], and the massless particle is called a *Goldstone boson*. The symmetry that is spontaneously broken is, of course, still a symmetry of the system. However, it is manifested not through the invariance of the lowest state, but in the "Goldstone mode"—through the existence of a Goldstone boson.

There are physical examples of Goldstone bosons in non-relativistic many-body systems: spin waves in a ferromagnet, phonons in a crystalline solid and in liquid helium.

3.3 Spontaneous Breaking of Local Gauge Invariance: Higgs Mode

When a *local* gauge symmetry is spontaneously broken, the symmetry is again manifested in a manner other than the invariance of the lowest state. In this case, however, no Goldstone boson occurs. Instead, the gauge field acquires mass. We say that the symmetry is manifested in the "Higgs mode".

Let us consider scalar electrodynamics, and choose V to have the form (3.20):

$$\mathcal{L}(x) = -\tfrac{1}{4}F^{\mu\nu}F_{\mu\nu} + (D^\mu\phi)^*(D_\mu\phi) - V(\phi^*\phi), \quad (3.25)$$
$$D^\mu\phi = (\partial^\mu + ieA^\mu)\phi,$$
$$V(\phi^*\phi) = \lambda(\phi^*\phi - \phi_0^2)^2, \quad (\phi_0 \neq 0).$$

The Lagrangian is invariant under a local gauge transformation

$$A^\mu(x) \to A^\mu(x) + \partial^\mu\omega(x),$$
$$\phi(x) \to e^{-ie\omega(x)}\phi(x), \quad (3.26)$$
$$\phi^*(x) \to e^{ie\omega(x)}\phi(x),$$

where $\omega(x)$ is an arbitrary real function. The local gauge symmetry is spontaneously broken because $\phi_0 \neq 0$. The field $\phi(x)$, which causes the breakdown, is referred to as a "Higgs field".

In canonical formalism the variables are as follows:

Field	Canonical conjugate	
A^μ	$-F_{0\mu}$	
ϕ	$\pi \equiv (D^0\phi)^*$	(3.27)
ϕ^*	$\pi^* \equiv D^0\phi$	

Since the canonical conjugate to A^0 is identically zero, A^0 is not an independent variable but can be eliminated in terms of the others through the equations of

---

[6] J. Goldstone, *N. Cim.* **19** 154 (1961); J. Goldstone, A. Salam, and S. Weinberg, *Phys. Rev.* **127**, 965 (1962).

motion. Introduce the electric field **E** and the magnetic field **B** by

$$E^k = F^{k0},$$
$$B^k = -\tfrac{1}{2}\varepsilon^{ijk}F^{ij}, \tag{3.28}$$

or

$$\mathbf{E} = -\partial \mathbf{A}/\partial t - \nabla A^0,$$
$$\mathbf{B} = \nabla \times \mathbf{A}. \tag{3.29}$$

The Hamiltonian can be brought to the form

$$H = \int d^3x [\tfrac{1}{2}(\mathbf{B}\cdot\mathbf{B} + \mathbf{E}\cdot\mathbf{E}) + |\pi|^2 + |\mathbf{D}\phi|^2 + V]. \tag{3.30}$$

In deriving (3.30), the equations of motion have been used, and a surface integral $\int d^3x \, \nabla\cdot(\mathbf{E}A^0)$ has been dropped[d]. The form (3.30) is manifestly gauge invariant, but it still contains A^0. The final form, after A^0 has been eliminated, will depend on the particular gauge one chooses. The numerical value of H is of course gauge invariant.

It is clear from (3.30) that a lowest-energy solution is

$$A^\mu(x) = 0,$$
$$\phi(x) = \phi_0 e^{i\alpha_0}. \tag{3.31}$$

To study the classical modes near this solution, it is convenient to go to "unitary gauge", in which $\phi(x)$ is real, by transforming away its phase through a continuous local gauge transformation. This can always be done because ϕ satisfies a second order differential equation, and hence its phase must have continuous derivatives. Thus, we can always write

$$\phi(x) = \rho(x) \quad \text{(real)}. \tag{3.32}$$

The equations of motion become

$$\partial_\mu F^{\mu\nu} = -2e^2\rho^2 A^\nu,$$
$$(\partial^\mu + ieA^\mu)(\partial_\mu + ieA^\mu)\rho = 2\lambda\rho(\phi_0^2 - \rho^2). \tag{3.33}$$

Since $\partial_\nu\partial_\mu F^{\mu\nu} \equiv 0$, we must have

$$\partial_\mu A^\mu(x) = 0 \quad \text{wherever } \rho(x) \neq 0. \tag{3.34}$$

Now we put

$$\rho(x) = \phi_0 + \eta(x), \tag{3.35}$$

and treat $\eta(x)$ and $A^\mu(x)$ as small quantities. The linearized equations of motion read

$$(\Box^2 + 2e^2\phi_0^2)A^\mu = 0 \quad (\partial_\mu A^\mu = 0),$$
$$(\Box^2 + 4\lambda\phi_0^2)\eta = 0. \tag{3.36}$$

---

[d] See the derivation of the more general expression (4.92) in chap. IV.

Maxwell Field: U(1) Gauge Theory

In the quantum theory, these lead to a spin 1 particle of mass $\sqrt{2}e\phi_0$, and a spin 0 particle of mass $2\sqrt{\lambda}\,\phi_0$. There is no massless particle. The original fields (in a fixed gauge) A^1, A^2, ϕ, ϕ^* are now replaced by A^1, A^2, A^3, η. (The subsidiary condition $\partial_\mu A^\mu = 0$ takes out the spin zero part of **A**). One could say that gauging the symmetry converts the "would-be Goldstone boson" into the longitudinal part of the resulting massive gauge field. If we had used some gauge other than the unitary gauge, the nature of the spectrum would be the same, but perhaps less obvious. The manner through which the photon mass comes about is called the "Higgs Mechanism".[7]

A physical example in which the Higgs mechanism actually takes place is superconductivity. The Lagrangian of the theory is invariant under local phase changes of the electron field, but the ground state is not, owing to a condensation of Cooper pairs made up of two electrons. As a consequence, the photon becomes massive inside a superconducting body. In particular, an externally applied magnetic field can penetrate the body only to a finite depth equal to the inverse mass (Meissner effect). A phenomenological way to describe the condensation phenomenon is to introduce an "order parameter" $\phi(x)$ to describe the condensate, as done in the Landau-Ginsberg theory[8]. In such an approach, (3.25) serves as a phenomenology Lagrangian.

Spontaneous symmetry breaking without a fundamental Higgs field, such as in the case of superconductivity, is sometimes called "dynamical symmetry breaking". This terminology carries the implication that Higgs fields are "normally" needed for symmetry breaking. This is a prejudice based not on physical fact, but solely on mathematical simplicity.

3.4 Classical Finite-Energy Solutions

We can see from (3.30) that a finite-energy solution must satisfy the requirements

$$\int d^3x \, (\mathbf{B} \cdot \mathbf{B} + \mathbf{E} \cdot \mathbf{E}) < \infty,$$
$$\int d^3x \, V(\phi^*\phi) < \infty. \tag{3.37}$$

The first of these implies that asymptotically $F^{\mu\nu} \to O(x^{-2})$, and hence $A^\mu \to \partial^\mu\omega + O(x^{-1})$. That is,

$$A^\mu(x) \xrightarrow[x \to \infty]{} \text{(pure gauge form)} + O(x^{-1}). \tag{3.38}$$

The second condition in (3.37) implies that

$$\phi(x) \xrightarrow[x \to \infty]{} \phi_0 e^{i\alpha(x)} + O(x^{-4}). \tag{3.39}$$

We now impose these conditions on the equations of motion (3.11), which may be rewritten in the form

$$\partial_\mu F^{\mu\nu} = -2e \, \text{Im}(\phi^* D^\nu \phi). \tag{3.40}$$

---

[7] P. W. Higgs, *Phys. Rev. Lett.* **12**, 132, (1964); F. Englert and R. Brout, *Ibid*, **13**, 321 (1964); G. S. Guralnik, C. R. Hagen, and T. W. B. Kibble, *Ibid*, **13**, 585 (1964).
[8] Ginsberg and L. Landau, *Zh. Teor. Fiz.* **47**, 2222 (1957). [*Sov. Phys. JETP* **5**, 1174 (1957)].

We know by (3.38) that $\partial_\mu F^{\mu\nu} \to O(x^{-3})$; the same must be true for the right hand side. From (3.38) and (3.39) we find that

$$D^\nu\phi \xrightarrow[x\to\infty]{} i\phi\partial^\nu(e\omega + \alpha) + O(x^{-2}). \tag{3.41}$$

Hence,

$$\text{Im}(\phi^* D^\nu\phi) \xrightarrow[x\to\infty]{} \phi_0^2 \partial^\nu(e\omega + \alpha) + O(x^{-2}). \tag{3.42}$$

For this to be $O(x^{-3})$, the first term must cancel the second, which requires $\partial^\nu(e\omega + \alpha) \to O(x^{-2})$, or

$$D^\nu\phi \xrightarrow[x\to\infty]{} O(x^{-2}). \tag{3.43}$$

Along any infinitesimal segment dx on a circle of very large radius R, the length of the segment is $R\,d\theta$, where $d\theta$ is the angular displacement. Hence, by (3.43), $dx^\mu D_\mu \phi \sim R^{-1}\,d\theta$. That is, a classical finite-energy solution must obey the boundary condition

$$dx^\mu D_\mu \phi(x) \xrightarrow[x\to\infty]{} 0. \tag{3.44}$$

Classical finite-energy solutions are interesting because they include "solitons" solutions that will be discussed later. This does not necessarily mean that classical infinite-energy solutions are irrelevant to physics. For example, classically, a plane wave has infinite energy, but leads to single particle states of finite energy in the quantum theory.

3.5 Magnetic Flux Quantization

Consider a static classical finite-energy solution. We can always choose $A^0 = 0$. At infinity, the boundary conditions (3.38) and (3.39) read

$$\mathbf{A}(\mathbf{x}) \xrightarrow[x\to\infty]{} \frac{1}{e} \nabla \alpha(\mathbf{x}), \tag{3.45}$$

$$\phi(\mathbf{x}) \xrightarrow[x\to\infty]{} \phi_0 e^{i\alpha(\mathbf{x})}.$$

We denote $\alpha(\mathbf{x})$ and $\phi(\mathbf{x})$ on a fixed circle C of very large radius by $\alpha(\theta)$ and $\phi(\theta)$ respectively, for they depend only on the angular position $\theta(0 \le \theta < 2\pi)$. Since $\phi(\theta)$ satisfies a differential equation, it must be continuous in space. Hence

$$\phi(2\pi) = \phi(0), \tag{3.46}$$

and this implies

$$\alpha(2\pi) - \alpha(0) = 2\pi n \quad (n = 0, \pm 1, \pm 2, \ldots). \tag{3.47}$$

The net magnetic flux linking C is

$$\Phi = \iint d\mathbf{S} \cdot \mathbf{B} = \oint_C d\mathbf{s} \cdot \mathbf{A}, \tag{3.48}$$

Maxwell Field: U(1) Gauge Theory

where **dS** is a surface element of a surface spanning C, and **ds** is an element of arc along C. Using (3.45) and (3.47) we find

$$\Phi = \frac{2\pi n}{e} \quad (n = 0, \pm 1, \pm 2, \ldots), \tag{3.49}$$

which is the statement of flux quantization, a necessary condition for a finite-energy solution.

In three spatial dimensions, we must choose $n = 0$; otherwise magnetic field lines would run off to infinity violating the condition (3.37) for finite energy. In two spatial dimensions, however, n can take on any integer value (with the magnetic field normal to the two-dimensional plane). The two-dimensional case may be looked upon as a three dimensional solution that is independent of the z coordinate, and has finite energy per unit length along the z axis.

Consider, then, a two dimensional problem in the x-y plane. Let us put $\alpha(\theta) = n\theta + \beta(\theta)$, where $\beta(2\pi) - \beta(0) = 0$. Thus, $\beta(\theta)$ is a continuous function, which can be transformed away through a continuous gauge transformation. Therefore there is a choice of gauge in which

$$\mathbf{A} \xrightarrow[x \to \infty]{} \frac{1}{e} \nabla(n\theta) = \hat{\boldsymbol{\theta}} \frac{n}{er} \quad \text{(pure gauge)}, \tag{3.50}$$

$$\phi \xrightarrow[x \to \infty]{} \phi_0 e^{in\theta}.$$

Note that **A** is multivalued, but $\oint \mathbf{ds} \cdot \mathbf{A}$ is single-valued. For any n, the magnetic field approached zero asymptotically, because **A** approaches a pure-gauge form. The integer n labels different "gauge types", which cannot be transformed into one another through continuous gauge transformations. This shows that magnetic flux confined to a finite portion of the x-y plane can make its presence known at infinity.

This fact underlies the Aharonov-Bohm effect[9], namely, an electron is scattered in different manners by an impenetrable cylinder depending on the magnetic flux trapped entirely inside, even though the electron never enters the field region. The reason is that the phase of the electron wave function has an angular dependence correlated with the non-vanishing pure-gauge vector potential outside. The phase depends only on the total flux inside, and is therefore gauge invariant.

The flux quantization condition (3.49) arises from energetic considerations. One can always prepare an initial state that does not obey this condition; but then surface currents will be induced to adjust the flux to quantized values eventually. It should also be noted that (3.49) applies only to cylindrical geometry and does not hold, for example, for magnetic flux trapped inside a torus.[10]

In the classical electrodynamics of charged particles, a knowledge of $F^{\mu\nu}$ completely determines the properties of the system. A knowledge of A^μ is redundant there, because it is determined only up to gauge transformations,

---

[9] Y. Aharonov and D. Bohm, *Phys. Rev.* **115**, 485 (1959).
[10] K. Huang and R. Tipton, *Phys. Rev.* **23**, 3050 (1981), Appendix.

which do not affect $F^{\mu\nu}$. As we have seen, such is not the case in quantum theory, in which charged fields are coupled directly to A^{μ}; a knowledge of $F^{\mu\nu}$ is not enough here. Although the continuous gauge transformations of A^{μ} remain physically irrelevant, discontinuous gauge transformations can generate different gauge types that give rise to different physical effects, the gauge invariance of $F^{\mu\nu}$ notwithstanding. The complete information that specifies the system, consists of $F^{\mu\nu}$ plus a specification of the gauge type.

3.6 Soliton Solutions: Vortex Lines

A soliton solution is a classical finite-energy solution whose energy density remains non-vanishing in a finite region of space. In the three-dimensional case, the only possible solitons correspond to solutions in which magnetic flux is trapped inside a close tube topologically equivalent to a torus[11]. In the two-dimensional case (taken in the sense of a z-independent three-dimensional case), the solitons are "vortex lines", configurations in which quantized magnetic flux is trapped inside a linear tube of finite radius, with the Higgs field assuming the normal value ϕ_0 outside the tube, but decreasing to zero towards the axis of the tube. Such vortex lines have been experimentally observed in superconducting bodies, with the ends of a vortex line attached to surfaces of the body.[12]

We set up cylindrical coordinates (r, θ) in the x-y plane, and seek static solutions to the equations of motion in Coulomb gauge ($\nabla \cdot \mathbf{A} = 0$) subject to the boundary conditions (3.50). The static equations of motion read

$$(\nabla^2 - 2e^2\phi^*\phi)\mathbf{A} = -i\phi^* \overset{\leftrightarrow}{\nabla} \phi,$$
$$(\nabla - ie\mathbf{A})^2\phi = 2\lambda\phi(\phi^*\phi - \phi_0^2), \quad (3.51)$$

in which we have set $A^0 = 0$. To look for cylindrically symmetric solutions, put

$$\mathbf{A}(r, \theta) = \hat{\boldsymbol{\theta}}A(r),$$
$$\phi(r, \theta) = \rho(r)e^{in\theta}. \quad (3.52)$$

The magnetic field is given by

$$\mathbf{B} = \nabla \times \mathbf{A} = \hat{\mathbf{z}}\frac{1}{r}\frac{d}{dr}[rA(r)]. \quad (3.53)$$

Now put

$$A(r) = \frac{n}{er}[1 - F(r)],$$
$$B(r) = \frac{n}{er}F'(r), \quad (3.54)$$

---

[11] K. Huang and R. Tipton, *op. cit.*
[12] K. Mendelsohn, *The Quest for Absolute Zero* (Taylor and Francis, London, 1977).

Maxwell Field: U(1) Gauge Theory

where a prime (') denotes differentiation with respect to r. Then (3.51) can be rewritten in the form

$$F'' - \frac{F'}{r} - 2e^2\rho^2 F = 0,$$

$$\rho'' + \frac{\rho'}{r} - \frac{n^2 F^2}{r^2}\rho - 2\lambda\rho(\rho^2 - \phi_0^2) = 0, \tag{3.55}$$

with the boundary conditions

$$F \xrightarrow[r\to\infty]{} 0,$$
$$\rho \xrightarrow[r\to\infty]{} \phi_0. \tag{3.56}$$

Flux quantization implies

$$2\pi \int_0^\infty dr\, r\, B(r) = \frac{2\pi n}{e}, \tag{3.57}$$

or

$$F(0) = 1 \quad (\text{for } n \neq 0).$$

Near $r = 0$, the solutions are

$$F \xrightarrow[r\to 0]{} 1 - O(r^2),$$
$$B \xrightarrow[r\to 0]{} \text{constant}, \tag{3.58}$$
$$\rho \xrightarrow[r\to 0]{} ar^\mu.$$

For $r \to \infty$, we may set $\rho = \phi_0$ in (3.55), and obtain the asymptotic forms

$$F(r) \xrightarrow[r\to\infty]{} \text{const.}\, r^{1/2} \exp(-\sqrt{2}e\phi_0 r),$$
$$A(r) \xrightarrow[r\to\infty]{} \frac{n}{er} + O(e^{-r}), \tag{3.59}$$
$$B(r) \xrightarrow[r\to\infty]{} O(e^{-r}).$$

That a solution exists can be shown by a variational principle. For static solutions, the action is proportional to the negative of the energy:

$$\mathscr{E} = \int_0^\infty dr\, r\left[\frac{1}{2}\left(\frac{n}{er}\right)^2 (F')^2 + \left(\frac{n}{r}\right)^2 F^2\rho^2 + (\rho')^2 + V(\rho)\right], \tag{3.60}$$

with the conditions

$$F(0) = 1, \quad F(\infty) = 0,$$
$$\rho(0) = 0, \quad \rho(\infty) = \phi_0. \tag{3.61}$$

Since every term in (3.60) is non-negative, $\mathscr{E}$ has a minimum with respect to variations of F and ρ. The F and ρ that minimize $\mathscr{E}$ are the solutions of lowest energy, with given n, and with the required boundary conditions. The qualitative nature of the solutions can be seen by inspection of (3.60). To minimize $\mathscr{E}$, $(F')^2$ wants to make F as smooth as possible, $(\rho')^2$ wants to make ρ as smooth as possible, and $F^2\rho^2$ wants to be as small as possible. The last condition means that where F is large, ρ wants to be small, and vice versa. Using these facts together with (3.58) and (3.59), we can make a rough sketch of F and ρ, as shown in Fig. 3.3., which shows the qualitative features of a vortex line.

The classical treatment given here may be viewed as the classical limit in which the trapped flux becomes very strong. A complete quantum mechanical treatment presents difficulties at present.

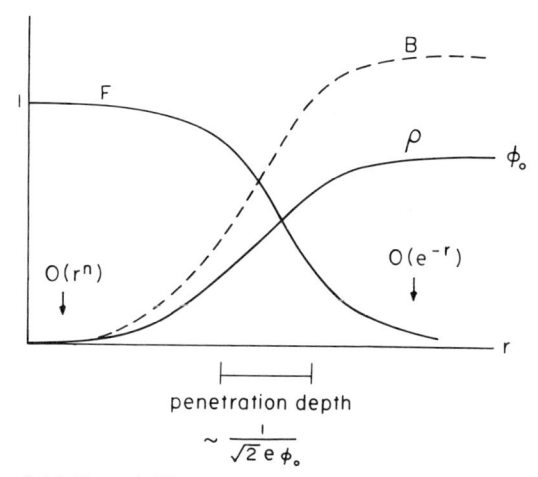

Fig. 3.3 Magnetic field B and Higgs field ρ in a vortex line, as function of normal distance r from axis of vortex.

CHAPTER IV
YANG-MILLS FIELDS: NON-ABELIAN GAUGE THEORIES

4.1 Introductory Note

The Maxwell field gives us the freedom to assign arbitrary signs to the charge of a particle at different space-time points. Yang-Mills fields play a similar role with respect to quantum numbers associated with higher symmetries. Historically, they were introduced with isospin in mind by Yang and Mills[1], who stated their motivation as follows:

> The conservation of isotopic spin[a] is identical with the requirement of invariance of all interactions under isotopic spin rotation. This means that when electromagnetic interactions can be neglected, as we shall hereafter assume to be the case, the orientation of the isotopic spin is of no physical significance. The differentiation between a neutron and a proton is then a purely arbitrary process. As usually conceived, however, once one chooses what to call a proton, what to call a neutron, at one space-time point, one is then not free to make any choices at other space-time points.
> It seems that this is not consistent with the localized field concept that underlies the usual physical theories . . .

We now believe that isospin is not a gauge symmetry, and cannot be, because it is not an exact symmetry in nature. But the idea of Yang and Mills, as applied to other internal symmetries, has led to the current gauge theories of interactions. The relevant internal symmetries are associated with Lie groups.

4.2 Lie Groups[2]

1 Structure Constants

For our purpose, a *Lie group* is a continuous group generated by a *Lie algebra*, which is a space whose basis consists of N generators L_a ($a = 1, \ldots,$

---

[a] Isotopic spin is the old name for isospin.
[1] C. N. Yang and R. L. Mills, *Phys. Rev.* **96**, 191 (1954).
[2] For general reference, see W. Miller, *Symmetry Groups and Their Applications* (Academic Press, New York, 1972); R. Gilmore, *Lie Groups, Lie Algebras, and Some of Their Applications* (Wiley-Interscience, New York, 1974).

N) that are closed under commutations:

$$[L_a, L_b] = iC^c_{ab}L_c. \tag{4.1}$$

Here we use the summation convention for repeated indices. The constants C^c_{ab} are real numbers called *structure constants*, which completely characterize the Lie algebra. An element of G is of the form[b]

$$U = e^{-i\omega_a L_a}, \tag{4.2}$$

where ω_a are arbitrary real numbers.[c]

The structure constants are not unique, for they change under a linear transformation of the generators, which does not change the group. Some properties, however, are invariant. First, it is obvious from (4.1) that

$$C^c_{ab} = -C^c_{ba}. \tag{4.3}$$

Secondly, the *Jacobi identity* for commutators

$$[[L_a, L_b], L_c] + [[L_b, L_c], L_a] + [[L_c, L_a], L_b] \equiv 0 \tag{4.4}$$

requires the identity

$$C^n_{ab}C^d_{nc} + C^n_{bc}C^d_{na} + C^n_{ca}C^d_{nb} \equiv 0, \tag{4.5}$$

which we shall also refer to as the Jacobi identity.

If all the structure constants vanish, G would be an Abelian group—the direct product of $NU(1)$ groups. If not all the structure constants are zero, G is non-Abelian. The smallest non-Abelian Lie group is $SU(2)$, with $N = 3$.

If, after making a linear transformation on $\{L_a\}$ if necessary, the index a can be divided into two sets, such that $C^c_{ab} = 0$ whenever a belongs to one set and b to the other, then the Lie algebra breaks up into two commuting subalgebras. In this case G is a direct product of two independent Lie groups. A non-Abelian Lie group that cannot be so factorized is called *simple*. [Note that $U(1)$ is excluded]. It is characterized by the property that any two generators can be connected to each other through a chain of commutations. A direct product of simple Lie groups is called *semi-simple*.

2 Matrix Representations

We represent the generators by matrices, thereby inducing a matrix representation of G. To avoid needless complications, we consider only finite-dimensional matrices[d]. If $\{L_a\}$ satisfies (4.1), so does the hermitian conjugate set $\{L_a^\dagger\}$. Hence, it is possible to represent $\{L_a\}$ by finite hermitian matrices. Group elements are then represented by finite unitary matrices.

---

[b] That these form a group is a non-trivial fact, which depends on the Baker-Hausdorf-Campbell theorem (see Ref. 2).
[c] We adopt a narrow definition of a Lie group that would suffice for later applications. To mathematicians, a Lie group is much more general, i.e., a continuous group that can be parametrized analytically.
[d] This excludes the Lorentz group from our discussions, because its unitary representations are all infinite-dimensional. On the other hand, its finite-dimensional representations are not unitary, and are thus excluded by (4.2).

Yang-Mills Fields: Non-Abelian Gauge Theories

The faithful representation of lowest dimensionality is called the *fundamental representation*. Other representations may be obtained by taking the repeated direct products of this representation with itself. An example of this procedure was discussed in Sec. 2.2 for the group $SU(3)$.

There is always one irreducible representation (not necessarily faithful) that is completely determined by the structure constants, namely, the *adjoint representation*. Its dimensionality is N (the number of generators), and the representative of L_a is given by

$$(L_a)_{bc} = -iC_{ab}^c. \tag{4.6}$$

This representation will play a central role in Yang-Mills gauge theory. To show that this is a representation, we calculate the commutator $[L_a, L_b]$:

$$\begin{aligned}[][L_a, L_b]_{cd} &= (L_a)_{cn}(L_b)_{nd} - (L_b)_{cn}(L_a)_{nd} \\ &= -C_{ac}^n C_{bn}^d + C_{bc}^n C_{an}^d = C_{ca}^n C_{bn}^d + C_{bc}^n C_{an}^d \\ &= -C_{ab}^n C_{nc}^d \quad \text{(by the Jacobi identity)} \\ &= C_{ab}^n i(L_n)_{cd}. \quad \blacksquare \end{aligned} \tag{4.7}$$

A simple Lie algebra can be made a metric space by defining the scalar product between L_a and L_b by

$$g_{ab} \equiv \text{Tr}(L_a L_b)_{\text{Adj. rep.}} = -C_{an}^m C_{bm}^n, \tag{4.8}$$

which is clearly a symmetric tensor. The norm of L_a is positive definite (i.e., $g_{aa} > 0$) because L_a is hermitian. Hence, by a linear transformation of the generators, g_{ab} can be diagonalized, and all the eigenvalues made unity:

$$g_{ab} = \delta_{ab}. \tag{4.9}$$

Defining new structure constants C_{abc} by

$$C_{abc} \equiv C_{ab}^n g_{nc}, \tag{4.10}$$

we can show that C_{abc} is completely antisymmetric in the indices a, b, c, by using the Jacobi identity. Using (4.9), we obtain

$$C_{ab}^c = C_{abc}. \tag{4.11}$$

Therefore, C_{ab}^c is completely antisymmetric in a, b, c. We shall write C_{abc} in place of C_{ab}^c from now on.

For any representation of a *simple* Lie group, the statements (4.8) and (4.9) can be generalized to

$$\text{Tr}(L_a L_b) = K\delta_{ab} \quad \text{(for simple Lie group)}, \tag{4.12}$$

where K depends on the representation, but not on a. The proof depends on (4.11). First, note that the tensor $\text{Tr}(L_a L_b)$ can always be diagonalized by a suitable choice of $\{L_a\}$, so that

$$\text{Tr}(L_a L_b) = \begin{cases} 0 & \text{if } a \neq b \\ K_a & \text{if } a = b. \end{cases} \tag{4.13}$$

We only need to show that K_a is independent of a. To do this, define

$$d_{abc} \equiv \text{Tr}\{[L_a, L_b]L_c\} = \text{Tr}(L_a L_b L_c) - \text{Tr}(L_b L_a L_c). \tag{4.14}$$

Obviously, this is completely antisymmetric in a, b, c. Using (4.1) and (4.13) successively, we can write

$$d_{abc} = iC_{abn}\,\text{Tr}(L_n L_c) = iC_{abc}K_c \quad \text{(no sum on } c\text{)}. \tag{4.15}$$

Interchanging the indices b and c, we have

$$d_{acb} = iC_{acb}K_b \quad \text{(no sum on } b\text{)}. \tag{4.16}$$

Since $d_{abc} = -d_{acb}$, $C_{abc} = -C_{acb}$, on comparing (4.15) and (4.16), we conclude that

$$K_c = K_b, \tag{4.17}$$

whenever $[L_c, L_b] \neq 0$. Since the group is simple, any two generators are connected by a chain of commutations. Therefore all K_a's are equal to one another, thus proving (4.11).

Finally, we note that the hermitian matrices $\{L_a\}$ can be replaced by real antisymmetric matrices $\{T_a\}$, defined by

$$T_a \equiv -iL_a. \tag{4.18}$$

The commutation relations read

$$[T_a, T_b] = -C_{abc}T_c, \tag{4.19}$$

and the adjoint representation is

$$(T_a)_{bc} = -C_{abc}. \tag{4.20}$$

Unitary matrices representing group elements take the form

$$U = e^{\omega_a T_a}, \quad U^{-1} = e^{-\omega_a T_a}. \tag{4.21}$$

3 Topological Properties

In a matrix representation, G is parametrized by $\{\omega_a\}$. In order that there be a one-to-one correspondence between $\{\omega_a\}$ and a group element, the possible values of $\{\omega_a\}$ must be suitably restricted. The space of the possible values of $\{\omega_a\}$ is called the *group manifold* of G. The group is called *compact* if the group manifold is a compact set. It is said to be *simply-connected* if every closed path in the group manifold can be continuously deformed to a point. Otherwise it is *multiply-connected*.

The correspondence between Lie group and Lie algebra is many-to-one. For example, $SU(2)$ and the rotation group $O(3)$ are different groups, but they share the same Lie algebra. The difference between the two groups is that $SU(2)$ is simply-connected, while $O(3)$ is doubly-connected.

Among all Lie groups sharing the same Lie algebra, only one is simply-connected, and this is called the *covering group*. Thus, $SU(2)$ is the covering group of $O(3)$.

Yang-Mills Fields: Non-Abelian Gauge Theories

To illustrate these ideas, we discuss $SU(2)$ and $O(3)$ in more detail. They are the only two groups sharing the Lie algebra characterized by $C_{abc} = \varepsilon_{abc}$, and both are compact groups. It is well-known that $O(3)$ admits only integer angular momentum representations, whereas $SU(2)$ admits integer and half-integer representations. The difference arises from the topology of the respective group manifolds, as determined by the definitions of these groups.

We can associate a rotation with the tip of a 3-dimensional vector that points along the axis of rotation, with length equal to the angle of rotation. Thus, the manifold of $O(3)$ is the volume enclosed by a sphere of radius π, with diametrically opposite points on the surface identified as the same point. The last condition comes from the fact that rotations of $\pm\pi$ about the same axis are one and the same. There are two classes of closed paths in the group manifold: closed loops drawn within the sphere, and diameters of the sphere. The former can be continuously deformed to a point, while the latter cannot. Hence $O(3)$ is doubly-connected.

On the other hand, $SU(2)$ is defined as the group generated by the fundamental representation of the Lie algebra, i.e., by 2×2 matrices. Its most general element is of the form

$$U = e^{i\theta \hat{n} \cdot \sigma/2} = b_0 + i\mathbf{b} \cdot \boldsymbol{\sigma},$$
$$b_0 \equiv \cos \theta/2, \tag{4.22}$$
$$\mathbf{b} \equiv \hat{n} \sin \theta/2.$$

Hence, the group is parametrized by the four numbers $b_0, \mathbf{b}$, with the condition

$$\sum_{i=1}^{4} b_i^2 = 1. \tag{4.23}$$

The manifold of $SU(2)$ is therefore the surface of a unit sphere in 4-dimensional Euclidean space, and is obviously simply-connected.

The difference between $SU(2)$ and $O(3)$ can be illustrated by the following experiment. Hang a sign from the ceiling with rubber bands, and anchor it to the floor with rubber bands, as shown in Fig. 4.1(a). Rotating the sign about a vertical axis through angle 2π produces a twisting of the rubber bands that cannot be unravelled by deforming the rubber bands continuously [see Fig. 4.1(b)]. However, the twisting produced by a 4π rotation can be so unravelled [see Fig. 4.1(c)]. Fig. 4.2 shows how.

This experiment illustrates the fact that a 2π rotation changes the relationship between the rotated object and its surroundings, even though the object is returned to its original orientation. Only a 4π rotation truly changes nothing. The group $SU(2)$ distinguishes between a 2π rotation and no rotation, whereas $O(3)$ identifies them by ignoring the rubber bands. Stated more precisely, $SU(2)$ contains a center Z_2, which is the subgroup (consisting of the two elements ± 1) that commutes with all group elements. The group $O(3)$ is the factor group with Z_2 taken to be the identity element, i.e., $O(3) = SU(2)/Z_2$.

As another example, consider $U(1)$. Since its elements have the form $e^{i\theta}$, its manifold is the unit circle, which is compact. However, the manifold is not simply connected. There are an infinite number of classes of closed paths labelled by the number of times the path winds around the circle, and paths with different winding numbers cannot be deformed into each other continuously. The covering group is of course simply-connected; but it is not compact because its manifold is the real line.

4 General Remarks

We have confined our discussion to groups of finite unitary matrices. Some of the properties we mentioned are in fact more general. For example, the complete antisymmetry of C_{ab}^c, which follows from (4.11), is true for any compact semi-simple Lie group. The general definition of a semi-simple Lie

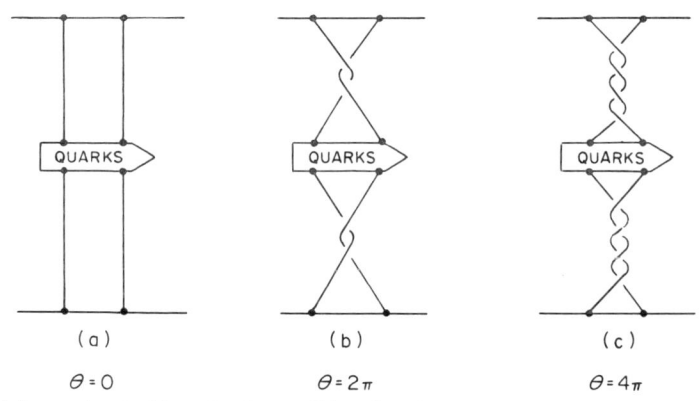

Fig. 4.1 (a) a sign held up by four rubber bands
      (b) The sign is rotated about a vertical axis through angle $\theta = 2\pi$. The twists in the rubber bands cannot be undone by continuous deformations of the rubber bands.
      (c) The sign is rotated through angle $\theta = 4\pi$. The situation is now topologically equivalent to (a), as shown in the next figure.

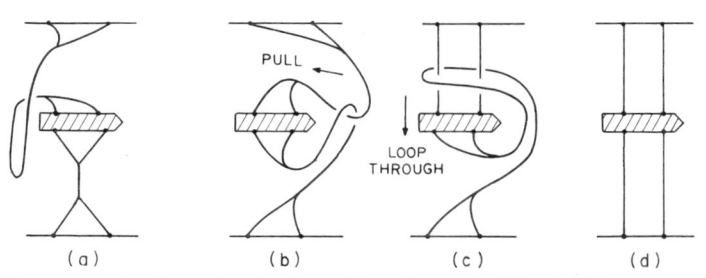

Fig. 4.2 How to undo the twists in the rubber bands in Fig. 4.1(c).

group is one whose Lie algebra does not contain an Abelian sub-algebra. Cartan's theorem states that a Lie group is semi-simple if and only if $\det\|g_{ab}\| \neq 0$, where g_{ab} is defined by (4.8). Hence, for a semi-simple Lie group, g_{ab} can be diagonalized, and all the eigenvalues can be made $+1$ or -1 by re-scaling L_a. If the group is compact, they can all be made $+1$, and (4.11) follows.

As we shall see in the next section, the necessary and sufficient conditions for the construction of a Yang-Mills gauge theory are the complete antisymmetry of C^c_{ab}, and the Jacobi identity. This means that the Lie group has to be a compact semi-simple group, and the Jacobi identity should hold. The Jacobi identity is automatic for a Lie algebra of finite matrices; but it can fail for infinite matrices.

4.3 The Yang-Mills Construction

1 Global Gauge Invariance

Yang-Mills fields are required when we enlarge global gauge symmetries to local ones. We begin with a description of globally gauge invariant systems.

A matrix representation of the Lie group element U is a generalization of the $U(1)$ element $e^{i\theta}$. We continue to refer to it as a gauge transformation (of the representational space on which it acts). If the matrix elements of U are space-time independent, U is called a global gauge transformation. Otherwise it is called a local gauge transformation.

A system of fields (called matter fields) that is globally gauge invariant generally contains multicomponent fields, which we denote collectively by $\Psi(x)$. The various components are grouped into multiplets that transform according to definite irreducible representations of the Lie group G. In general, these irreducible representations are different for different multiplets.

A multiplet may be boson or fermion; but a boson and a fermion field cannot be included in the same multiplet, because they respectively obey commutation and anticommutation rules in the quantum theory, and so cannot transform into each other.

Let the Lagrangian density be $\mathscr{L}_0(\Psi(x), \partial^\mu \Psi(x))$. Then, global gauge invariance states that

$$\mathscr{L}_0(U\Psi, \partial^\mu U\Psi) = \mathscr{L}_0(\Psi, \partial^\mu \Psi). \tag{4.24}$$

There are requirements arising from Lorentz invariance:

(a) Fermion fields ψ (of spin 1/2) are complex Dirac spinor fields. The canonical conjugate of ψ is $i\psi^\dagger$, i.e. ψ appears in L_0 in the form $i\psi^\dagger\psi$.

(b) Boson fields ϕ may be real or complex. The canonical conjugate to ϕ is $\dot\phi^*$, i.e., $\dot\phi$ appears in L_0 in the form $\dot\phi^*\dot\phi$.

The two independent parts ϕ and ϕ^* of a complex boson field may be replaced by a pair of real fields, the real and imaginary parts of ϕ. Correspondingly, the representative of an element of G may be expressed either in complex form or real form. As an example, consider a complex boson field forming an n-dimensional irreducible representation of G. There are $2n$ independent fields, which can be represented as two complex column vectors:

$$\phi = \begin{pmatrix} \phi_1 \\ \vdots \\ \phi_n \end{pmatrix}, \qquad \phi^* = \begin{pmatrix} \phi_1^* \\ \vdots \\ \phi_n^* \end{pmatrix}. \tag{4.25}$$

An infinitesimal gauge transformation is of the form

$$\begin{aligned} \phi &\to \phi + \delta\phi, & \delta\phi &= -i\omega\phi, \\ \phi^* &\to \phi^* + \delta\phi^*, & \delta\phi^* &= i\omega^T\phi, \end{aligned} \tag{4.26}$$

where ω is an element of the Lie algebra:

$$\omega \equiv \omega_a L_a, \tag{4.27}$$

where ω_a are infinitesimal real numbers, and L_a is an $n \times n$ hermitian matrix representing a generator of G. ω^T denotes the transpose of ω.

Alternatively, we may consider the independent fields to be the real and imaginary parts of ϕ:

$$\phi = 2^{-1/2}(A + iB), \tag{4.28}$$

where A and B are both n-component real fields. From (4.26), we find easily that under an infinitesimal gauge transformation,

$$A \to A + \delta A, \qquad \delta A = -\frac{i}{2}(\omega - \omega^T)A + \frac{1}{2}(\omega + \omega^T)B,$$

$$B \to B + \delta B, \qquad \delta B = -\frac{1}{2}(\omega + \omega^T)A - \frac{i}{2}(\omega - \omega^T)B. \tag{4.29}$$

These can be stated in terms of one real field of $2n$ components:

$$\Phi = \begin{pmatrix} A \\ B \end{pmatrix}$$

$$\Phi \to \Phi + \delta\Phi, \qquad \delta\Phi = \omega_a T_a \Phi, \tag{4.30}$$

$$T_a = -\begin{pmatrix} \operatorname{Im} L_a & \operatorname{Re} L_a \\ -\operatorname{Re} L_a & \operatorname{Im} L_a \end{pmatrix} \quad (2n \times 2n \text{ matrix}).$$

As an illustrative example of how the various fields appear in the Lagrangian density, let us take G to be the isospin group $SU(2)$, and choose the matter fields to consist of the following multiplets:

$$\Psi = \{\pi, K, N, \bar{N}, \Sigma, \bar{\Sigma}\}, \tag{4.31}$$

with transformation properties given in Table 4.1. A free Lagrangian density with global isospin invariance is

$$\begin{aligned} \mathcal{L}_0(\Psi, \partial^\mu \Psi) &= \tfrac{1}{2}\partial_\mu \pi_a \partial^\mu \pi_a + \partial_\mu K^* \partial^\mu K \\ &\quad + \bar{N}(i\gamma_\mu \partial^\mu - m)N + \bar{\Sigma}(i\gamma_\mu \partial^\mu - M)\Sigma. \end{aligned} \tag{4.32}$$

Yang-Mills Fields: Non-Abelian Gauge Theories 69

2 Local Gauge Invariance

Under a local gauge transformation,

$$\Psi(x) \to U(x)\Psi(x),$$
$$\partial^\mu \Psi(x) \to U(x)\partial^\mu \Psi(x) + [\partial^\mu U(x)]\Psi(x). \tag{4.33}$$

Since $\partial^\mu U(x) \neq 0$, $\partial^\mu \Psi$ does not transform in the same manner as Ψ, and this spoils the invariance of $\mathcal{L}_0$. To learn how to cancel the unwanted term $(\partial^\mu U)\Psi$, it suffices to consider infinitesimal local gauge transformations:

$$\delta\Psi(x) = -i\omega(x)\Psi(x),$$
$$\delta[\partial^\mu \Psi(x)] = -i\omega(x)\partial^\mu \Psi(x) - i[\partial^\mu \omega(x)]\Psi(x), \tag{4.34}$$

where $\omega(x)$ is an element of the Lie algebra. We define the covariant derivative by

$$D^\mu \Psi(x) \equiv [\partial^\mu + igA^\mu(x)]\Psi(x), \tag{4.35}$$

where $A^\mu(x)$ is an element of the Lie algebra:

$$A^\mu(x) \equiv A_a^\mu(x) L_a. \tag{4.36}$$

Thus, we need N gauge fields $A_a^\mu(x)$ $(a = 1, \ldots, N)$ called Yang-Mills fields. Under an infinitesimal local gauge transformation,

$$D^\mu \Psi \to [\partial^\mu + ig(A^\mu + \delta A^\mu)](\Psi + \delta\Psi). \tag{4.37}$$

Note that $\delta\Psi = -i\omega\Psi$ does not commute with $A^\mu + \delta A^\mu$, because ω and

Table 4.1 EXAMPLES OF SU(2) MULTIPLETS

	Boson		Fermion	
Fields	$\pi = \begin{pmatrix} \pi_1 \\ \pi_2 \\ \pi_3 \end{pmatrix}$	$K = \begin{pmatrix} K^0 \\ \bar{K}^0 \end{pmatrix}$	$N = \begin{pmatrix} p \\ n \end{pmatrix}$	$\Sigma = \begin{pmatrix} \Sigma^+ \\ \Sigma^0 \\ \Sigma^- \end{pmatrix}$
			$\bar{N} = \begin{pmatrix} \bar{p} \\ \bar{n} \end{pmatrix}$	$\bar{\Sigma} = \begin{pmatrix} \bar{\Sigma}^+ \\ \bar{\Sigma}^0 \\ \bar{\Sigma}^- \end{pmatrix}$
Isospin	1	1/2	1/2	1
Generators L_a $(a = 1, 2, 3)$	$(L_a)_{bc} = -i\varepsilon_{abc}$ Adjoint representation	$L_a = \tfrac{1}{2}\tau_a$ Fundamental representation	$L_a = \tfrac{1}{2}\tau_a$	$(L_a)_{bc} = -i\varepsilon_{abc}$
Remarks	π_a $(a = 1, 2, 3)$ are real numbers	$K^0, \bar{K}^0$ are complex numbers	Each entry in the column vector is a Dirac spinor with 4 complex components	

$A^\mu + \delta A^\mu$ are elements of the Lie algebra. The change in $D^\mu \Psi$ is given by

$$\delta(D^\mu \Psi) = -i\omega D^\mu \Psi + ig\left\{\delta A^\mu - \frac{1}{g}\partial^\mu \omega + i[\omega, A^\mu]\right\}. \tag{4.38}$$

To make $D^\mu \Psi$ transform in the same manner as Ψ, we require that the last term vanish, namely,

$$\delta A^\mu(x) = \frac{1}{g}\partial^\mu \omega(x) - i[\omega(x), A^\mu(x)]. \tag{4.39}$$

Multiplying both sides by L_a, taking the trace, and using (4.12), we obtain

$$\delta A_a^\mu(x) = \frac{1}{g}\partial^\mu \omega_a(x) + C_{abc}\omega_b(x)A_c^\mu(x), \tag{4.40}$$

where we have made use of the antisymmetry of C_{abc}. This makes $\mathcal{L}_0$ invariant under local gauge transformations.

To make the fields A_a^μ dynamical objects, we still have to add to $\mathcal{L}_0$ their free Lagrangian density, which should be quadratic in space-time derivatives of A_a^μ, and should be both gauge invariant and Lorentz invariant. In analogy with the Maxwell case, we may try to define a field tensor $\partial^\mu A_a^\nu - \partial^\nu A_a^\mu$; but this transforms in a complicated way under (4.40). We seek a field tensor that transforms according to an irreducible representation of G that is solely determined by C_{abc}. Hence, the representation must be the adjoint representation. The following satisfies our requirements:

$$F_a^{\mu\nu}(x) \equiv \partial^\mu A_a^\nu(x) - \partial^\nu A_a^\mu(x) - gC_{abc}A_b^\mu(x)A_c^\nu(x). \tag{4.41}$$

The corresponding element of the Lie algebra, $F^{\mu\nu} \equiv F_a^{\mu\nu} L_a$, is given by

$$F^{\mu\nu}(x) = \partial^\mu A^\nu(x) - \partial^\nu A^\mu(x) + ig[A^\mu(x), A^\nu(x)]. \tag{4.42}$$

Under an infinitesimal local gauge transformation, it can be verified from (4.42), using the Jacobi identity, that

$$\begin{aligned}\delta F^{\mu\nu}(x) &= -i[\omega(x), F^{\mu\nu}(x)], \\ \delta F_a^{\mu\nu}(x) &= C_{abc}\omega_b(x)F_c^{\mu\nu}(x).\end{aligned} \tag{4.43}$$

From (4.6) and the antisymmetry of C_{abc}, we see that $C_{abc} = -i(L_b)_{ac}$. Hence

$$\delta F_a^{\mu\nu} = -i(\omega_b L_b)_{ac} F_c^{\mu\nu}. \tag{4.44}$$

This shows that $F_a^{\mu\nu}$ transforms according to the adjoint representation.

We have shown that the Jacobi identity and the complete antisymmetry of C_{abc} are necessary for (4.42) to lead to (4.43). Glashow and Gell-Mann[3] have shown that the same conditions are also necessary for the converse, i.e., that (4.43) leads to (4.42). Therefore, $F_a^{\mu\nu}$ is uniquely given by (4.42).

---

[3] S. L. Glashow and M. Gell-Mann, *Annals of Phys.*, **15**, 437 (1961).

Yang-Mills Fields: Non-Abelian Gauge Theories

We now take the free Lagrangian density of the gauge field to be $-\frac{1}{4}F_a{}^{\mu\nu}F_{a\mu\nu}$, which is gauge invariant (see Table 4.2). The factor 1/4 is conventional.

The complete locally gauge invariant Lagrangian density is

$$\mathscr{L} = -\tfrac{1}{4}F_a{}^{\mu\nu}F_{a\mu\nu} + \mathscr{L}_0(\Psi, D^\mu\Psi). \qquad (4.45)$$

This is the Yang-Mills Lagrangian density[e]. The necessary and sufficient conditions for this construction to be possible are that.
 (a) the Jacobi identity holds,
 (b) C_{abc} is completely antisymmetric.
Condition (b) requires G to be a compact semi-simple Lie group (see Sec. 4.2). Condition (a) is automatic for finite matrices, but may not hold for infinite matrices.

In the definition (4.41), the only arbitrary parameters are g, and the relative normalization constants between subsets of C_{abc} corresponding to the various simple Lie groups contained in G, the arbitrariness of C_{abc} for a simple Lie group having been fixed by the convention (4.12). The parameter g is called a *gauge coupling constant*. It fixes the relative scale between the gauge fields and the

Table 4.2 LOCAL GAUGE TRANSFORMATIONS

Infinitesimal Transformations

$\omega(x) \equiv \omega_a(x)L_a$

$\delta\Psi_i(x) = -i\omega_a(x)(L_a)_{ij}\Psi_j(x),$

$\delta A_a{}^\mu(x) = \dfrac{1}{g}\partial^\mu\omega_a(x) + C_{abc}\omega_b(x)A_c{}^\mu(x)$

$\delta F_a{}^{\mu\nu}(x) = C_{abc}\omega_b(x)F_c{}^{\mu\nu}(x)$

$\delta\Psi(x) = -i\omega(x)\Psi(x)$

$\delta A^\mu(x) = \dfrac{1}{g}\partial^\mu\omega(x) - i[\omega(x), A^\mu(x)]$

$\delta F^{\mu\nu}(x) = -i[\omega(x), F^{\mu\nu}(x)]$

Finite Transformations

$U(x) \equiv e^{-i\omega(x)}$

$\Psi(x) \to U(x)\Psi(x)$

$A^\mu(x) \to U(x)A^\mu(x)U^{-1}(x) - \dfrac{i}{g}U(x)\partial^\mu U^{-1}(x)$

$F^{\mu\nu}(x) \to U(x)F^{\mu\nu}(x)U^{-1}(x)$

---

[e] This does not include Einstein's theory of gravitation, which requires tensor gauge fields instead of vector gauge fields. The reason is that the Lorentz group is a group of spacetime transformations, so that the index labelling a group generator is also a space-time index.

matter fields. For a simple G, it can be absorbed by redefining the gauge fields to be

$$\tilde{A}_a{}^\mu(x) \equiv g A_a{}^\mu(x). \tag{4.46}$$

If G is not simple, there would be one independent gauge coupling constant for each simple sub-group, and each $U(1)$ sub-group. In general, not all of these can be absorbed simultaneously by re-scaling the gauge fields. From now, we assume that G is simple [and not $U(1)$], for generalizations are straightforward.

The finite forms of the local gauge transformations can be obtained easily from the infinitesimal forms. We give the results in Table 4.2. Note $F_a{}^{\mu\nu}F_{a\mu\nu}$ is proportional to $\text{Tr}(F^{\mu\nu}F_{\mu\nu})$, which is obviously gauge invariant.

The pure-gauge form of $A^\mu(x)$ is

$$A^\mu(x) = -\frac{i}{g} U(x) \partial^\mu U^{-1}(x) \quad \text{(pure gauge)}, \tag{4.47}$$

which is the generalization of $\partial^\mu \omega$ in the $U(1)$ case. It can be easily shown that (4.47) gives $F^{\mu\nu} = 0$. The matrix $U(x)$ is not necessarily obtainable from $U(x) = 1$ by continuous changes. As in the $U(1)$ case, this gives rise to different pure-gauge types.

To express (4.47) in terms of $\omega = i \ln U$, we note that an infinitesimal change in $\omega(x)$ leads to a change in $U(x)$ given by[f]

$$dU \equiv e^{-i(\omega + d\omega)} - e^{-i\omega} = \int_0^1 dt \, e^{-(1-t)i\omega}(-i\,d\omega)e^{-it\omega}. \tag{4.48}$$

Using this in conjunction with (4.47) leads to

$$A^\mu(x) = -\frac{1}{g} \int_0^1 dt \, e^{-it\omega(x)} [\partial^\mu \omega(x)] e^{it\omega(x)} \quad \text{(pure gauge)}. \tag{4.49}$$

4.4 Properties of Yang-Mills Fields

1 Electric and Magnetic Fields

We define electric and magnetic fields $\mathbf{E} \equiv \mathbf{E}_a L_a$ and $\mathbf{B} \equiv \mathbf{B}_a L_a$, which are elements of the Lie algebra, by designating the components of the antisymmetric tensor (or 6-vector) $F^{\mu\nu}$ as follows:

$$F^{\mu\nu} = \begin{pmatrix} 0 & -E^1 & -E^2 & -E^3 \\ E^1 & 0 & -B^3 & B^2 \\ E^2 & B^3 & 0 & -B^1 \\ E^3 & -B^2 & B^1 & 0 \end{pmatrix}, \tag{4.50}$$

---

[f] Owing to the well-known formula

$$e^{A+B} = e^A + \int_0^1 dt \, e^{(1-t)A} B \, e^{tA} + \cdots$$

Yang-Mills Fields: Non-Abelian Gauge Theories

or, (with $k = 1, 2, 3$),
$$E^k = F^{k0},$$
$$B^k = -\tfrac{1}{2}\varepsilon^{kij}F^{ij}, \quad F^{ij} = -\varepsilon^{ijk}B^k. \tag{4.51}$$

In terms of $A^\mu = (A^0, \mathbf{A})$:
$$E^k = \partial^k A^0 - \partial^0 A^k + ig[A^k, A^0],$$
$$B^k = (\partial^l A^m - \partial^m A^l) - ig[A^l, A^m] \quad (k, l, m \text{ cyclic}). \tag{4.52}$$

These matrices are not gauge-invariant, but undergo unitary transformations under a group element U (see Table 4.2). We can also write

$$\mathbf{E}_a = -\nabla A_a - \frac{\partial \mathbf{A}_a}{\partial t} - g C_{abc} \mathbf{A}_b A_c^0,$$
$$\mathbf{B}_a = \nabla \times \mathbf{A}_a + \tfrac{1}{2} g C_{abc} \mathbf{A}_b \times \mathbf{A}_c. \tag{4.53}$$

Note that
$$\nabla \cdot \mathbf{B}_a = \tfrac{1}{2} g C_{abc} \nabla \cdot (\mathbf{A}_b \times \mathbf{A}_c), \tag{4.54}$$

which shows that magnetic charge density can exist. The total magnetic charge is a surface integral at spatial infinity, and there are special solutions (monopole solutions) for which it does not vanish, as we shall discuss in chapter V.

2 Dual Tensor

Define the dual tensor by
$$\tilde{F}^{\mu\nu} \equiv \tfrac{1}{2} \varepsilon^{\mu\nu\alpha\beta} F_{\alpha\beta}, \tag{4.55}$$

which reads, as an antisymmetric tensor in μ, ν:

$$\tilde{F}^{\mu\nu} = \begin{pmatrix} 0 & -B^1 & -B^2 & -B^3 \\ B^1 & 0 & -E^3 & E^2 \\ B^2 & E^3 & 0 & -E^1 \\ B^3 & -E^2 & E^1 & 0 \end{pmatrix}. \tag{4.56}$$

It is obtainable from $F^{\mu\nu}$ by interchanging $\mathbf{B}$ and $\mathbf{E}$. The following are Lorentz invariant quantities:
$$\tfrac{1}{4} F_a{}^{\mu\nu} F_{a\mu\nu} = \tfrac{1}{2}(\mathbf{B}_a \cdot \mathbf{B}_a - \mathbf{E}_a \cdot \mathbf{E}_a) \quad (\text{scalar}),$$
$$\tfrac{1}{4} \tilde{F}_a{}^{\mu\nu} F_{a\mu\nu} = -\mathbf{E}_a \cdot \mathbf{B}_a \quad (\text{pseudoscalar}). \tag{4.57}$$

In the Abelian case, the absence of magnetic current is implied by the *identity* $\partial_\mu \tilde{F}^{\mu\nu} \equiv 0$, which has no dynamical content, but is a direct consequence of the definition $F^{\mu\nu} = \partial^\mu A^\nu - \partial^\nu A^\mu$. In the non-Abelian case here, we can verify from (4.55), using the Jacobi identity, that
$$\partial^\mu \tilde{F}_{\mu\nu} + ig[A^\mu, \tilde{F}_{\mu\nu}] \equiv 0,$$
$$\partial^\mu \tilde{F}_{a\mu\nu} - g C_{abc} A_{b\mu} \tilde{F}_c{}^{\mu\nu} \equiv 0. \tag{4.58}$$

We can also write
$$D^\mu \tilde{F}_{\mu\nu} \equiv 0, \tag{4.59}$$

where

$$D^\mu = \partial^\mu + ig[A^\mu, \ ] \qquad (4.60)$$

is the covariant differentiation appropriate for the adjoint representation.

3 Path Representation of the Gauge Group

In the presence of Yang-Mills fields, each space-time point is associated with an independent choice of coordinate frames in the internal symmetry space of the matter fields. A change in the local frame between the space-time points x and $x + dx$ is correlated with a local gauge transformation of $A^\mu(x)$. The matter-field system does not care about the local frame, i.e., $\Psi(x)$ is locally physically indistinguishable from $U(x)\Psi(x)$. Similarly, local gauge transformations of $A^\mu(x)$ have no physical significance.

When there are no internal symmetries, $\Psi(x)$ is a single real field. The statement that $\Psi(x)$ does not change between x and $x + dx$ is then simply $dx_\mu \partial^\mu \Psi(x) = 0$. If $\Psi(x) = e^{-i\alpha(x)}|\Psi(x)|$ is a complex field coupled to a $U(1)$ gauge field, so that its phase has no physical significance, then for all physical purposes $\Psi(x)$ is constant over dx if $dx_\mu \partial^\mu |\Psi(x)| = 0$, or, $dx_\mu[\partial^\mu + i\partial^\mu \alpha(x)]\Psi(x) = 0$. In this case, we say that $\Psi(x)$ undergoes parallel displacement from x to $x + dx$, and we can restate that condition in terms of the $U(1)$ gauge field:

$$dx_\mu D^\mu \Psi(x) \equiv dx_\mu[\partial^\mu + igA^\mu(x)]\Psi(x) = 0. \qquad (4.61)$$

In the Yang-Mills case, we take over this definition of *parallel displacement*, with $A^\mu(x) = A_a^\mu(x)L_a$. The local frame in internal space is specified by the matrix representing the Lie algebra element $A^\mu(x)$, i.e. by $A_a^\mu(x)$, when the matrix L_a is fixed.

Suppose $\Psi(x)$ undergoes parallel displacement along a path P, which is parametrized by $0 \le s \le 1$. Then at any point x on P,

$$\frac{dx^\mu}{ds}[\partial^\mu + igA^\mu(s)]\Psi(s) = 0, \qquad (4.62)$$

where $\Psi(s) \equiv \Psi(x(s))$, $A^\mu(s) \equiv A^\mu(x(s))$. The solution to the equation is

$$\Psi(s) = T\left[\exp\left(-ig \int_0^s ds'(dx^\mu/ds')A_\mu(s')\right)\right]\Psi(0), \qquad (4.63)$$

where T is a "path-ordering operator" which instructs us to order the matrices $A_\mu(s')$ in increasing order of s', in every term of the power series expansion of the exponential function. Thus, associated with every directed path P is the matrix

$$\Omega(P) \equiv T\exp\left(-ig \int_P dx^\mu A_\mu(x)\right) \qquad (4.64)$$

which is a path-dependent representation of an element of G, the particular representation being determined by the representation of L_a in the Lie algebra

Yang-Mills Fields: Non-Abelian Gauge Theories

element $A^\mu(x) = A_a{}^\mu(x)L_a$. The significance of $\Omega(P)$ is that it is the gauge transformation that takes $\Psi(x_0)$ to $\Psi(x_1)$ along P:

$$\Psi(x_1) = \Omega(P)\Psi(x_0), \tag{4.65}$$

where x_0 and x_1 are the endpoints of $P: x(0) = x_0$, $x(1) = x_1$.

Under a point-wise gauge transformation $U(x)$, $\Omega(P)$ transform as follows:

$$\Omega(P) \to \Omega'(P),$$
$$\Omega'(P) = U(x_1)\Omega(P)U^{-1}(x_0). \tag{4.66}$$

It follows that for a closed path C, Tr $\Omega(C)$ is gauge invariant.

Proof: Let $\Psi'(x) = U(x)\Psi(x)$. The definition of $\Omega'(P)$ is given by

$$\Psi'(x_1) = \Omega'(P)\Psi'(x_0) \tag{4.67}$$

or,

$$U(x_1)\Psi(x_1) = \Omega'(P)U(x_0)\Psi(x_0),$$
$$U(x_1)\Omega(P)\Psi(x_0) = \Omega'(P)U(x_0)\Psi(x_0),$$
$$U(x_1)\Omega(P) = \Omega'(P)U(x_0) \quad \text{(since } \Psi(x_0) \text{ is arbitrary)}.$$
$$\therefore \quad U(x_1)\Omega(P)U^{-1}(x_0) = \Omega'(P). \quad \blacksquare$$

For an infinitesimal closed path at x, of area $dx^\mu\, dy^\nu$,

$$\Omega(dx\, dy) = 1 - ig\, dx^\mu\, dx^\nu F_{\mu\nu}(x). \tag{4.68}$$

Proof: Number the sides of the rectangle $n = 1, 2, 3, 4$ as shown in Fig. 4.3. Along any infinitesimal dx located at x, we have (to second order)

$$\Omega(dx) = 1 - ig\, dx_\mu A^\mu(x) - \tfrac{1}{2}g^2\, dx_\mu\, dx_\nu A^\mu(x)A^\nu(x).$$

Let

$$\lambda_n \equiv \int_{\text{side } n} dx^\mu A_\mu(x).$$

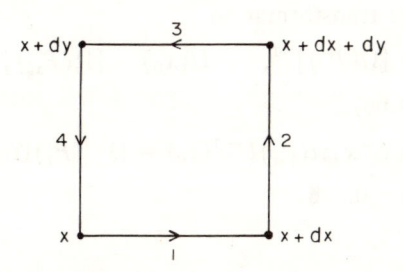

Fig. 4.3 An infinitesimal rectangle in space-time.

By the group property of Ω,

$$\Omega(dx\,dy) = \Omega_1\Omega_2\Omega_3\Omega_4 = 1 - ig(\lambda_1 + \lambda_2 + \lambda_3 + \lambda_4)$$
$$-g^2[\lambda_1\lambda_2 + \lambda_1\lambda_3 + \lambda_1\lambda_4 + \lambda_2\lambda_3 + \lambda_2\lambda_4 + \lambda_3\lambda_4$$
$$+ \tfrac{1}{2}(\lambda_1^2 + \lambda_2^2 + \lambda_3^2 + \lambda_4^2)].$$

Sample side calculations are as follows:

$$\lambda_1 + \lambda_3 = \left(\int_x^{x+dx} - \int_{x+dy}^{x+dx+dy}\right) dz^\mu A_\mu(z)$$
$$= dy^\mu \frac{\partial}{\partial x^\mu} \int_x^{x+dx} dz^\mu A_\mu(z) = dy^\mu\,dx^\nu A_\nu(x).$$

$$\lambda_1 + \lambda_2 + \lambda_3 + \lambda_4 = dx^\mu\,dy^\nu(\partial_\mu A_\nu - \partial_\nu A_\mu).$$

$$g^2[\lambda_1\lambda_2 + \cdots] = -g^2\,dx^\mu\,dy^\nu[A^\mu(x), A^\nu(x)].$$

Hence,

$$\Omega(dx\,dy) = 1 - ig\,dx^\mu\,dy^\nu F_{\mu\nu}(x). \quad\blacksquare$$

A knowledge of $\Omega(P)$ for all paths P determines $A_a{}^\mu(x)$, by differentiation with respect to an endpoint of P. The classical gauge theory can be formulated entirely in terms of $\Omega(P)$;[4] but the canonical quantization is still based on the quantization of $A_a{}^\mu(x)$, which will be done in Chapter VIII. A knowledge of $\Omega(C)$ for all closed loops C contains all the physical information about the classical gauge field without redundancy.[5]

To illustrate a practical use for $\Omega(P)$, we show that *any space-time component of $A_a{}^\mu(x)$ can be reduced to zero by means of a continuous local gauge transformation*. When a spatial component is zero, $A_a{}^\mu(x)$ is said to be in the *axial gauge*. When the time component is zero, it is said to be in the *temporal gauge*. Suppose we wish to make $A_a{}^0(x) = 0$. Let P_x be the space-time path directing linearly from $x_0 = (0, \mathbf{x})$ to $x = (t, \mathbf{x})$. Consider

$$\Omega(P_x) = T\exp\left(-ig\int_0^t dt'A^0(t', \mathbf{x})\right). \tag{4.69}$$

Clearly, $\Omega(P_x) = 1$ for $x = x_0$. If $A_a{}^0(x)$ is not already zero, make the continuous local gauge transformation

$$U(x) = [\Omega(P_x)]^{-1}, \qquad U(x_0) = [\Omega(P_{x_0})]^{-1} = 1. \tag{4.70}$$

Then, according to (4.66),

$$\Omega(P_x) \to U(x)\Omega(P_x)U^{-1}(x_0) = \Omega^{-1}(P_x)\Omega(P_x) = 1, \tag{4.71}$$

which implies $A_a{}^0(x) = 0$. $\blacksquare$

---

[4] C. N. Yang, *Phys. Rev. Lett.*, **33**, 445 (1974).
[5] T. T. Wu and C. N. Yang, *Phys. Rev.* **D12**, 3845 (1975).

4.5 Canonical Formalism

1 Equations of Motion

We consider the classical Lagrangian density

$$\mathcal{L} = -\tfrac{1}{4}F_a{}^{\mu\nu}F_{a\mu\nu} + (D^\mu\phi)^*(D_\mu\phi) - V(\phi)$$
$$+ \bar{\psi}(i\gamma_\mu D^\mu - m)\psi, \qquad (4.72)$$

where ϕ is a set of complex boson fields, ψ is a set of fermion fields, and m is a constant mass matrix. The fields form sets of irreducible representations of the gauge group G. The boson self-interacting term is gauge-invariant:

$$V(\phi) = V(U\phi). \qquad (4.73)$$

For renormalizability of the quantized theory, $V(\phi)$ must be a polynomial in ϕ and ϕ^* of degree no greater than 4. The same consideration rules out fermion self-interactions. Apart from $V(\phi)$, all interactions are mediated by the gauge fields, $A_a{}^\mu$, which are coupled to ϕ and ψ solely through $D^\mu\phi$ and $D^\mu\psi$. We confine our discussions here to the classical theory, leaving quantization to later chapters.

The fields ϕ, ϕ^* can be replaced by the equivalent set Re ϕ, Im ϕ. The only case not covered by (4.72) is that of N real fields ϕ_a transforming according to the adjoint representation, whose free Lagrangian density $\tfrac{1}{2}D^\mu\phi_a D^\mu\phi_a$ can be added on if desired. We also leave out possible Yukawa couplings of the general form $\bar{\psi}_i\phi_j\psi_k$, where i, j, k are indices labelling field components. The allowed combinations of i, j, k depend on the group G, and on the representations to which the fields belong. These Yukawa couplings can contribute to the fermion mass, as we shall see in Chap. VI.

It should be noted that $\mathcal{L}$ may be invariant under a global group larger than G, but that only G is gauged. For example, with $G = SU(2)$, take the boson fields to consist of a triplet $\{\phi_1, \phi_2, \phi_3\}$ and a doublet $\{K_1, K_2\}$, and take

$$V = a(\phi_1{}^2 + \phi_2{}^2 + \phi_3{}^2) + b(K_1{}^*K_1 + K_2{}^*K_2).$$

This is invariant under independent transformations of ϕ and K under $SU(2)$. Hence, the global symmetry group is $SU(2) \times SU(2)$; but the gauged symmetry is the $SU(2)$ group of simultaneous transformations of ϕ and K.

Table 4.3 FIELDS AND CANONICAL CONJUGATES

	Field	Canonical Conjugate
gauge fields	$A_a{}^\nu(x)$	$-F_a{}^{\nu 0}(x) = \begin{cases} -E_a{}^k(x) & (\nu = k = 1, 2, 3) \\ 0 & (\nu = 0) \end{cases}$
scalar fields	$\phi(x)$ $\phi^*(x)$	$[D^0\phi(x)]^* \equiv \pi(x)$ $D^0\phi(x) \equiv \pi^*(x)$
spinor fields	$\psi(x)$	$i\psi^\dagger(x)$

The independent fields and their canonical conjugates for our system are listed in Table 4.3. The equations of motion are

$$D_\mu D^\mu \phi = -\frac{\partial V}{\partial \phi^*}, \quad (D_\mu D^\mu \phi)^* = -\frac{\partial V}{\partial \phi}, \tag{4.74}$$

$$(i\gamma^\mu D_\mu - m)\psi = 0, \quad i(D_\mu \bar\psi)\gamma^\mu + m\bar\psi = 0, \tag{4.75}$$

$$\partial_\mu F_a^{\mu\nu} - gC_{abc}A_{b\mu}F_c^{\mu\nu} = j_a^\nu, \tag{4.76}$$

where j_a^ν is the matter-field current:

$$\begin{aligned}j_a^\nu &= -ig[(D^\nu\phi)^* L_a \phi - \phi^* L_a (D^\nu\phi)] + (\bar\psi \gamma^\nu L_a \psi) \\ &= -ig(\phi^* \overleftrightarrow{\partial}^\nu L_a \phi) - g^2 A_b^\nu \phi^*\{L_a, L_b\}\phi + (\bar\psi\gamma^\nu L_a\psi).\end{aligned} \tag{4.77}$$

The current j_a^ν is gauge-covariant but not conserved. On the other hand, $j_a^\nu + gC_{abc}A_{b\mu}F_c^{\mu\nu}$ is conserved but not gauge-covariant.

We can rewrite (4.76) in the form

$$\partial_\mu F^{\mu\nu} + ig[A_\mu, F^{\mu\nu}] = j^\nu, \tag{4.78}$$

or

$$D_\mu F^{\mu\nu} = j^\nu, \tag{4.79}$$

where D^μ is defined in (4.60), and

$$j^\nu = j_a^\nu L_a. \tag{4.80}$$

We concentrate on the gauge fields, since the behavior of the matter fields is familiar, except for possible spontaneous symmetry breaking due to $V(\phi)$, which we discuss separately later.

The three-vector form of (4.76) is

$$\nabla \cdot \mathbf{E}_a + gC_{abc}\mathbf{A}_b \cdot \mathbf{E}_c = j_a^0, \tag{4.81}$$

$$\nabla \times \mathbf{B}_a - \frac{\partial \mathbf{E}_a}{\partial t} + gC_{abc}(A_b^0 \mathbf{E}_c + \mathbf{A}_b \times \mathbf{B}_c) = \mathbf{j}_a, \tag{4.82}$$

where $\mathbf{E}_a$ and $\mathbf{B}_a$ are defined in (4.53). These are generalizations of Maxwell's equations, but they cannot be expressed solely in terms of $\mathbf{E}_a$ and $\mathbf{B}_a$.

Equation (4.81), which is the generalization of Gauss' Law, does not involve $\partial A_a^0/\partial t$. Thus we can use it to eliminate A_a^0 in terms of the other fields. That this should be possible is indicated by the fact that the canonical conjugate of A_a^0 is identically zero (Table 4.3). It is also evident from the fact that A_a^0 can always be reduced to zero by means of a continuous gauge transformation (Sec. 4.4). If one imposes some initial gauge condition, then (4.81) is a constraint on the initial values of A_a^0, so that the gauge condition is respected by the dynamics. In terms of A_a^0, (4.81) can be rewritten in the form

$$\nabla^2 A_a^0 + M_{ab} A_b^0 + N_a = -\frac{\partial}{\partial t}(\nabla \cdot \mathbf{A}_a) + gC_{abc}(\nabla \cdot \mathbf{A}_b) A_c^0, \tag{4.83}$$

Yang-Mills Fields: Non-Abelian Gauge Theories

where
$$M_{ab} = 2gC_{abc}\mathbf{A}_c \cdot \nabla + g^2 C_{amn}C_{nlb}\mathbf{A}_m \cdot \mathbf{A}_l,$$
$$N_a = gC_{abc}\mathbf{A}_b \cdot \frac{\partial \mathbf{A}_c}{\partial t} - j_a^0. \tag{4.84}$$

If we choose the Coulomb gauge $\nabla \cdot \mathbf{A}_a = 0$ at a particular time, then the condition
$$\nabla^2 A_a^0 + M_{ab}A_b^0 + N_a = 0 \tag{4.85}$$
ensures $\partial(\nabla \cdot \mathbf{A}_a)/\partial t = 0$, so that the Coulomb gauge will be maintained at all times. In the $U(1)$ case, the analogue of (4.85) is $\nabla^2 A^0 = -j^0$, whose general solution leads to the Coulomb potential. In the present case, it is coupled to (4.82), because M_{ab} and N_a depend on $\mathbf{A}_a$. This makes it considerably more complicated, and no general solution is known.

2 Hamiltonian

The Hamiltonian density can be obtained by the standard recipe $\mathcal{H} = p\dot{q} - \mathcal{L}$, where q is an independent field, and p its canonical conjugate:
$$\mathcal{H} = \tfrac{1}{2}(\mathbf{B}_a \cdot \mathbf{B}_a - \mathbf{E}_a \cdot \mathbf{E}_a) - \mathbf{E}_a \cdot \dot{\mathbf{A}}_a + \pi\dot{\phi} + \pi^*\dot{\phi}^*$$
$$- (D^\mu \phi)^*(D_\mu \phi) + V(\phi) + \bar{\psi}\left(\frac{1}{i}\boldsymbol{\alpha} \cdot \mathbf{D} + m\beta\right)\psi. \tag{4.86}$$

The second term can be re-expressed as follows:
$$-\mathbf{E}_a \cdot \dot{\mathbf{A}}_a = \mathbf{E}_a \cdot (\mathbf{E}_a + \nabla A_a^0 + gC_{abc}\mathbf{A}_b A_c^0)$$
$$= \mathbf{E}_a \cdot \mathbf{E}_a + \nabla \cdot (\mathbf{E}_a A_a^0) - (\nabla \cdot \mathbf{E}_a)A_a^0$$
$$+ gC_{abc}(\mathbf{E}_a \cdot \mathbf{A}_b)A_c^0. \tag{4.87}$$

The next term can be re-expressed as follows:
$$\pi\dot{\phi} + \pi^*\dot{\phi}^* - (D^\mu\phi)^*(D_\mu\phi) = \pi(\pi^* - igA_0\phi)$$
$$+ \pi^*(\pi + igA_0\phi^*) - \pi^*\pi + (D\phi)^* \cdot (D\phi)$$
$$= \pi^*\pi + (D\phi)^* \cdot (D\phi) + ig(\pi^*A_0\phi^* - \pi A_0\phi). \tag{4.88}$$

Hence,
$$\mathcal{H} = \tfrac{1}{2}(\mathbf{B}_a \cdot \mathbf{B}_a + \mathbf{E}_a \cdot \mathbf{E}_a) + \pi^*\pi + (D\phi)^* \cdot (D\phi)$$
$$+ V(\phi) + \bar{\psi}\left(\frac{1}{i}\boldsymbol{\alpha} \cdot \mathbf{D} + m\beta\right)\psi + X, \tag{4.89}$$

where
$$X = \nabla \cdot (\mathbf{E}_a A_a^0) - (\nabla \cdot \mathbf{E}_a)A_a^0 + gC_{abc}\mathbf{E}_a \cdot \mathbf{E}_b A_c^0$$
$$+ ig(\pi^*A_0\phi^* - \pi A_0\phi). \tag{4.90}$$

Using (4.81) and (4.77), we find that all the terms after the first cancel one another. Therefore,

$$X = \nabla \cdot (\mathbf{E}_a A_a{}^0), \qquad (4.91)$$

which we can ignore upon integration over all space. Hence, the total Hamiltonian, which is the total energy, is given by

$$H = \int d^3x \left[\frac{1}{2} (\mathbf{B}_a \cdot \mathbf{B}_a + \mathbf{E}_a \cdot \mathbf{E}_a) + \pi^*\pi + (\mathbf{D}\phi)^* \cdot (\mathbf{D}\phi) \right.$$
$$\left. + V(\phi) + \bar{\psi} \left(\frac{1}{i} \boldsymbol{\alpha} \cdot \mathbf{D} + m\beta \right) \psi \right]. \qquad (4.92)$$

The independent gauge fields are $\mathbf{A}_a$, with canonical conjugates $\mathbf{E}_a$, and $\mathbf{B}_a$ is a function of $\mathbf{A}_a$ given by (4.53). All dependencies on $A_a{}^0$ have been eliminated. The Hamiltonian is clearly locally gauge invariant. One can fix the gauge completely so that for each a, only two of the three fields $\mathbf{A}_a$ are independent. The form of H after gauge fixing will depend on the gauge choice.

Quantization of ϕ and ψ is done by imposing the usual commutation or anticommutation relations, which we shall not go into. Quantization of $A_a{}^\mu$, however, is complicated by the need for gauge fixing, and we postpone it to Chap. VIII.

Finally, we mention that the Poincaré invariance of the quantized theory has to be proven by constructing from the field operators infinite-dimensional unitary representations of the Poincaré group, which is a Lie group with an algebra consisting of the ten generators: energy, linear momenta, angular momenta, and Lorentz boosts. This problem has been partially solved by Schwinger[6]. However, a complete solution taking full account of the non-Abelian nature is still lacking. In particular, there are novel features concerning the representation of the angular momentum operators that remain to be fully understood.[7]

4.6 Spontaneous Symmetry Breaking

1 The Little Group

Spontaneous symmetry breaking in the present model means that some spin 0 fields (Higgs fields) have non-vanishing vacuum values; how this comes about depends entirely on the form of $V(\phi)$. We assume that higher-spin fields have zero vacuum values, for otherwise Lorentz invariance would be spontaneously broken, in apparent contradiction to experimental evidence. As we mentioned in Chap. III, the treatment of spontaneous symmetry in terms of a Higgs field provides us with a convenient mathematical description; but in reality the Higgs fields might turn out to be merely phenomenological devices. Here, we treat the problem classically, leaving a discussion of quantum corrections to Chap. X.

---

[6] J. Schwinger, *Phys. Rev.* **127**, 324 (1962).
[7] N. Christ, A. Guth, and E. Weinberg, *Nucl. Phys.* **B114**, 61 (1976); K. Huang and D. Stump, *Phys. Rev.* **D15**, 3660 (1977).

Let the Higgs fields be denoted collectively by ϕ, and suppose that $V(\phi)$ has a lowest minimum at $\phi = \rho$, with the minimum value taken to be zero:

$$V(\rho) = 0,$$
$$V'(\rho) = 0, \qquad (4.93)$$
$$V''(\rho) > 0.$$

Spontaneous symmetry breaking occurs if $\rho \neq 0$. In that event, a vacuum solution (lowest-energy solution) is

$$\phi(x) = \rho,$$
$$A_a{}^\mu(x) = 0, \qquad (4.94)$$

for it clearly satisfies the equations of motion, and has the lowest possible energy.

Just as in the $U(1)$ case, ρ is not unique, because $U(x)\rho$ is equivalent to ρ. But, in contrast to the $U(1)$ case, not all $U(x)\rho$ are independent of one another here. To emphasize this point, let us look at an example.

Take $G = SU(2)$, and $\boldsymbol{\phi} = (\phi_1, \phi_2, \phi_3)$ transforming according to the adjoint representation. Take

$$V(\phi) = \lambda(\phi_1{}^2 + \phi_2{}^2 + \phi_3{}^2 - a^2)^2.$$

Then, $\boldsymbol{\rho} = (\rho_1, \rho_2, \rho_3)$ is any 3-vector satisfying

$$\rho_1{}^2 + \rho_2{}^2 + \rho_3{}^2 = a^2.$$

That is, $\boldsymbol{\rho}$ is a 3-vector whose tip lies on a sphere of radius a. Therefore, $U\boldsymbol{\rho} = \boldsymbol{\rho}$, for any rotation U of the sphere about $\boldsymbol{\rho}$. Choose the x_3 axis to lie along $\boldsymbol{\rho}$. Then the most general rotation that leaves $\boldsymbol{\rho}$ invariant is $[\exp(-i\omega L_3)]\boldsymbol{\rho} = \boldsymbol{\rho}$, or $L_3\boldsymbol{\rho} = 0$. Thus, the vacuum solution is invariant under the $U(1)$ subgroup of $SU(2)$ generated by L_3. In this case, the $SU(2)$ symmetry is said to be spontaneously broken down to $U(1)$.

Returning to the general case, let us specify that ρ is constant (i.e., all components of ρ are independent of x). This is possible because $V(\rho)$ has no explicit x dependence. Still, ρ is not unique, because $U\rho$ will serve equally well as ρ, where U is a global gauge transformation. The set of all elements U of G that leaves ρ invariant forms a subgroup of G (obvious). Since $\rho \neq 0$ by assumption, this subgroup is not G itself, but a proper subgroup. The largest subgroup H that leaves ρ invariant is called the *little group* with respect to ρ. The Lie algebra of the little group consists of a subset $\{l_\alpha\}$ of the Lie algebra $\{L_a\}$ of G, with

$$[l_\alpha, l_\beta] = iC_{\alpha\beta\gamma}l_\gamma. \qquad (4.95)$$

Under an infinitesimal element of H,

$$\delta\rho = -i\omega_\alpha l_\alpha \rho = 0. \qquad (4.96)$$

Since $\{\omega_\alpha\}$ is arbitrary, we have

$$l_\alpha \rho = 0. \qquad (4.97)$$

We say that the symmetry G is spontaneously broken down to H.

The elements of G fall into equivalence classes that are the distinct cosets of G with respect to H, namely H, U_1H, U_2H, ..., where U_iH is the coset $\{U_i u | U_i \in G, u \in H\}$. The independent vectors among $U\rho$ are $\rho, C_1\rho, C_2\rho, \ldots$, where $C_i \in U_iH$. The collection of all cosets, denoted by G/H, is in general not a group. It is a group if and only if H is a normal subgroup of G (i.e., $UH = HU$). In this case, G/H is called the factor group, with the multiplication rule $(U_1H)(U_2H) = (U_1U_2)H$.

The generators of G not in the set $\{l_\alpha\}$ cannot annihilate ρ, by definition. Thus, we can divide $\{L_a\}$ into two disjoint subsets:

$$\{L_a\} = \{L_j, l_\alpha\},$$
$$L_j\rho \neq 0 \quad (j = 1, \ldots, K), \tag{4.98}$$
$$l_\alpha\rho = 0 \quad (\alpha = 1, \ldots, N - K).$$

The choice of $\{L_j\}$ and $\{l_\alpha\}$ depends on the particular gauge for ρ, and hence is non-unique; but the number of generators in each set is gauge-invariant, because

$$l_\alpha\rho = 0 \Rightarrow [U(x) l_\alpha U^{-1}(x)] \, U(x)\rho = 0. \tag{4.99}$$

The generators $\{L_j\}$ ($j = 1, \ldots, k$) generate a group if and only if $\{l_a\}$ is a normal subalgebra.

To proceed further, it will be convenient to represent the generators by real antisymmetric matrices in accordance with (4.18):

$$T_j = -iL_j \quad (j = 1, \ldots, K),$$
$$t_\alpha = -il_\alpha \quad (\alpha = 1, \ldots, N - K). \tag{4.100}$$

We also take ρ to have R real components, so that the representational vector space is a real vector space of dimensionality R. Scalar products in this space are denoted by

$$(f, Og) = \sum_{n=1}^{R} \sum_{m=1}^{R} f_n O_{nm} g_m. \tag{4.101}$$

The real symmetric matrix $(T_i\rho, T_j\rho)$ has positive-definite eigenvalues. Therefore, $T_i\rho$ ($i = 1, \ldots, K$) are independent vectors that span a K-dimensional subspace of the R-dimensional representational vector space. A necessary condition for the little group H to be non-empty is therefore

$$R - K > 0. \tag{4.102}$$

We call the K-dimensional space spanned by $T_i\rho$ ($i = 1, \ldots, K$), the *Goldstone space*, and its complement, of dimensionality $R - K$, the *Higgs space*.

For any vector ϕ in the representational vector space, and for compact gauge group G, there is always a gauge transformation U_0 such that $U_0\phi$ is orthogonal to the Goldstone space:

$$(T_j\rho, U_0\phi) = 0 \quad (j = 1, \ldots, K). \tag{4.103}$$

We say that $U_0\phi$ is in *unitary gauge*. That U_0 exists can be shown as follows.[8]

For given ρ and ϕ, consider a mapping of G to the real line defined by

$$f(U) = (\rho, U\phi).$$

---

[8] S. Weinberg, *Phys. Rev.* **D7**, 1068 (1973).

Yang-Mills Fields: Non-Abelian Gauge Theories

Since G is compact, the values of $f(U)$ lie in a real compact set. Hence $f(U)$ has extrema. Let $f(U_0)$ be an extremum. A small variation of U near U_0 gives

$$\delta f \equiv f(U_0 + \delta U) - f(U_0) = 0.$$

Now any change in a group element U can be written as a left multiplication by another group element. Hence we can write

$$\delta U = \omega_a T_a U_0,$$

where ω_a is arbitrary. Therefore,

$$0 = \delta f = (\rho, \omega_a T_a U_0 \phi) = \omega_j(\rho, T_j U_0 \phi) = -\omega_j(T_j \rho, U_0 \phi).$$

Since ω_j is arbitrary, we have $(T_j \rho, U_0 \phi) = 0$. ∎

The above result holds at each space-time point. If $\phi(x)$ is a solution, it must be a continuous function of x. Therefore, there exists $U_0(x)$, continuous in x, such that $U_0(x)\phi(x)$ is in unitary gauge:

$$\phi(x) \to U_0(x)\phi(x),$$

$$U_0(x)\phi(x) = \begin{pmatrix} 0 \\ \tilde{\phi}(x) \end{pmatrix} \begin{matrix} \text{Goldstone space, } K\text{-dimensional} \\ \text{Higgs space, } (R-K)\text{-dimensional.} \end{matrix} \qquad (4.103)$$

2 Higgs Mechanism

In unitary gauge, the vacuum solution is

$$\phi(x) = \rho = \begin{pmatrix} 0 \\ \tilde{\rho} \end{pmatrix},$$

$$A_a{}^\mu(x) = 0. \qquad (4.104)$$

Solutions with energy near the vacuum solution are of the form

$$\phi(x) = \begin{pmatrix} 0 \\ \tilde{\rho} + \eta(x) \end{pmatrix}, \qquad (4.105)$$

$$A_a{}^\mu(x) \quad \text{small},$$

where $\eta(x)$ and $A_a{}^\mu(x)$ are both small because the energy is a continuous functional of η and $A_a{}^\mu$.

To first order in these small quantities, we have

$$V(\phi) = \tfrac{1}{2}(\eta, V''(\rho)\eta),$$
$$j_a{}^\mu = -g^2(T_a\rho, T_b\rho)A_b{}^\mu. \qquad (4.106)$$

We define the following matrices, which will turn out to be mass matrices:

$$(\mu^2)_{rs} \equiv \begin{array}{|c|c|l} \hline 0 & 0 & \text{Goldstone space} \\ \hline 0 & V''(\rho) & \text{Higgs space} \\ \hline \end{array}$$

$$(M^2)_{ab} \equiv g^2(T_a\rho, T_b\rho) = \begin{array}{|c|c|l} \hline (M^2)_{ij} & 0 & \text{Goldstone space} \\ \hline 0 & 0 & \text{Higgs space} \\ \hline \end{array} \qquad (4.107)$$

Using these, the linearized equations can be written in the form

$$\Box^2 \eta_r + (\mu^2)_{rs}\eta_s = 0 \quad (r = 1, \ldots, R - K),$$
$$\Box^2 A_i{}^\nu + (M^2)_{ij} A_j{}^\nu = 0 \quad (\partial_\mu A_i{}^\mu = 0) \quad (i = 1, \ldots, K), \quad (4.108)$$
$$\Box^2 A_\alpha{}^\nu - \partial^\nu(\partial_\mu A_\alpha{}^\mu) = 0 \quad (\alpha = 1, \ldots, N - K).$$

These respectively describe massive Higgs bosons, massive vector bosons, and massless vector bosons, as summarized in Table 4.4. The massless vector bosons are the gauge particles associated with the unbroken symmetry H.

The number of massive vector bosons is equal to the number of Goldstone bosons that would be present if there were no gauge coupling. As it is, there are no Goldstone bosons; they have been "eaten up" by the massive vector bosons.

If the field ϕ is complex, and we choose not to put it in real form, then the vector meson mass matrix could be represented in the alternative form

$$(M^2)_{ab} = \tfrac{1}{2}g^2 \rho^\dagger \{L_a, L_b\}\rho. \quad (4.109)$$

Examples

We illustrate the group-theoretic aspects of spontaneous symmetry breaking by two examples.

(i) First, consider $G = O(n)$, $\phi = \{\phi_1, \ldots, \phi_n\}$ (fundamental representation), and

$$V(\phi) = \lambda[(\phi_1^2 + \cdots + \phi_n^2) - a^2]^2.$$

We can choose $\rho = \{0, \ldots 0, a\}$. Clearly $H = O(n-1)$. The group $O(n)$ has $\tfrac{1}{2}n(n-1)$ generators, and $O(n-1)$ has $\tfrac{1}{2}(n-1)(n-2)$ generators. Therefore:

no. of massless vector bosons = $\tfrac{1}{2}(n-1)(n-2)$,
no. of massive vector bosons = $\tfrac{1}{2}n(n-1) - \tfrac{1}{2}(n-1)(n-2) = n - 1$,
no. of Higgs bosons = 1.

Table 4.4 HIGGS MECHANISM

G is spontaneously broken down to H
N = No. of generators of G
$N - K$ = No. of generators of H
R = Dimensionality of real representation of G

Field	No. of Fields	No. of indep. Components
η_r (Higgs, massive)	$R - K$	$R - K$
$A_i{}^\mu$ (gauge, massive)	K	$3K$
$A_\alpha{}^\mu$ (gauge, massless)	$N - K$	$2(N - K)$
	Total	$N + R$

(ii) As a second example, consider $G = SU(2)$, with $N = 3$. Suppose the scalar field ϕ consists of a real triplet π and a complex doublet K:

$$\phi = \begin{pmatrix} \pi \\ K \end{pmatrix}, \quad \pi = \begin{pmatrix} \pi_1 \\ \pi_2 \\ \pi_3 \end{pmatrix}, \quad K = \begin{pmatrix} K_1 \\ K_2 \end{pmatrix}.$$

Choose

$$V = \lambda(\pi_1^2 + \pi_2^2 + \pi_3^2 - a^2)^2 + \lambda'(K_1^*K_1 + K_2^*K_2 - b^2)^2.$$

Then,

$$\rho = \begin{pmatrix} \phi_0 \\ K_0 \end{pmatrix}, \quad \phi_0 = \begin{pmatrix} 0 \\ 0 \\ a \end{pmatrix}, \quad K_0 = K_0^* = \begin{pmatrix} 0 \\ b \end{pmatrix}.$$

The generators are represented by

$$L_a = \begin{pmatrix} I_a & 0 \\ 0 & \tfrac{1}{2}\tau_a \end{pmatrix},$$

where I_a are 3×3 matrices for the adjoint representation, and τ_a are 2×2 Pauli matrices. Thus,

$$L_a\rho = \begin{pmatrix} I_a\phi_0 \\ \tfrac{1}{2}\tau_a K_0 \end{pmatrix}.$$

We distinguish the two cases $b \neq 0$, and $b = 0$.

If $b \neq 0$, then $K_0 \neq 0$. In this case no generator can annihilate ρ, because no τ_a can annihilate K_0. The proof is as follows:

Assume the contrary, e.g., $\tau_3 K_0 = 0$. Then $-i[\tau_1, \tau_2]K_0 = 0$, or $\tau_1 \tau_2 K_0 = \tau_2 \tau_1 K_0$. But $\tau_1 \tau_2 = -\tau_2 \tau_1$. Hence $K_0 = 0$ (contradiction). Therefore, in this case the symmetry is completely broken, i.e., H is empty, and there are:
— no massless vector bosons,
— three massive vector bosons,
— three Higgs bosons (one real and one complex field).

If $b = 0$, then $H = U(1)$.

CHAPTER V
TOPOLOGICAL SOLITONS

5.1 Solitons

As we have mentioned in Chapter III, a classical soliton solution (soliton) is a solution of the classical equations of motion whose energy density is non-zero only in a finite region of space. Its total energy is therefore finite. Quantization leads to a corresponding quantum soliton. We discuss solitons in Yang-Mills field theory and deal mainly with the classical case here.

A topological soliton can occur only when the vacuum is degenerate, with the physical vacuum identified with one of the vacua. The topological soliton solution approaches a vacuum solution at spatial infinity, with different topological properties from those of the physical vacuum. The difference in topology makes the soliton stable. There are also non-topological soliton solutions,[1] which approach the physical vacuum asymptotically; but we shall not go into those here.

Solitons can be static or time-dependent. A static soliton is one for which $A^\mu(x)$ can be made time-independent through a continuous gauge transformation. This definition is equivalent to the statement that the time evolution of $A^\mu(x)$ is a continuous gauge transformation, i.e.,

$$A^\mu(\mathbf{x}, t + dt) = A^\mu(\mathbf{x}, t) + \frac{1}{g} \partial^\mu d\omega(\mathbf{x}, t) - i[d\omega(\mathbf{x}, t), A^\mu(\mathbf{x}, t)], \tag{5.1}$$

or

$$\dot{A}^\mu(x) = \frac{1}{g} \partial^\mu \dot{\omega}(x) - i[\dot{\omega}(x), A^\mu(x)], \tag{5.2}$$

where $\omega(x)$ is an arbitrary continuous function. To show the equivalence, we note that the solution to (5.2) can be made time-independent through a gauge transformation $-\omega(x)$, with $\omega(x)$ satisfying

$$\partial^\mu \dot{\omega}(x) = ig[\dot{\omega}(x), A^\mu(x)]. \tag{5.3}$$

The solution is

$$\dot{\omega}(x) = \Omega(P)\dot{\omega}(x_0)\Omega^{-1}(P), \tag{5.4}$$

where $\Omega(P)$ is defined in (4.64), with the path P beginning at x_0 and ending at x. From (5.4) we can obtain $\omega(x)$ by integrating over time. Although $\omega(x)$ depends

---
[1] T. D. Lee, *Phys. Reports*, **23**, 254 (1976).

on the choice of the path P, its time-dependence is independent of P, because the endpoints of P are fixed. The freedom in the choice of P merely reflects the fact that a static $A^\mu(x)$ still has the freedom of time-independent gauge transformations. The gauge in which a static solution is time-independent will be referred to as the *static gauge*.

A static solution of finite energy is either the vacuum or a static soliton. For a pure Yang-Mills field (i.e., one without coupling to matter fields), we can show the following property:[2]

There are no static solitons in pure Yang-Mills theory except in 4 spatial dimensions.

Proof: In $(n + 1)$-dimensional Minkowski space, let the space-time index be $\mu = 0, 1, \ldots, n$, and the spatial index be $k = 1, \ldots, n$. Consider static solutions of finite energy other than the vacuum. The canonical energy-momentum tensor is given by

$$\theta^{\mu\nu} = -F_a{}^{\mu\lambda}F_{a\lambda}{}^\nu + \tfrac{1}{4}g^{\mu\nu}F_a{}^{\alpha\beta}F_{a\alpha\beta}, \quad (\partial_\mu \theta^{\mu\nu} = 0).$$

This is gauge-invariant, and hence independent of x^0. The following quantities will be relevant:

$$\theta^\mu{}_\mu = \tfrac{1}{4}(n-3)F_a{}^{\alpha\beta}F_{a\alpha\beta},$$
$$\theta^{00} = \tfrac{1}{2}F_a{}^{k0}F_a{}^{k0} + \tfrac{1}{4}F_a{}^{ij}F_a{}^{ij},$$
$$\theta^{0k} = F_a{}^{j0}F_a{}^{jk}.$$

The requirement of finite energy means

$$\int d^n x (\tfrac{1}{2}F_a{}^{k0}F_a{}^{k0} + \tfrac{1}{4}F_a{}^{ij}F_a{}^{ij}) < \infty,$$

which leads to the condition

$$F_a{}^{\mu\nu} \xrightarrow[r \to \infty]{} O(r^{-n/2-\varepsilon}), \quad (\varepsilon > 0),$$

$$r^2 \equiv \sum_{k=1}^n x_k{}^2.$$

We can show that $F_a{}^{k0} = 0$ for a static soliton as follows. In the static gauge, $\partial^0 A^k = 0$. Hence, by (4.42)

$$F^{k0} = \partial^k A^0 + ig[A^k, A^0].$$

The relevant equation of motion in (4.78) reads

$$\partial_k F^{k0} + ig[A_k, F^{k0}] = 0.$$

Hence,

$$A_0 \{\partial_k F^{k0} + ig[A_k, F^{k0}]\} = 0,$$
$$\partial_k (A_0 F^{k0}) - (\partial_k A^0) F^{k0} + ig(A_0 A_k F^{k0} - A_0 F^{k0} A_k) = 0,$$

---

[2] S. Deser, *Phys. Letters*, **64**, B463 (1976).

$$\text{Tr} \int d^n x \{\partial_k A^0 + ig[A_k, A_0]\} F^{k0} = 0,$$
$$\text{Tr} \int d^n x F_{k0} F^{k0} = 0,$$
$$\int d^n x F_a^{k0} F_a^{k0} = 0.$$

Since $F_a^{k0} F_a^{k0} = |\mathbf{E}_a|^2$ is non-negative, it must be zero. Hence, $E_a^{\ k} = F_a^{\ k0} = 0$ in the static gauge. Since $F^{k0} \to U F^{k0} U^{-1}$ under a gauge transformation, we conclude that $F_a^{\ 0k} = 0$ for a static solution in any gauge.

Now consider

$$\partial_j (x^k \theta^{jk}) = \theta^k_{\ k} + x^k \partial_j \theta^{jk}.$$
$$\partial_\mu \theta^{\mu k} = 0 \Rightarrow \partial_0 \theta^{0k} + \partial_j \theta^{jk} = 0.$$
$$\therefore \quad \partial_j (x^k \theta^{jk}) = \theta^k_{\ k} - x^k \partial_0 \theta^{0k}.$$

The last term vanishes for static solutions. Hence,

$$0 = \int d^n x \, \partial_j (x^k \theta^{kj}) = \int d^n x \, \theta^k_{\ k}.$$

Using $\theta^k_{\ k} = \theta^\mu_{\ \mu} - \theta^{00}$, we obtain

$$\int d^n x [(2-n) F_a^{\ k0} F_a^{\ k0} - \tfrac{1}{2}(4-n) F_a^{\ jk} F_a^{\ jk}] = 0.$$

As shown earlier, $F_a^{\ k0} = 0$. Hence

$$(4-n) \int d^n x F_a^{\ ij} F_a^{\ ij} = 0.$$

Therefore $F_a^{\ ij} = 0$ unless $n = 4$. ∎

For $n = 4$, a pure Yang-Mills static soliton can be constructed explicitly. An example is the instanton discussed in Sec. 5.2. For $n \neq 4$, there can be static solitons only if there are matter fields. An example is the monopole in 3 spatial dimensions, described in Sec. 5.3.

Quantization of a soliton solution requires appropriate handling of the translational motion of the soliton as a whole, and is quite involved[3]. We will not discuss it here.

5.2 The Instanton

1 Topological Charge

As we have shown, a pure Yang-Mills theory can have static soliton solutions only in 4 Euclidean dimensions. We now construct an example, the instanton solution, which is characterized by a "topological charge" different from that of the vacuum. One might wonder why a phenomenon that exists only in Euclidean

---

[3] J. Goldstone and R. Jackiw, *Phys. Rev.* **D11**, 1486 (1975); N. H. Christ and T. D. Lee, *Phys. Rev.* **D12**, 1606 (1975).

Topological Solitons

4-space should concern us. The answer lies in the fact that a quantum field theory in Minkowski 4-space is equivalent to a classical field theory in Euclidean 4-space, as we shall show in Chap. VII. The instanton can have physical applications, as will be discussed in Chap. VIII.

In this section, we consider $G = SU(2)$. A vector in Euclidean 4-space is denoted by $x^\mu (\mu = 1, 2, 3, 4)$. There is no distinction between upper and lower indices. We may look upon x^μ either as the spatial components of a vector in Minkowski 5-space with metric $(1, -1, -1, -1, -1)$, or as a 4-vector in the usual Minkowski 4-space with the time-component continued to imaginary values: $x^0 \to -ix^4$. The topological charge is defined by

$$q \equiv \frac{g^2}{16\pi^2} \int d^4x \, \text{Tr} \, \tilde{F}^{\mu\nu} F_{\mu\nu}. \tag{5.5}$$

First we show that the integrand is a total 4-divergence:

$$\tfrac{1}{4} \text{Tr} \, \tilde{F}^{\mu\nu} F_{\mu\nu} = \partial_\mu X^\mu,$$

$$X^\mu = \varepsilon^{\mu\alpha\beta\gamma} \text{Tr} \left[\tfrac{1}{2} A_\alpha \partial_\beta A_\gamma + \frac{i}{3} g A_\alpha A_\beta A_\gamma \right]. \tag{5.6}$$

Proof:

$$\tfrac{1}{4} \text{Tr} \, \tilde{F}^{\mu\nu} F_{\mu\nu} = \tfrac{1}{2} \varepsilon^{\mu\nu\alpha\beta} \text{Tr}\{[(\partial_\alpha + igA_\alpha)A_\beta][(\partial_\mu + igA_\mu)A_\nu]\}$$
$$= \tfrac{1}{2} \varepsilon^{\mu\nu\alpha\beta} \text{Tr}[(\partial_\alpha A_\beta)(\partial_\mu A_\nu) + 2ig(\partial_\alpha A_\beta)A_\mu A_\nu - g^2 A_\alpha A_\beta A_\mu A_\nu].$$

The last term will not contribute because it is symmetric in α, β. Side calculations:

$$(\partial_\alpha A_\beta)(\partial_\mu A_\nu) = \partial_\alpha(A_\beta \partial_\mu A_\nu) - A_\beta(\partial_\alpha \partial_\mu A_\nu).$$

$$\varepsilon^{\mu\nu\alpha\beta} \text{Tr}[(\partial_\alpha A_\beta) A_\mu A_\nu]$$
$$= \varepsilon^{\mu\nu\alpha\beta} \text{Tr}[\partial_\alpha(A_\beta A_\mu A_\nu) - (\partial_\alpha A_\mu)A_\nu A_\beta - (\partial_\alpha A_\nu) A_\beta A_\mu].$$

The last two terms are the same, and equal to the left-hand side. Hence,

$$\varepsilon^{\mu\nu\alpha\beta} \text{Tr}[(\partial_\alpha A_\beta) A_\mu A_\nu] = \tfrac{1}{3} \varepsilon^{\mu\nu\alpha\beta} \partial_\alpha \text{Tr}(A_\beta A_\mu A_\nu).$$

Using the above, we obtain

$$\tfrac{1}{4} \text{Tr} \, \tilde{F}^{\mu\nu} F_{\mu\nu} = \varepsilon^{\mu\nu\alpha\beta} \partial_\alpha \text{Tr}(\tfrac{1}{2} A_\beta \partial_\mu A_\nu + \tfrac{1}{3} ig A_\beta A_\mu A_\nu). \blacksquare$$

Note that the proof goes through for Euclidean as well as Minkowski 4-space. Using (5.6) we can write

$$q = \frac{1}{4\pi^2} \int d^4x \, \partial_\mu X^\mu = \frac{1}{4\pi^2} \int dS_\mu X^\mu, \tag{5.7}$$

where dS_μ is an element of a 3-dimensional spherical hyper-surface S^3, with radius $R \to \infty$.

Now impose the condition of finite energy:

$$F^{\mu\nu} \xrightarrow[r \to \infty]{} O(r^{-3}), \qquad r^2 \equiv x_4^2 + |\mathbf{x}|^2.$$

$$A^\mu \xrightarrow[r \to \infty]{} -\frac{i}{g} U \partial^\mu U^{-1} + O(r^{-2}), \tag{5.8}$$

where $U \in SU(2)$. Let

$$\lambda^\mu \equiv U \partial^\mu U^{-1}. \tag{5.9}$$

Then $A^\mu \xrightarrow[r \to \infty]{} -(i/g)\lambda^\mu$. Using this form in (5.6) and (5.7) gives

$$X^\mu = \frac{1}{6g^2} \varepsilon^{\mu\alpha\beta\gamma} \mathrm{Tr}(\lambda^\alpha \lambda^\beta \lambda^\gamma),$$

$$q = \frac{1}{24\pi^2} \int_{S^3} \mathrm{d}S_\mu \varepsilon^{\mu\alpha\beta\gamma} \mathrm{Tr}(\lambda^\alpha \lambda^\beta \lambda^\gamma). \tag{5.10}$$

To do the surface integral, we parametrize S^3 by three angles θ_1, θ_2, θ_3. The associated unit vectors (4-vectors) are denoted by $\hat{\theta}_1{}^\alpha$, $\hat{\theta}_2{}^\alpha$, $\hat{\theta}_3{}^\alpha$ ($\alpha = 1, \ldots, 4$), with

$$\hat{\theta}_i{}^\alpha \hat{\theta}_j{}^\alpha = \delta_{ij}. \tag{5.11}$$

Then the unit normal vector to $\mathrm{d}S_\mu$ can be written as

$$\begin{aligned} n_\mu &= \varepsilon_{\mu\alpha\beta\gamma} \hat{\theta}_1{}^\alpha \hat{\theta}_2{}^\beta \hat{\theta}_3{}^\gamma, \\ n_\mu n^\mu &= 1, \\ n_\mu \hat{\theta}_i{}^\mu &= 0. \end{aligned} \tag{5.12}$$

With these, we can write $\mathrm{d}S_\mu = n_\mu \, \mathrm{d}S$, and hence

$$\begin{aligned} \int \mathrm{d}S_\mu X^\mu &= \int \mathrm{d}S \varepsilon_{\mu\alpha\beta\gamma} \hat{\theta}_1{}^\alpha \hat{\theta}_2{}^\beta \hat{\theta}_3{}^\gamma X^\mu \\ &= \frac{1}{6g^2} \int \mathrm{d}S \varepsilon_{\mu\alpha\beta\gamma} \varepsilon^{\mu\rho\sigma\tau} \hat{\theta}_1{}^\alpha \hat{\theta}_2{}^\beta \hat{\theta}_3{}^\gamma \mathrm{Tr}(\lambda^\rho \lambda^\sigma \lambda^\tau) \\ &= \frac{1}{6g^2} \int \mathrm{d}S \hat{\theta}_1{}^\alpha \hat{\theta}_2{}^\beta \hat{\theta}_3{}^\gamma \mathrm{Tr}(\lambda^\alpha \lambda^\beta \lambda^\gamma) \\ &= \frac{1}{6g^2} \int \mathrm{d}S \varepsilon^{ijk} \mathrm{Tr}(\lambda^i \lambda^j \lambda^k), \end{aligned}$$

where

$$\lambda^i \equiv \hat{\theta}_i{}^\alpha \lambda^\alpha \quad (i = 1, 2, 3). \tag{5.13}$$

Topological Solitons

Therefore,
$$q = \frac{1}{24\pi^2} \int_{S^3} dS \varepsilon^{ijk} \text{Tr}(\lambda^i \lambda^j \lambda^k),$$
$$\lambda^k = U \partial^k U^{-1}, \qquad (5.14)$$
$$\partial^k \equiv \hat{\theta}_k{}^\mu \partial_\mu.$$

Clearly, $q = 0$ when $U = 1$, and, $q \to -q$ when $\lambda^k \to -\lambda^k$.

We now show that q is invariant under a continuous gauge transformation.

Proof: Any infinitesimal change of U can be represented as the transformation
$$U \to U(1 - i\omega),$$
$$U^{-1} \to (1 + i\omega)U,$$
where $\omega = \omega_a(x) L_a$ is an arbitrary infinitesimal element of the Lie algebra. Thus,
$$\delta U = -iU\omega, \qquad \delta U^{-1} = i\omega U^{-1}.$$
$$\delta \lambda^k = \delta(U \partial^k U^{-1}) = (\delta U)(\partial^k U^{-1}) + U \partial^k (\delta U^{-1}) = iU(\partial^k \omega) U^{-1}.$$
$$\delta q = \frac{1}{24\pi^2} \int dS \varepsilon^{ijk} \text{Tr}[(\delta\lambda^i)\lambda^j\lambda^k + \lambda^i(\delta\lambda^j)\lambda^k + \lambda^i\lambda^j(\delta\lambda^k)]$$
$$= \frac{1}{8\pi^2} \int dS \varepsilon^{ijk} \text{Tr}(\lambda^i \lambda^j \delta\lambda^k)$$
$$= \frac{1}{8\pi^2} \int dS \varepsilon^{ijk} \text{Tr}[(U\partial^i U^{-1})(U\partial^j U^{-1}) U(\partial^k \omega) U^{-1}]$$
$$= \varepsilon^{ijk} \int dS \partial^k \text{Tr}[(\partial^i U^{-1})(\partial^j U^{-1})\omega]$$
$$- \int dS \varepsilon^{ijk} \text{Tr}[(\partial^i \partial^k U^{-1})(\partial^j U)\omega + (\partial^i U^{-1})(\partial^k \partial^j U^{-1})\omega].$$

The first term vanishes because S^3 has no boundary, and the second vanishes because of ε^{ijk}. Therefore $\delta q = 0$. ∎

One cannot conclude from the above that $q = 0$, because U falls into equivalence classes that are not related by continuous transformations. This may be seen as follows:

For $x \in S^3$, $U(x)$ represents a mapping of S^3 to the group manifold of $SU(2)$, which is topologically the same as S^3. Therefore, $U(x)$ is a mapping $S^3 \to S^3$. These mappings fall into homotopy classes characterized by "winding numbers" n, the number of times the spatial S^3 is covered by the group manifold S^3. A representative of the nth class is
$$U(x) = [u(x)]^n \quad (n = 0, \pm 1, \pm 2, \ldots,),$$
$$u(x) = \frac{1}{r}(x_4 + \mathbf{x} \cdot \boldsymbol{\tau}) \quad \left(r^2 \equiv \sum_{i=1}^{4} x_i^2\right). \qquad (5.15)$$

Within each class, q is invariant, and we denote it by q_n. It is straightforward to show that

$$q_n = nq_1. \tag{5.16}$$

It is only necessary to calculate q_1:

$$q_1 = \frac{1}{24\pi^2} \int_{S^3} dS \varepsilon^{ijk} \mathrm{Tr}(\lambda^i \lambda^j \lambda^k), \tag{5.17}$$

$$\lambda^k = u\partial^k u.$$

To calculate q_1, we note that the surface integral extends over values of x on S^3. Going from one value of x to another on S^3 is a rotation of S^3, and such a rotation can always be undone by a continuous change of u, which does not affect q. Under a continuous change of u, the integrand of (5.17) changes by a 4-divergence $\partial_\mu X^\mu$. Therefore, we can write the integrand as its value at any fixed x, plus $\partial_\mu X^\mu$, and the latter gives no contribution upon integration. Therefore, to calculate q_1, it suffices to evaluate the integrand at any fixed x, and multiply the result by $2\pi^2$, the volume of S^3. It is most convenient to evaluate the integrand at $x_4 = 1$, $\mathbf{x} = 0$. In the neighborhood of this point, we can set up 3-dimensional Cartesian axes $\hat{\mathbf{x}}_1, \hat{\mathbf{x}}_2, \hat{\mathbf{x}}_3$ lying in S^3, and take $\hat{\boldsymbol{\theta}}_i = \hat{\mathbf{x}}_i$ ($i = 1, 2, 3$). Then we have, at this point,

$$u = 1,$$

$$\partial^k u^{-1} = \partial^k \left(\frac{x_4 + i\mathbf{x} \cdot \boldsymbol{\tau}}{r} \right) = \frac{i\tau^k}{r} - \frac{x^k}{r^3}(x_4 + i\mathbf{x} \cdot \boldsymbol{\tau}) = i\tau^k.$$

Therefore,

$$\lambda^k = i\tau^k,$$

from which follows

$$q_1 = (2\pi^2)\left(-\frac{i}{24\pi^2}\right) \varepsilon^{ijk} \mathrm{Tr}(\tau^i \tau^j \tau^k) = 1.$$

Thus, the possible values of the topological charge for a finite-energy configuration of the Yang-Mills field are

$$q_n = n \quad (n = 0, \pm 1, \pm 2, \ldots). \tag{5.18}$$

2 Explicit Solution[4]

We now construct an explicit solution with $q = 1$. Take $L_a = \frac{1}{2}\tau_a$, and use the shorthand $F^2 = F^{\mu\nu}F_{\mu\nu}$, $\tilde{F} \cdot F = \tilde{F}^{\mu\nu}F_{\mu\nu}$. We note that $\tilde{F}^2 = F^2$. The

---

[4] A. A. Belavin, A. M. Polyakov, A. S. Schwartz, and Yu. S. Tyupkin, *Phys. Letters* **59** B 85 (1975); G. 't Hooft, *Phys. Rev. Lett.* **37**, 8 (1976).

Topological Solitons

Lagrangian density and the Euclidean action are given respectively by

$$\mathcal{L} = -\tfrac{1}{2}\text{Tr } F^2, \tag{5.19}$$

$$S = -\tfrac{1}{2}\int d^4x \text{ Tr } F^2.$$

A solution corresponds to an extremum of S. Consider the inequality

$$\int d^4x \text{ Tr}(F \pm \tilde{F})^2 \geq 0,$$

or

$$\int d^4x (F^2 \pm \tilde{F} \cdot F) \geq 0.$$

We choose the $+$ sign if $\text{Tr } \tilde{F} \cdot F > 0$, and the $-$ sign if $\text{Tr } \tilde{F} \cdot F < 0$. Hence

$$\int d^4x \text{ Tr } F^2 \geq \left| \int d^4x \text{ Tr } \tilde{F} \cdot F \right|$$

or,

$$S \leq -\tfrac{1}{2} \left| \int d^4x \text{ Tr } \tilde{F} \cdot F \right|. \tag{5.20}$$

The equality holds if $F = \pm \tilde{F}$, and when this is true we say that F is self-dual (anti-self-dual). Since $F = \pm \tilde{F}$ corresponds to an extremum S, any self-dual or anti-self-dual F is a solution. An instanton solution is a self-dual F having $q = 1$. Its contribution to the (Euclidean) action is

$$S_{\text{instanton}} = -8\pi^2/g^2. \tag{5.21}$$

Let

$$\begin{aligned}
\tau_\mu &= (\tau, i) \quad (\mu = 1, 2, 3, 4), \\
\tau_\mu^\dagger &= (\tau, -i), \\
\tau_{\mu\nu} &= i(\tau_\mu \tau_\nu^\dagger - \delta_{\mu\nu}) \\
\tilde{\tau}_{\mu\nu} &\equiv \tfrac{1}{2}\varepsilon_{\mu\nu\alpha\beta}\tau_{\alpha\beta} = \tau_{\mu\nu}.
\end{aligned} \tag{5.22}$$

Note that $\tau_{\mu\nu}$ is self-dual. Using the above notation we can write

$$(x \cdot \tau)(x \cdot \tau^\dagger) = r^2,$$

$$u = i\frac{x \cdot \tau^\dagger}{r},$$

$$u^{-1} = -i\frac{x \cdot \tau^\dagger}{r}, \tag{5.23}$$

$$A^\mu_{\text{pure-gauge}} = -\frac{i}{g} u \partial^\mu u^{-1} = \frac{1}{g}\frac{\tau_{\mu\nu} x_\nu}{r^2}.$$

For a finite-energy solution with $q = 1$, put

$$A_\mu = \frac{1}{g} \frac{\tau_{\mu\nu} x_\nu}{r^2} f(r^2),$$

$$f(0) = 0 \text{ (regularity at } r = 0\text{)},\qquad(5.24)$$

$$f(\infty) = 1 \text{ (finite energy, } q = 1\text{)}.$$

To make this a solution, we only need to choose f such that $F_{\mu\nu} = \tilde{F}_{\mu\nu}$. From (5.24) we obtain

$$F_{\mu\nu} = \frac{2}{g} \left\{ \frac{f(1-f)}{r^2} \tau_{\mu\nu} + \left[f' - \frac{f(1-f)}{r^2} \right] (\tau_{\mu\lambda} x_\lambda x_\nu - \tau_{\nu\lambda} x_\lambda x_\mu) \right\},\qquad(5.25)$$

$$\tilde{F}_{\mu\nu} = \frac{2}{g} \left\{ \frac{f(1-f)}{r^2} \tau_{\mu\nu} + \left[f' - \frac{f(1-f)}{r^2} \right] \varepsilon_{\mu\nu\alpha\beta} (\tau_{\alpha\lambda} x_\lambda x_\beta - \tau_{\beta\lambda} x_\lambda x_\alpha) \right\}.\qquad(5.26)$$

Thus, $F_{\mu\nu}$ is self-dual if the second term vanishes;

$$f' - \frac{f(1-f)}{r^2} = 0.\qquad(5.27)$$

The general solution satisfying the required boundary conditions is

$$f(r^2) = \frac{r^2}{\rho^2 + r^2},\qquad(5.28)$$

where ρ is an arbitrary scale parameter. With this, we have

$$F_{\mu\nu} = \tilde{F}_{\mu\nu} = \frac{2}{g} \frac{\rho^2}{(r^2 + \rho^2)^2} \tau_{\mu\nu},\qquad(5.29)$$

which is the field tensor for an instanton with $q = 1$. An instanton with $q = -1$ is called an anti-instanton, and may be obtained from (5.29) by replacing $\tau_{\mu\nu}$ by $\bar{\tau}_{\mu\nu}$, with

$$\bar{\tau}_{ij} = \tau_{ij},$$
$$\bar{\tau}_{i4} = -\tau_{i4}.\qquad(5.30)$$

The physical relevance of the instanton will be discussed in Chap. VIII.

5.3 The Monopole

1 Topological Stability

There are no solitons in pure Yang-Mills theory in 3 spatial dimensions; but they can exist when matter fields are present. We now discuss such an example. Consider any simple gauge group G, and introduce scalar matter fields $\phi(x) = \{\phi_1(x), \phi_2(x), \ldots\}$, which form a (generally reducible) representation of G. Using a real representation, we write the total energy as

$$\mathscr{E} = \int d^3x [\tfrac{1}{2}(\mathbf{B}_a \cdot \mathbf{B}_a + \mathbf{E}_a \cdot \mathbf{E}_a + \tfrac{1}{2}(D^0\phi, D^0\phi) \\ + \tfrac{1}{2}(D^k\phi, D^k\phi) + V(\phi)]. \tag{5.31}$$

We assume that the lowest minimum of $V(\phi)$ occurs at $\phi(x) = \rho(x)$:

$$V(\rho) = 0, \quad V'(\rho) = 0, \quad V''(\rho) > 0. \tag{5.32}$$

The conditions for a finite-energy solution are

(a) $F^{\mu\nu}(x) \xrightarrow[r\to\infty]{} O(r^{-2}), \quad r \equiv |\mathbf{x}|,$

$$A^\mu(x) \xrightarrow[r\to\infty]{} -\frac{i}{g} U\partial^\mu U^{-1} + O(r^{-1}), \tag{5.33}$$

(b) $D^\mu\phi(x) \xrightarrow[r\to\infty]{} O(r^{-2}),$

(c) $\phi(x) \xrightarrow[r\to\infty]{} \rho(x) + O(r^{-2})$

In (b), the first term of A^μ is of pure-gauge form, but not the $O(r^{-1})$ term. It is the latter that will be of interest.

We begin by considering the implications of (c). As noted in Sec. 4.6, $\rho(x)$ is not unique. The independent ρ's are of the form $U(x)\rho_0$, where ρ_0 is a constant satisfying (5.32), and $U(x)$ belongs to a coset in G/H (H is the little group with respect to ρ, the largest subgroup of G that leaves ρ invariant). Conversely, a given $\rho(x)$ can always be written in the form $U(x)\rho_0$. Since $\rho(x)$ is the value of $\phi(x)$ at a point on S^2 (a 3-sphere of radius $r \to \infty$), the function $\rho(x)$ is a mapping of S^2 to G/H:

$$\rho(x): S^2 \to G/H. \tag{5.34}$$

The identity map corresponds to the physical vacuum, by definition. A topological soliton can exist only if there exists a non-trivial map. The topology is then different from that of the physical vacuum, making the soliton stable.

The mappings (5.34) fall into homotopy classes which form a group, namely, the second homotopy group of G/H, denoted by $\pi_2(G/H)$. In general, the nth homotopy group $\pi_n(X)$ is the group of inequivalent mappings $S^n \to X$. If $\pi_2(G/H) = 0$ (i.e. it contains only the identity map and its continuous deformations), then no topological solitons exist. Some general properties of $\pi_2(G/H)$ are as follows[5]:

(a) For simply-connected G,

$$\pi_2(G/H) = \pi_1(H), \\ \pi_1(G/H) = 0. \tag{5.35}$$

---

[5] M. I. Monastyrskii and A. M. Perelomov, *JETP Letters*, **21**, 43 (1975).

(b) If $G = \tilde{G}/C$, where $\tilde{G}$ is simply-connected with a finite center (i.e. a subgroup that commutes with all elements of $\tilde{G}$), and C is a subgroup of the center of $\tilde{G}$, then

$$\pi_2(G/H) = \pi_2(\tilde{G}/H) = \pi_1(H),$$
$$\pi_1(G/H) = C. \tag{5.36}$$

We give some examples of these properties, assuming that H is non-empty:
Ex. 1. $G = SU(2)$.
(a) For half-integer representations, $H = U(1)$, $\pi_1(H) = 0$.
(b) For integer representations, $H = U(1)$, $\pi_1(H) = Z$ (the set of all integers).
Ex. 2. $G = SU(3)$.
(a) For the triplet representation, $H = SU(2)$, $\pi_1(H) = 0$.
(b) For the octet representation, ρ can be represented as a 3×3 traceless matrix. If all eigenvalues are distinct, then $H = U(1) \times U(1)$, $\pi_1(H) = Z + Z$. If two of the eigenvalues are the same, then $H = U(2)$, $\pi_1(H) = Z$.
Ex. 3. $G = SU(2) \times U(1)$. $H = U(1)$, $\pi_2(G/H) = 0$. This shows that the Weinberg-Salam model has no topological solitons. We shall discuss this in more detail in Chap. VI.

2 Flux Quantization[6]

Assume $\pi_2(G/H) \neq 0$. Then a topological soliton is possible, and condition (b) of (5.33) imposes boundary conditions on S^2: to order $O(r^{-2})$,

$$D^\mu \rho(x) = \partial^\mu \rho(x) + ig A_a{}^\mu(x) L_a \rho(x) = 0. \tag{5.37}$$

The $O(r^{-1})$ terms in $A_a{}^\mu$ must be cancelled by $\partial^\mu \rho$. Thus, $A_a{}^\mu$ and ρ are related. In particular, if ρ is not constant, there will be a Coulombic potential. For static solutions, this cannot be an electric potential. Hence, there will be a magnetic potential corresponding to a magnetic monopole.

The condition (5.37) can always be fulfilled, if $V(\phi)$ is such that the lowest minima are all related by continuous gauge transformations. We assume this is true.

Consider three great circles $C_k (k = 1, 2, 3)$ on S^2, whose normals are respectively $\hat{\mathbf{x}}_k$, as shown in Fig. 5.1. Let ΔC_k be a finite arc of C_k; with endpoints y and z. According to (5.37), $\rho(x)$ undergoes parallel displacement from y and z to order r^{-2}:

$$\rho(z) = \Omega(\Delta C_k) \rho(y),$$
$$\Omega(\Delta C_k) = T \exp\left(ig \int_{\Delta C_k} \mathbf{ds} \cdot \mathbf{A}(x) \right), \tag{5.38}$$

where $\mathbf{ds}$ is an element of arc on S^2. On the other hand, $\rho(z)$ can be obtained from $\rho(y)$ by a rotation through some angle $\Delta \theta$ about $\hat{\mathbf{x}}_k$, generated by a linear

---
[6] K. Huang and D. R. Stump, *Phys. Rev.* D**15**, 3660 (1977).

Topological Solitons

combination of the generator $\{L_a\}$ of G. Since these rotations form an Abelian group, there must be a choice of $\{L_a\}$ such that

$$[\Omega(\Delta C_k), \Omega(\Delta C_k')] = 0, \tag{5.39}$$

where ΔC_k and $\Delta C_k'$ are two arcs of C_k. Therefore, the operation T in (5.38) can be dropped. Thus

$$\Delta\Omega(C_1) = \exp\left(ig \int_{\Delta C_1} \mathbf{ds} \cdot \mathbf{A}\right) = \exp\left(-i(\Delta\theta_1)\mathcal{J}^1\right),$$

$$\Delta\Omega(C_2) = \exp\left(ig \int_{\Delta C_2} \mathbf{ds} \cdot \mathbf{A}\right) = \exp\left(-i(\Delta\theta_2)\mathcal{J}^2\right), \tag{5.40}$$

$$\Delta\Omega(C_3) = \exp\left(ig \int_{\Delta C_3} \mathbf{ds} \cdot \mathbf{A}\right) = \exp\left(-i(\Delta\theta_3)\mathcal{J}^3\right).$$

It is possible to make all these rotations with the same choice of axes $\hat{\mathbf{x}}_1, \hat{\mathbf{x}}_2, \hat{\mathbf{x}}_3$; for example the closed circuit OPQ in Fig. 5.1. Therefore, with the same choice of $\{L_a\}$, $\mathcal{J}^k$ can be constructed such that

$$[\mathcal{J}^1, \mathcal{J}^2] = i\mathcal{J}^3 \quad (1, 2, 3 \text{ cyclic}). \tag{5.41}$$

Since the eigenvalues of $\mathcal{J}^k$ are integers or half-integers, the angles $\Delta\theta_k$ are defined only mod (4π). This corresponds to the fact that a rotation through angle 4π leaves the system truly unchanged, as we pointed out in Sec. 4.2.

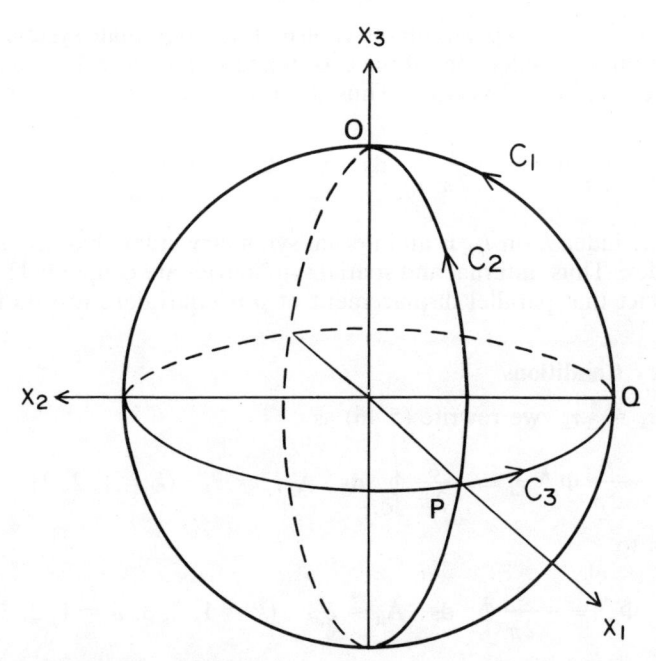

Fig. 5.1 Great circles on a sphere

Now take ΔC_k to be a complete great circle, with $\theta_k = 2\pi$:

$$\Omega(C_k) = \exp\left(ig \oint_{C_k} \mathbf{ds} \cdot \mathbf{A}\right) = \exp(-2\pi i \mathcal{J}^k). \qquad (5.42)$$

Let the flux matrix Φ^k be defined by

$$\Phi^k \equiv \oint_{C_k} \mathbf{ds} \cdot \mathbf{A}, \quad \Phi_k \equiv -\Phi^k. \qquad (5.43)$$

This refers to the flux of $\nabla \times \mathbf{A}$, which is divergenceless, and is in general not the flux of $\mathbf{B}$. Its eigenvalues are defined only mod $(4\pi/g)$. By (5.42) and (5.41) we have

$$\left[\frac{g}{2\pi}\Phi_1, \frac{g}{2\pi}\Phi_2\right] = \frac{ig}{2\pi}\Phi_3 \quad (1, 2, 3 \text{ cyclic}). \qquad (5.44)$$

This is the generalization of the flux quantization condition (3.49) in the $U(1)$ case. We have assumed that $\{L_a\}$ have been chosen in a special way. This is equivalent to saying that (5.42) is true in a gauge in which $\mathbf{A}(x)$ has no singularities on S^2.

For a non-trivial solution to (5.44) to be possible, G must contain $SU(2)$. Choose $\{L_a\}$ such that the first three are generators of $SU(2)$:

$$\begin{aligned} \{L_a\} &= \{L_1, L_2, L_3, L_\alpha\} \quad (\alpha = 4, 5, \ldots, N), \\ [L_1, L_2] &= iL_3 \quad (1, 2, 3 \text{ cyclic}). \end{aligned} \qquad (5.45)$$

The choice, of course, depends on the choice of the coordinate system $\hat{\mathbf{x}}_1, \hat{\mathbf{x}}_2, \hat{\mathbf{x}}_3$. Under rotations and reflections of the coordinate system both Φ_k and L_k ($k = 1, 2, 3$) must change like 3-vectors. Thus the following is a 3-vector equation:

$$\frac{g}{2\pi}\Phi_k = -\frac{g}{2\pi}\oint_{C_k} \mathbf{ds} \cdot \mathbf{A} = L_k, \quad (k = 1, 2, 3). \qquad (5.46)$$

Note that the index k on L_k is an internal symmetry index that also serves as a 3-vector index. Thus, internal and spatial symmetries are coupled. The origin of this is the fact that parallel displacement of ρ is equivalent to rotation.

3 Boundary Conditions

Taking $L_k = \tfrac{1}{2}\tau_k$, we rewrite (5.46) as

$$-\frac{g}{2\pi}\Phi_a{}^k \tau_a \equiv -\frac{g}{2\pi}\oint_{C_k} \mathbf{ds} \cdot \mathbf{A}_a \tau_a = \tau_k \quad (k = 1, 2, 3), \qquad (5.47)$$

which leads to

$$-\frac{g}{2\pi}\Phi_a{}^k \equiv -\frac{g}{2\pi}\oint_{C_k} \mathbf{ds} \cdot \mathbf{A}_a = \delta_{ak} \quad (k = 1, 2, 3; \ a = 1, 2, 3). \qquad (5.48)$$

This requires $A_a{}^k$ ($k = 1, 2, 3; \ a = 1, 2, 3$) to be a pseudotensor that falls off

Topological Solitons

like r^{-1} asymptotically:

$$A_\alpha{}^k = 0 \quad (\alpha = 4, 5, \ldots, N),$$

$$A_a{}^k = g' \varepsilon^{akj} \frac{x^j}{r^2} + O(r^{-2}), \tag{5.49}$$

where g' can be determined by substituting this into (5.48):

$$gg' = 1 \mod(2). \tag{5.50}$$

The case $gg' = \pm 2$ is gauge-equivalent to the trivial case $gg' = 0$, and $gg' = \pm 1$ are gauge-equivalent. Hence, the boundary conditions for $A_a{}^\mu(x)$ are[7]

$$A_a{}^k(x) \xrightarrow[r\to\infty]{} \frac{1}{g} \varepsilon^{akj} \frac{x^j}{r^2} + O(r^{-2}) \quad (k = 1, 2, 3; a = 1, 2, 3). \tag{5.51}$$

This describes the only possible topological soliton with "spherical symmetry" (i.e., for which **A** has no singularity on S^2).

Note that g is the gauge-coupling constant, or the "charge" of the gauge field. The charge of an irreducible multiplet ϕ is

$$e = \kappa g, \tag{5.52}$$

where κ is the smallest positive eigenvalue of the matrix representing a generator of G. For example, for fundamental representations, $e = g$ for $O(3)$, $e = g/2$ for $SU(2)$, and $e = g/3$ for $SU(3)$.

We now examine the boundary conditions for $\rho(x)$. Substituting (5.51) into (5.37), we obtain

$$d\mathbf{s} \cdot \left(\nabla \rho + i \frac{\mathbf{r} \times \mathbf{L}}{r^2} \rho \right) = 0, \tag{5.53}$$

where $\mathbf{L} = (L_1, L_2, L_3)$ and $d\mathbf{s}$ is an infinitesimal arc element on S^2. Writing $d\mathbf{s} = \hat{\mathbf{n}} \times \mathbf{r} \, d\theta$, with the symbols defined in Fig. 5.2, we obtain, after a little algebra,

$$\frac{\partial \rho}{\partial \theta} = i \hat{\mathbf{n}} \cdot \mathbf{L} \rho. \tag{5.54}$$

In the representation for which $\hat{\mathbf{n}} \cdot \mathbf{L}$ is diagonal with eigenvalue λ, the solution is $\rho = \rho_0 e^{i\lambda\theta}$. Since ρ must be a continuous function on S^2, λ must be an integer. Therefore, a topological soliton can exist only if ρ belongs to an integer representation of the $SU(2)$ subgroup of G. Let us assume the simplest case of the adjoint representation. Then ρ is a triplet of fields ρ_a ($a = 1, 2, 3$), with $(L_a)_{bc} = -i\varepsilon^{abc}$, and (5.53) can be written as

$$\partial^k \rho_b = -\varepsilon^{akj} \frac{x^j}{r^2} \varepsilon^{abc} \rho_c$$

$$= -\frac{1}{r^2} (\delta_{kb} x^c \rho_c - x^b \rho_k). \tag{5.55}$$

---

[7] T. T. Wu and C. N. Yang, in *Properties of Matter Under Unusual Conditions*, edited by H. Mark and S. Fernbach (Interscience, New York, 1969), p. 349.

The solution is

$$\rho_a(x) = \frac{x^a}{r}\rho_0, \tag{5.56}$$

where ρ_0 is any constant satisfying (5.32). In summary, the boundary conditions for a spherical symmetric topological soliton are

$$A_a^k(x) \xrightarrow[r\to\infty]{} \frac{1}{g}\varepsilon^{akj}\frac{x^j}{r^2},$$

$$\phi_a(x) \xrightarrow[r\to\infty]{} \frac{x^a}{r}\rho_0. \tag{5.57}$$

The magnetic field can be shown to have the behavior

$$B_a^k(x) \xrightarrow[r\to\infty]{} \frac{1}{g}\frac{x^a x^k}{r^4}, \tag{5.58}$$

while the electric field vanishes to order r^{-2}.

4 Explicit Solution[8]

To show a solution actually exists, take $G = SU(2)$, with a triplet of Higgs fields $\phi_a(x)$. According to the Higgs mechanism (see Sec. 4.6), there should be one massless vector boson, two massive vector bosons, and one massive Higgs boson.

For a static solution, take $A_a^0 = 0$, and

$$A_a^k(x) = \frac{1}{g}\varepsilon^{akj}\frac{x^j}{r^2}F(r),$$

$$\phi_a(x) = \frac{x^a}{r}\rho_0\eta(r), \tag{5.59}$$

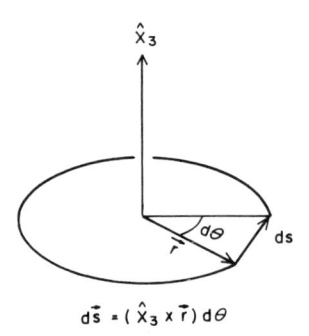

Fig. 5.2 Definition of symbols in Eq. (5.54).

---

[8] G. 't Hooft, *Nucl. Phys.* **B79**, 276 (1974); A. M. Polyakov, *JETP Lett.* **20**. 194 (1974).

with the boundary conditions
$$F(0) = 0, \quad F(\infty) = 1,$$
$$\eta(0) = 0, \quad \eta(\infty) = 1, \tag{5.60}$$

where F and η are required to vanish at $r = 0$, in order to render the solution regular. The electric and magnetic fields are given by

$$E_a^k = 0,$$
$$B_a^k = \frac{1}{g} \frac{x^a x^k}{r^4} F(1 - F) + \frac{1}{g} \left(\delta_{ak} - \frac{x^a x^k}{r^2} \right) \frac{F'}{r}. \tag{5.61}$$

The total energy is given by

$$\mathscr{E} = \int d^3x \left\{ \frac{1}{g^2} \left(\frac{F'}{r} \right)^2 + \frac{1}{2g^2} \left[\frac{F(1-F)}{r^2} \right]^2 \right.$$
$$\left. + \rho_0^2 \left[\frac{\eta(1-F)}{r} \right]^2 + \frac{\rho_0^2}{2} \left[\frac{\eta(1-F)}{r} + \eta'F \right]^2 + V \right\}. \tag{5.62}$$

All terms are positive-definite, and neither $F = 0$ nor $F = 1$ give the lowest minimum. Hence there is a solution, by the variational principle. Explicit numerical calculations with a quartic form for V show that the mass of the topological soliton is of the order of magnitude

$$M \sim \frac{m_V}{g^2/4\pi}, \tag{5.63}$$

where m_V is a vector boson mass.

We can see from (5.62) that for the pure Yang-Mills case ($\eta \equiv 0$), the F that minimizes $\mathscr{E}$ would be $F = 1$, which leads to an unacceptable singularity in (5.59) at $r = 0$. This is why Higgs fields are needed to give a static soliton solution.

5 Physical Fields

The boundary conditions (5.57) are valid only in a special gauge, the Coulomb gauge (we can verify that $\partial_k A_a^k = 0$). In this gauge most fields are not physical. It is difficult to see that there is only one massless vector field, and which combination of A_a^k it corresponds to. To display the physical fields, we go to unitary gauge, which is defined as the gauge in which $\phi(x)$ has only one non-vanishing component, as indicated in the following comparison:

$$\text{Coulomb gauge: } \phi(x) \xrightarrow[r \to \infty]{} \rho_0 \begin{pmatrix} x/r \\ y/r \\ z/r \end{pmatrix},$$
$$\tag{5.64}$$
$$\text{Unitary gauge: } \phi(x) \xrightarrow[r \to \infty]{} \rho_0 \begin{pmatrix} 0 \\ 0 \\ 1 \end{pmatrix}$$

In the Coulomb gauge, the axis along which $\phi(x)$ points, in internal space (i.e. the 3-axis that defines the diagonal generator L_3), coincides with the radial vector $\mathbf{r}$ in ordinary space. In unitary gauge, the direction of $\phi(x)$ in internal space is independent of $\mathbf{r}$. To transform from Coulomb gauge to unitary gauge, we make a gauge transformation to rotate $\hat{\mathbf{r}}$ into $\hat{\mathbf{x}}_3$ at every point in space:

$$\phi \to U\phi$$
$$U = e^{-i\theta L_2} e^{-i\varphi L_3} \quad (5.65)$$

Where θ, φ are the polar angles of $\mathbf{r}$ in a spherical coordinate system. It is clear that this transformation is ill-defined along the negative z-axis, because the downward vector $\hat{\mathbf{r}}$ has to be rotated to its opposite, and the way to do this is not unique. Therefore, in unitary gauge, $A_a{}^k$ is not defined along the negative z-axis. The flux quantization condition (5.44) does not hold in this gauge.

The massless vector field tensor $B^{\mu\nu}$ must be the projection of $F_a{}^{\mu\nu}$ into the Higgs space (see Sec. 4.6.):

$$B^{\mu\nu} = F_a{}^{\mu\nu} \phi_a / \rho_0. \quad (5.66)$$

This is gauge-invariant, for under an infinitesimal gauge transformation,

$$\delta F_a{}^{\mu\nu} = \varepsilon_{abc} \omega_b F_c{}^{\mu\nu},$$
$$\delta \phi_a = \varepsilon_{abc} \omega_b \phi_c,$$

and hence

$$\delta(F_a{}^{\mu\nu} \phi_a) = \varepsilon_{abc} \omega_b (F_c{}^{\mu\nu} \phi_a + F_a{}^{\mu\nu} \phi_c) = 0.$$

Therefore, $F_a{}^{\mu\nu} \phi_a$ is the same in unitary gauge as in the Coulomb gauge, and we can calculate it conveniently in the latter. We calculate the asymptotic form only:

$$B^k \xrightarrow[r\to\infty]{} \frac{1}{g} \frac{x^a x^k}{r^4} \frac{x^a}{r} = \frac{1}{g} \frac{x^k}{r^3}. \quad (5.67)$$

Note that this is well-defined all over S^2, although A^k is not defined along the negative z-axis. To summarize, in unitary gauge we have

$$\mathbf{B} \xrightarrow[r\to\infty]{} \frac{1}{g} \frac{\hat{\mathbf{r}}}{r^2} \quad \text{(Massless gauge field)},$$

$$\phi \xrightarrow[r\to\infty]{} \rho_0 \begin{pmatrix} 0 \\ 0 \\ 1 \end{pmatrix} \quad \text{(Higgs field)}. \quad (5.68)$$

The other 2 vector fields are massive, and hence fall off exponentially as $r \to \infty$.

Far away from the origin, $\mathbf{B}$ appears to be the field produced by a magnetic monopole of magnetic charge $1/g$ located at $\mathbf{r} = 0$, hence the name monopole solution. However, there is no singularity at $\mathbf{r} = 0$.

As a comparison, we give the vector potential and magnetic field for the Dirac monopole[9], with a string along the negative z-axis:

$$\mathbf{A}_D = g' \frac{\hat{\mathbf{z}} \times \hat{\mathbf{r}}}{r + z}, \quad A_0 = 0, \tag{5.69}$$

where g' is such that $2g'e =$ integer, e being the electronic charge. The components of $\mathbf{A}_D$ are, in rectangular coordinates:

$$A^2 = -\frac{g'}{r+z}\frac{y}{r},$$
$$A^2 = -\frac{g'}{r+z}\frac{x}{r}, \tag{5.70}$$
$$A^3 = 0.$$

and in polar coordinates:

$$A_r = 0,$$
$$A_\theta = 0, \tag{5.71}$$
$$A_\varphi = \frac{g'}{r}\frac{1 - \cos\theta}{\sin\theta}.$$

The magnetic field is given by

$$\mathbf{B}_D = \frac{\hat{\mathbf{r}} g'}{r^2} + \hat{\mathbf{z}} b_0 \delta(x)\delta(y)\theta(1 - z). \tag{5.72}$$

This is the same as (5.68), except that this is supposed to hold everywhere, including $\mathbf{r} = 0$, except for the string. In the Dirac case, which is a construction not based on a complete theory, the string is a necessary artifact. In our case, there is really no string because it does not appear in $\mathbf{B}$, and even in $\mathbf{A}$ it can be transformed away by a gauge transformation.

So far, there is no experimental evidence for the magnetic monopole, possibly because even if it exists, the mass would have to be so large that it cannot be produced in present-day accelerators.

6 Spin from Isospin[10]

A monopole can trap a boson of "isospin" 1/2 [i.e. a boson $SU(2)$ doublet], resulting in a topological soliton with intrinsic angular momentum $\hbar/2$, although Dirac fields are not present. The monopole converts isospin into spin. This phenomenon occurs only in the quantum theory.

The point is that the flux matrices $(g/2\pi)\Phi_k$ ($k = 1, 2, 3$) are generators of rotations in space, according to (5.44). Hence, they should be added to the total angular momentum of the system. Their eigenvalues are half-integers, if there are fields of isospin 1/2.

---

[9] P. A. M. Dirac, *Proc. Roy. Soc.* **A133**, 60 (1931).
[10] R. Jackiw and C. Rebbi, *Phys. Rev. Lett.*, **36**, 1116 (1976), P. Hasenfratz and G. 't Hooft, *Phys. Rev. Lett.*, **36**, 1119 (1976).

As an example, consider $G = SU(2)$, and take the matter fields to consist of a triplet of Higgs fields ϕ and a doublet K, with

$$\phi(x) \xrightarrow[r \to \infty]{} \rho \neq 0, \qquad K(x) \xrightarrow[r \to \infty]{} 0. \tag{5.73}$$

The flux matrices can be represented by

$$\frac{g}{2\pi} \Phi_k = L_k = \begin{pmatrix} C_k & 0 \\ 0 & \frac{1}{2}\tau_k \end{pmatrix} \quad (k = 1, 2, 3), \tag{5.74}$$

where C_k are 3×3 matrices appropriate for the adjoint representation. We take this to be the extra angular momentum of the system, to be added to the "normal" total angular momentum, which has only integer eigenvalues. Note that the gauge field makes no contribution to it. In quantum theory, we put $\phi(x) = \rho + \eta(x)$, and represent $\eta(\mathbf{r})$ and $K(\mathbf{r})$ by quantum fields in the Schrödinger picture:

$$\eta(\mathbf{r}) = \sum_{\mathbf{p}} \frac{1}{\sqrt{2\omega_p}} [e^{i\mathbf{p} \cdot \mathbf{r}} a(\mathbf{p}) + e^{-i\mathbf{p} \cdot \mathbf{r}} a^\dagger(\mathbf{p})],$$

$$K(\mathbf{r}) = \sum_{\mathbf{p}} \frac{1}{\sqrt{2E_p}} [e^{-i\mathbf{p} \cdot \mathbf{r}} b(\mathbf{p}) + e^{-i\mathbf{p} \cdot \mathbf{r}} c^\dagger(\mathbf{p})], \tag{5.75}$$

where $a(\mathbf{p})$ is an isovector, and $b(\mathbf{p})$, $c(\mathbf{p})$ are isospinors. The diagonal component of the extra angular momentum is

$$\mathcal{J}^3 = \sum_{\mathbf{p}} (a(\mathbf{p}), C_3 a(\mathbf{p})) + \sum_{\mathbf{p}} \left[b^\dagger(\mathbf{p}) \frac{\tau_3}{2} b^\dagger(\mathbf{p}) - c^\dagger(\mathbf{p}) \frac{\tau_3}{2} c(\mathbf{p}) \right], \tag{5.76}$$

where

$$C_3 = \begin{pmatrix} 1 & 0 & 0 \\ 0 & 0 & 0 \\ 0 & 0 & -1 \end{pmatrix}, \quad \frac{\tau_3}{2} = \begin{pmatrix} \frac{1}{2} & 0 \\ 0 & -\frac{1}{2} \end{pmatrix}. \tag{5.77}$$

The second term in (5.76) has half-integer eigenvalues. The reasons why $\mathcal{J}^3$ must be included in the total angular momentum are Poincaré invariance, and the need for gauge-fixing in quantum theory[11]. Goldhaber[12] has given an argument showing that monopoles with half-integer spin obey Fermi statistics.

A related phenomenon is that, in the presence of a monopole, a massless Dirac field possesses states of fermion number $1/2$.[13] A simpler version of the phenomenon in one spatial dimension (where the topological soliton is known as a "kink"), can be applied to polymers, and predicts states of electron number $1/2$ in polyacetylene.[14]

---

[11] K. Huang and D. R. Stump. *op. cit.*
[12] A. Goldhaber, *Phys. Rev. Lett.*, **19**, 1122 (1976).
[13] R. Jackiw and C. Rebbi, *Phys. Rev.* **D13**, 3398 (1976).
[14] W. P. Su., J. R. Schrieffer, and A. J. Heeger, *Phys. Rev.* **B22**, 2099 (1980).

CHAPTER VI
WEINBERG-SALAM MODEL

6.1 The Matter Fields

Glashow advanced the idea that the electromagnetic and weak interactions may be unified in a gauge theory based on the group $SU(2) \times U(1)$. The problem of generating masses in a manner consistent with gauge invariance was solved later by Weinberg and Salam, using the idea of spontaneous symmetry breaking[1]. The resulting theory, known as the Weinberg-Salam model, was shown by 't Hooft[2] to be a renormalizable quantum field theory. The inclusion of quarks in this theory was achieved by Glashow, Iliopoulos and Maiani[3]. It gained wide acceptance after experiments verified some of its predictions, chief among these being the structure of the neutral currents.

The basic idea of the Weinberg-Salam model has been discussed in Chapter I, and basic experimental facts concerning the electromagnetic and weak interactions have been reviewed in Chapter II. We reiterate some of these in a different form for emphasis.

Let us first review the definition of chirality, which is the eigenvalue of γ_5, with $\gamma_5 = +1$ corresponding to right-handedness, and $\gamma_5 = -1$ to left-handedness:

$$\gamma_5 R = R, \qquad \bar{R}\gamma_5 = -\bar{R},$$
$$\gamma_5 L = -L, \qquad \bar{L}\gamma_5 = \bar{L}, \tag{6.1}$$

where R, L are Dirac spinors with only two independent components. They may be obtained from a 4-component Dirac spinor ψ by the following projections:

$$R = \tfrac{1}{2}(1+\gamma_5)\psi, \qquad \bar{R} = \tfrac{1}{2}\bar{\psi}(1-\gamma_5),$$
$$L = \tfrac{1}{2}(1-\gamma_5)\psi, \qquad \bar{L} = \tfrac{1}{2}\bar{\psi}(1+\gamma_5). \tag{6.2}$$

For later applications, it is important to note that

$$\bar{\psi}\psi = \bar{L}R + \bar{R}L,$$
$$\bar{\psi}\gamma^\mu\psi = \bar{L}\gamma^\mu L + \bar{R}\gamma^\mu R. \tag{6.3}$$

The Dirac equation for a massive particle of 4-momentum $p^\mu = (E, \mathbf{p})$ may be

---

[1] For historical accounts and references see the Nobel lectures of Glashow, Weinberg, and Salam: S. L. Glashow, *Rev. Mod. Phys.* **53**, 539 (1980); S. Weinberg, *ibid*, **52**, 515 (1980); A. Salam, *ibid*, **52**, 525 (1980).
[2] G. 't Hooft, *Nuclear Physics*, B33, 173 (1971); B35, 167 (1971).
[3] S. L. Glashow, J. Iliopoulos, and L. Maiani, *Phys. Rev.* D3, 1043 (1981).

written as
$$(\boldsymbol{\alpha}\cdot\mathbf{p}+\beta m)\psi = E\psi, \qquad E=(p^2+m^2)^{1/2}. \tag{6.4}$$

Using the identity $\boldsymbol{\alpha}=\gamma_5\boldsymbol{\sigma}$, and the fact that γ_5 and $\boldsymbol{\sigma}$ commute, we can rewrite this in the form

$$\boldsymbol{\sigma}\cdot\hat{\mathbf{p}}R = \frac{E}{p}R - \frac{m}{p}\beta L, \qquad p\equiv|\mathbf{p}|,$$
$$\boldsymbol{\sigma}\cdot\hat{\mathbf{p}}L = -\frac{E}{p}L - \frac{m}{p}\beta R. \tag{6.5}$$

These equations become decoupled if $m=0$:
$$\boldsymbol{\sigma}\cdot\hat{\mathbf{p}}R = R,$$
$$\boldsymbol{\sigma}\cdot\hat{\mathbf{p}}L = -L. \tag{6.6}$$

Therefore, for massless Dirac particles, chirality is the same as helicity; for antiparticles, chirality is the opposite of helicity. [An antiparticle has the same chirality as the particle, by definition; but it has the opposite helicity due to a change in the sign of E in (6.4)].

The electromagnetic interaction Lagrangian density is given by
$$\mathscr{L}^{\text{em}} = e\bar{\psi}Q\slashed{A}\psi, \tag{6.7}$$

where Q is the charge matrix, and $\slashed{A}=A_\mu\gamma^\mu$. The charge-changing weak interaction Lagrangian density can be written in the form

$$\mathscr{L}^{\text{ch}} = \frac{g}{\sqrt{2}}\bar{L}(\slashed{W}_+\tau_- + \slashed{W}_-\tau_+)L = \frac{g}{2}\bar{L}(\slashed{W}_1\tau_1 + \slashed{W}_2\tau_2)L, \tag{6.8}$$

where $W_\pm^\mu$ are the vector boson fields that mediate this interaction, with

$$W_\pm^\mu = \frac{1}{\sqrt{2}}(W_1^\mu \pm iW_2^\mu),$$
$$\tau_\pm = \frac{1}{2}(\tau_1 \pm i\tau_2). \tag{6.9}$$

Adding (6.7) and (6.8), we have

$$\mathscr{L}^{\text{em}} + \mathscr{L}^{\text{ch}} = \bar{L}\left[g\left(\slashed{W}_1\frac{\tau_1}{2}+\slashed{W}_2\frac{\tau_2}{2}\right)+e\slashed{A}Q\right]L + \bar{R}e\slashed{A}QR. \tag{6.10}$$

The form suggests that the basic spinor fermion fields are not 4-component Dirac spinors, but rather their left and right-handed projections, and that A^μ may be combined with W_1^μ and W_2^μ to form a multiplet of some kind.

To illustrate the multiplet structure of the matter fields, and the need for spontaneous symmetry breaking, we consider only one lepton doublet consisting of the electron and its neutrino:

$$L = \begin{pmatrix} \nu_L \\ e_L \end{pmatrix}, \qquad R = e_R. \tag{6.11}$$

Weinberg-Salam Model

The neutrino is assumed to be massless, and hence ν_R is absent. The theory is assumed to be invariant under an $SU(2)$ group, under which L transforms as a doublet, and R transforms as a singlet.

A conventional mechanical mass term in the Lagrangian density cannot be invariant under $SU(2)$, because it is proportional to $\bar{\psi}\psi = \bar{L}R + \bar{R}L$. We cannot violate this symmetry (to however small a degree), because it is to be gauged. Therefore, in this theory the electron mass can arise only by virtue of a spontaneous breakdown of $SU(2)$. A convenient way to implement this is to introduce a doublet Higgs field

$$\phi = \begin{pmatrix} \phi_+ \\ \phi_0 \end{pmatrix}, \tag{6.12}$$

where the subscripts refer to the electric charge, and write the mass term as

$$\mathcal{L}^{\text{mass}} \propto \bar{L}\phi R + \bar{R}\phi^\dagger L, \tag{6.13}$$

where $\bar{L}\phi$ is an $SU(2)$ singlet, and a Dirac spinor. Thus, (6.13) is Lorentz invariant, and invariant under $SU(2)$. If ϕ has non-zero vacuum value, then for low excitations, (6.13) is indistinguishable from a conventional mass term. Writing out (6.13) in detail, we have

$$\mathcal{L}^{\text{mass}} \propto (\bar{\nu}_L \ \bar{e}_L)\begin{pmatrix} \phi_+ \\ \phi_0 \end{pmatrix} e_R + \bar{e}_R(\phi_- \ \phi_0)\begin{pmatrix} \nu_L \\ e_L \end{pmatrix}$$
$$= (\bar{e}_\nu e_R)\phi_+ + (\bar{e}_R e_L)\phi_- + (\bar{e}e)\phi_0. \tag{6.14}$$

The first two terms can be transformed away by a gauge transformation (unitary gauge). The last term gives mass to the electron if $\phi_0 \neq 0$ in the vacuum state. There is no mass term for the neutrino.

If one wishes to give the neutrino mass, one can make use of the conjugate Higgs doublet

$$\widetilde{\phi} \equiv \begin{pmatrix} \phi_0 \\ -\phi_- \end{pmatrix}, \quad \phi_- \equiv \phi_+^*, \tag{6.15}$$

which also transforms as an $SU(2)$ doublet, and add to (6.13) a term proportional to

$$\bar{L}\widetilde{\phi}\nu_R + \bar{\nu}_R\widetilde{\phi}^\dagger L = (\bar{\nu}_L \bar{e}_L)\begin{pmatrix} \phi_0 \\ -\phi_- \end{pmatrix}\nu_R + \bar{\nu}_R(\phi_0 \ -\phi_+)\begin{pmatrix} \nu_L \\ e_R \end{pmatrix}$$
$$= (\bar{\nu}\nu)\phi_0 - (\bar{e}_L\nu_R)\phi_- - (\bar{\nu}_R e_R)\phi_+. \tag{6.16}$$

The last two terms can be transformed away in unitary gauge, and the first term gives mass to the neutrino. However, we exclude neutrino mass for simplicity.

The Lagrangian density for the matter fields is taken to be

$$\mathcal{L}_0 = \bar{L}i\slashed{\partial}L + \bar{R}i\slashed{\partial}R + (\partial\phi)^\dagger \cdot (\partial\phi) - V(\phi^\dagger\phi) - \frac{m}{\rho_0}(\bar{L}\phi R + \bar{R}\phi^\dagger L), \tag{6.17}$$

$$V(\phi^\dagger\phi) = \lambda(\phi^\dagger\phi - \rho_0)^2,$$

where ρ_0, λ are real positive parameters, and m is the physical mass of the electron.

6.2 The Gauge Fields

1 Gauging $SU(2) \times U(1)$

The Lagrangian density (6.17) is globally invariant under $SU(2)$, whose generators will be denoted by $\mathbf{t}$ in general, and by $\boldsymbol{\tau}/2$ in the fundamental representation. The global gauge transformations are

$$L \to e^{-i\boldsymbol{\omega}\cdot\boldsymbol{\tau}/2}L, \quad \phi \to e^{-i\boldsymbol{\omega}\cdot\boldsymbol{\tau}/2}\phi,$$
$$R \to R. \tag{6.18}$$

In addition to $SU(2)$, the Lagrangian density is also invariant under independent phase changes of L and R:

$$L \to e^{-i\theta}L,$$
$$R \to e^{-i\theta'}R, \tag{6.19}$$
$$\phi \to e^{-i(\theta-\theta')}\phi.$$

These transformations form a group $U(1) \times U(1)$. We can identify the generator of one of the $U(1)$ groups as lepton number N, and call the generator of the other $U(1)$ group, weak hypercharge t_0:

$$L \to e^{-i(\alpha t_0 + \beta N)}L \quad (t_0 = -\tfrac{1}{2}, N = 1),$$
$$R \to e^{-i(\alpha t_0 + \beta N)}R \quad (t_0 = -1, N = 1), \tag{6.20}$$
$$\phi \to e^{-i(\alpha t_0 + \beta N)}\phi \quad (t_0 = \tfrac{1}{2}, N = 0).$$

The assignments for N are conventional, and t_0 satisfies the rule

$$Q = t_3 + t_0, \tag{6.21}$$

where Q is the electric charge, in units of the magnitude of the electronic charge. There is evidence that the $U(1)$ corresponding to lepton number is not a local gauge symmetry, because it is experimentally observed to be an unbroken symmetry; if it were gauged there would have to be a massless gauge field coupled to lepton number, which has not been observed.

The Weinberg-Salam model is obtained by gauging $SU(2) \times U(1)$, where the $U(1)$ is generated by weak hypercharge. To study the properties of the gauge fields, we continue to consider only one lepton doublet, since adding more doublets will not alter the gauge fields.

To gauge the symmetry $SU(2) \times U(1)$, we associate a gauge field with each of the generators, with notations given in Table 6.1. The locally gauge invariant

Table 6.1 GAUGE FIELDS OF THE WEINBERG-SALAM MODEL

Group	Generators	Gauge-Fields	Field Tensors
$SU(2)$	$\mathbf{t}$	$\mathbf{W}^\mu$	$\mathbf{G}^{\mu\nu} = \partial^\mu \mathbf{W}^\nu - \partial^\nu \mathbf{W}^\mu - g\mathbf{W}^\mu \times \mathbf{W}^\nu$
$U(1)$	t_0	W_0	$H^{\mu\nu} = \partial^\mu W_0{}^\nu - \partial^\nu W_0{}^\mu$

Weinberg-Salam Model

Lagrangian density is

$$\mathcal{L} = -\tfrac{1}{4}(G^{\mu\nu}\cdot G_{\mu\nu} + H^{\mu\nu}H_{\mu\nu}) + \bar{L}i\slashed{D}L + \bar{R}i\slashed{D}R \qquad (6.22)$$
$$+ (D\phi)^{\dagger}\cdot(D\phi) - V(\phi^{\dagger}\phi) - \frac{m}{\rho_0}(\bar{L}\phi R + \bar{R}\phi^{\dagger}L).$$

The covariant derivative D^{μ} is defined by

$$D^{\mu} = \partial^{\mu} + ig\mathbf{W}^{\mu}\cdot\mathbf{t} + ig'W_0{}^{\mu}t_0, \qquad (6.23)$$

where g and g' are two independent gauge coupling constants.

We require that there be only one massless neutral gauge field, the electromagnetic field A^{μ}, which is coupled to the charge eQ, with Q given by (6.21). Generally, it is a linear combination of $W_3{}^{\mu}$ and $W_0{}^{\mu}$. Accordingly we put

$$W_3{}^{\mu} = Z^{\mu}\cos\theta_W + A^{\mu}\sin\theta_W,$$
$$W_0{}^{\mu} = -Z^{\mu}\sin\theta_W + A^{\mu}\cos\theta_W. \qquad (6.24)$$

Solving this for A^{μ} and Z^{μ} gives

$$A^{\mu} = W_0{}^{\mu}\cos\theta_W + W_3{}^{\mu}\sin\theta_W,$$
$$Z^{\mu} = -W_0{}^{\mu}\sin\theta_W + W_3{}^{\mu}\cos\theta_W, \qquad (6.25)$$

where θ_W is called the Weinberg angle, a free parameter to be determined by experiments.

Given θ_W, the requirement that A^{μ} be the electromagnetic field imposes relations between g and g', as follows. Rewrite (6.23) as

$$D^{\mu} = \partial^{\mu} + ig(W_1{}^{\mu}t_1 + W_2{}^{\mu}t_2)$$
$$+ i(gt_3\sin\theta_W + g't_0\cos\theta_W)A^{\mu} \qquad (6.26)$$
$$+ i(gt_3\cos\theta_W - g't_0\sin\theta_W)Z^{\mu}$$

Requiring the coefficient of A^{μ} be eQ, we obtain the condition

$$gt_3\sin\theta_W + g't_0\cos\theta_W = e(t_3 + t_0), \qquad (6.27)$$

where $-e$ is the charge of the electron ($e^2/4\pi\hbar c \cong 1/137$). This leads to

$$e = g\sin\theta_W = g'\cos\theta_W, \qquad (6.28)$$

or

$$g'/g = \tan\theta_W,$$
$$e = gg'/(g^2 + g'^2)^{1/2}. \qquad (6.29)$$

With these, the covariant derivative can be written in the form

$$D^{\mu} = \partial^{\mu} + ig(W_1{}^{\mu}t_1 + W_2{}^{\mu}t_2) + ieQA^{\mu} + ieQ'Z\mu, \qquad (6.30)$$

where the neutral charge matrix Q' is defined by

$$Q' = t_3\cos\theta_W - t_0\tan\theta_W. \qquad (6.31)$$

To study the masses of the gauge fields, it is convenient to go to unitary gauge, in which

$$\phi = \begin{pmatrix} 0 \\ \rho \end{pmatrix}, \tag{6.32}$$

where ρ is a real field[a]. We can write

$$D^\mu \phi = \{\partial^\mu + ig[\tfrac{1}{2}(W_1^\mu - iW_2^\mu)\tau_+ + \tfrac{1}{2}(W_1^\mu + iW_2^\mu)\tau_- + W_3^\mu \tau_3]$$
$$+ ig'W_0^\mu t_0\} \begin{pmatrix} 0 \\ \rho \end{pmatrix} \tag{6.33}$$
$$= \begin{pmatrix} \tfrac{1}{2}ig(W_1^\mu - iW_2^\mu)\rho \\ \partial^\mu \rho - \dfrac{ig}{2\cos\theta_W} Z^\mu \rho \end{pmatrix}.$$

Hence, the kinetic term for the Higgs field in the Lagrangian density, which generates the masses, takes the form

$$(D^\mu\phi)^\dagger(D_\mu\phi) = \tfrac{1}{4}g^2\rho^2\left[(W_1^\mu W_{1\mu} + W_2^\mu W_{2\mu}) + \frac{Z^\mu Z_\mu}{\cos^2\theta_W}\right] \tag{6.34}$$
$$+ \partial^\mu\rho\,\partial_\mu\rho.$$

Note that A^μ does not appear, because the charged component of ϕ has been transformed away in (6.32). Therefore, A^μ is a massless field as desired. In terms of fields in the unitary gauge, the Lagrangian density is

$$\mathcal{L} = -\tfrac{1}{4}(\mathbf{G}\cdot\mathbf{G} + \mathbf{H}\cdot\mathbf{H}) + \tfrac{1}{4}g^2\rho^2\left(W_1^2 + W_2^2 + \frac{Z^2}{\cos^2\theta_W}\right)$$
$$- \bar{\nu}_L i\slashed{D}\nu_L + \bar{e}\left(i\slashed{D} - \frac{\rho}{\rho_0}m\right)e \tag{6.35}$$
$$+ \partial\rho\cdot\partial\rho - \lambda(\rho^2 - \rho_0^2)^2,$$

in an abbreviated notation that should be obvious. The masses m_W, m_Z, m_H of the fields $W_\pm^\mu$, Z^μ, and the Higgs fields $\eta = \rho - \rho_0$, are given by

$$m_W^2 = \tfrac{1}{2}g^2\rho_0^2,$$
$$m_Z/m_W = 1/\cos\theta_W, \tag{6.36}$$
$$m_H = 2\lambda^{1/2}\rho_0.$$

The motivation for constructing a gauge theory, with all masses generated by spontaneous symmetry breaking, is to have a renormalizable quantum field theory.

---

[a] It is more customary to write $\rho/\sqrt{2}$ in place of ρ in (6.32). We simply write ρ to avoid too many factors of 2.

Weinberg-Salam Model

2 Determination of Constants

The Fermi constant is given by

$$G = \frac{g^2}{2^{5/2} m_W^2} = 1.165 \times 10^{-5} \text{ (GeV)}^{-2}. \tag{6.37}$$

Substituting g^2 from (6.36) into (6.37), we have

$$m_W^2 = 2^{3/2} \rho_0^2 m_W^2 G. \tag{6.38}$$

Hence[b]

$$\rho_0^2 = 2^{-3/2} G^{-1}$$
$$\rho_0 = 174 \text{ GeV}. \tag{6.39}$$

Substituting (6.28) into (6.37), we obtain

$$m_W^2 = \frac{e^2}{2^{5/2} G \sin^2 \theta_W},$$
$$m_W \cong \frac{40}{\sin \theta_W} \text{ GeV}/c^2. \tag{6.40}$$

The Weinberg angle has been measured experimentally[4]:

$$\sin^2 \theta_W = 0.218 \pm 0.020. \tag{6.41}$$

Using this, we obtain

$$m_W \cong 80 \text{ GeV}/c^2,$$
$$m_Z \cong 90 \text{ GeV}/c^2. \tag{6.42}$$

The Higgs-electron coupling is extremely weak:

$$\frac{m_e}{\rho_0} = 3.86 \times 10^{-6}. \tag{6.43}$$

The Higgs mass is not determined, and it depends on the unknown dimensionless parameter λ.

3 Interactions

Rewrite (6.35) in unitary gauge as follows:

$$\mathscr{L} = \mathscr{L}_V + \mathscr{L}_F + \mathscr{L}_H + \mathscr{L}' + \mathscr{L}'', \tag{6.44}$$

where the first three terms denote the "free" Lagrangian density of the vector field, Fermi field, and Higgs field respectively:

$$\mathscr{L}_V = -\tfrac{1}{4}(\mathbf{G} \cdot \mathbf{G} + \mathbf{H} \cdot \mathbf{H}) + \tfrac{1}{2} m_W^2 (W_1^2 + W_2^2) + \tfrac{1}{2} m_Z^2 Z^2,$$
$$\mathscr{L}_F = \bar{L} i \partial\!\!\!/ L + \bar{R} i \partial\!\!\!/ R - m_e(\bar{L}R + \bar{R}L), \tag{6.45}$$
$$\mathscr{L}_H = \partial_\mu \eta \partial^\mu \eta - m_H^2 \eta \left(1 + \frac{\eta}{2\rho_0}\right)^2,$$

[b] The vacuum expectation value of the Higgs field is often quoted as $\langle \phi \rangle = \sqrt{2} \rho_0 = 247$ Gev.
[4] I. Liede and M. Roos, *Nucl. Phys.* **B167**, 397 (1980).

where η is the real Higgs field in unitary gauge, in which $\phi(x)$ has the form (6.32), and

$$\rho(x) = \rho_0 + \eta(x). \tag{6.46}$$

The terms $\mathcal{L}'$ and $\mathcal{L}''$ are interaction terms, with $\mathcal{L}'$ containing the electromagnetic and weak currents, and $\mathcal{L}''$ containing the interactions between the Higgs field and other fields:

$$\mathcal{L}'' = [m_W^2(W_1^2 + W_2^2) + m_Z^2 Z^2] \frac{\eta}{\rho_0}\left(1 + \frac{\eta}{2\rho_0}\right) + m(\bar{e}e)\frac{\eta}{\rho_0}, \tag{6.47}$$

$$\mathcal{L}' = g\bar{\psi}(W_1 t_1 + W_2 t_2)\psi + e\bar{\psi}QA\psi + e\bar{\psi}Q'Z\psi$$

$$= \frac{g}{2}(W_1^\mu J_{1\mu}^{\text{ch}} + W_2^\mu J_{2\mu}^{\text{ch}}) + eA^\mu J_\mu^{\text{em}} + eZ^\mu J_\mu^{\text{neut}}, \tag{6.48}$$

where

$$\begin{aligned} J_{i\mu}^{\text{ch}} &= \bar{L}\tau_i\gamma_\mu L \quad (i = 1, 2), \\ J_\mu^{\text{em}} &= \bar{\psi}Q\gamma_\mu\psi, \\ J_\mu^{\text{neut}} &= \bar{\psi}Q'\gamma_\mu\psi. \end{aligned} \tag{6.49}$$

The "free" Lagrangian for the gauge fields is, again,

$$\mathcal{L}_V = -\tfrac{1}{4}(\mathbf{G}\cdot\mathbf{G} + \mathbf{H}\cdot\mathbf{H}) + \tfrac{1}{2}m_W^2(W_1^2 + W_2^2) + \tfrac{1}{2}m_Z^2 Z^2. \tag{6.50}$$

To write this out more explicitly, we use the following notation:

$$\begin{aligned} (A \times B)^{\mu\nu} &\equiv A^\mu B^\nu - A^\nu B^\mu, \\ (A \times B)^2 &\equiv (A \times B)^{\mu\nu}(A \times B)_{\mu\nu}, \\ (A \times B)\cdot(C \times D) &\equiv (A \times B)^{\mu\nu}(C \times D)_{\mu\nu}. \end{aligned} \tag{6.51}$$

Then

$$\begin{aligned} \mathcal{L}_V = &-\tfrac{1}{4}[(\partial \times W_1)^2 + (\partial \times W_2)^2 + (\partial \times A)^2 + (\partial \times Z)^2] \\ &+ \tfrac{1}{2}m_W^2(W_1^2 + W_2^2) + \tfrac{1}{2}m_Z^2 Z^2 \\ &+ \tfrac{1}{2}g[(\partial \times W_1)\cdot(W_2 \times W_3) + (\partial \times W_2)\cdot(W_3 \times W_1) \\ &+ (\partial \times W_3)\cdot(W_1 \times W_2)] \\ &- \tfrac{1}{4}g^2[(W_1 \times W_2)^2 + (W_2 \times W_3)^2 + (W_3 \times W_1)^2], \end{aligned} \tag{6.52}$$

where W_3^μ is given by (6.24). The terms of orders g and g^2 describe interactions among the gauge fields, which arise from the non-Abelian structure of the gauge group. Note that Z^μ does not interact directly with A^μ, as we would expect of a neutral field.

The charged gauge fields W_1^μ, W_2^μ have electromagnetic interactions. The part of (6.52) containing these interactions can be written in the form

$$\begin{aligned} \mathcal{L}_W^{\text{em}} = &\, eW_1^\mu W_2^\nu F_{\mu\nu} + e[(\partial \times W_1)^{\mu\nu}W_{2\mu} - (\partial \times W_2)^{\mu\nu}W_{1\mu}]A_\nu \\ &+ \tfrac{1}{2}e^2[W_1^\mu W_2^\nu + W_2^\mu W_1^\nu - g^{\mu\nu}(W_1^2 + W_2^2)]A_\mu A_\nu, \end{aligned} \tag{6.53}$$

where $F^{\mu\nu} \equiv \partial^\mu A^\nu - \partial^\nu A^\mu$. Rewriting this in terms of the fields with definite

Weinberg-Salam Model

charge:
$$W = 2^{-1/2}(W_1 + iW_2),$$
$$W^* = 2^{-1/2}(W_1 - iW_2), \qquad (6.54)$$

we can rewrite (6.53) as
$$\begin{aligned}\mathscr{L}_W^{\text{em}} &= -ie(W_\nu^* \overleftrightarrow{\partial}_\mu W^\nu)A^\mu - ieW^{*\mu}W^\nu F_{\mu\nu} \\ &\quad - ie(W^{*\mu}W^\nu - \text{c.c.})A_\nu \\ &\quad + e^2(W^{*\mu}W^\nu - g^{\mu\nu}W^{*\lambda}W_\lambda)A_\mu A_\nu.\end{aligned} \qquad (6.55)$$

The second term gives rise to magnetic moments and electric quadrupole moments[5]. In general, a term of the form $\mathscr{L}_W^{\text{em}} = -ie\kappa W_\mu^* W_\nu F^{\mu\nu}$ gives rise to

$$\text{Magnetic dipole moment} = (1+\kappa)\frac{e}{2m_W}\mathbf{s},$$

$$\text{Electric quadrupole moment} = \int d^3x \rho(3z^2 - r^2) = -\frac{e\kappa}{m_W^2}, \qquad (6.56)$$

where $\mathbf{s}$ is the spin vector, and ρ is the static electric charge density in the state $s_z = 1$. Since $\kappa = 1$, the g-factor of the W boson is equal to 2.

The complicated structure of (6.55) actually serves a simple purpose; when (6.55) is added to the kinetic term for the W field, we obtain the Lagrangian density for a charged vector theory:

$$\begin{aligned}&-\tfrac{1}{4}[(\partial \times W_1)^2 + (\partial \times W_2)^2] + \mathscr{L}_W^{\text{em}} \\ &= -\tfrac{1}{2}(D \times W)^* \cdot (D \times W) - ieW^{*\mu}W^\nu F_{\mu\nu},\end{aligned}$$
$$D^\mu \equiv \partial^\mu - ieA^\mu.$$

6.3 The General Theory

1 Mass Terms

To include all the quarks and leptons in the three families as described in Chapter I, we make two modifications in (6.22).

First, the definitions of R and L are extended to include all quarks and leptons:
$$\begin{aligned}L &= \{l_{L\alpha}, q_{L\alpha}\} \quad (\alpha = 1, 2, 3), \\ R &= \{l_{R\alpha}, q_{R\alpha}\} \quad (\alpha = 1, 2, 3).\end{aligned} \qquad (6.57)$$

where the symbols are defined in Table 6.2. Secondly, the mass terms are modified to enable us to assign arbitrary masses. The locally gauge invariant Lagrangian is

$$\begin{aligned}\mathscr{L} &= -\tfrac{1}{4}(\mathbf{G}\cdot\mathbf{G} + \mathbf{H}\cdot\mathbf{H}) + \bar{L}i\slashed{D}L + \bar{R}i\slashed{D}R + (D\phi)^\dagger \cdot (D\phi) \\ &\quad - V(\phi^*\phi) + \mathscr{L}_{\text{mass}}.\end{aligned} \qquad (6.58)$$

The mass term is further split into lepton and quark contributions:
$$\mathscr{L}_{\text{mass}} = \mathscr{L}_{\text{mass}}^{\text{lept}} + \mathscr{L}_{\text{mass}}^{\text{quark}}, \qquad (6.59)$$

[5] T. D. Lee and C. N. Yang, *Phys. Rev.* **128**, 885 (1962).

Table 6.2 QUANTUM NUMBERS

Subscripts L, R denote left- and right-handed fields, respectively. A tilde over quark symbols denote generalized Cabbibo mixed states.

The Weinberg angle is denoted by θ ($\sin^2\theta \cong 1/4$).

Electric charge = eQ, $Q = t_3 + t_0$

Neutral charge = eQ', $Q' = t_3 \cot\theta - t_0 \tan\theta$

	Particles			t	t_3	t_0	Q	$Q' \sin 2\theta$	$\dfrac{Q'}{(\text{for } \sin^2\theta = 1/4)}$
Leptons $l_{L\alpha}$:	$\begin{pmatrix}\nu_L\\ e_L\end{pmatrix}$	$\begin{pmatrix}\nu'_L\\ \mu_L\end{pmatrix}$	$\begin{pmatrix}\nu''_L\\ \tau_L\end{pmatrix}$	$\dfrac{1}{2}$ $\dfrac{1}{2}$	$\dfrac{1}{2}$ $-\dfrac{1}{2}$	$-\dfrac{1}{2}$ $-\dfrac{1}{2}$	0 -1	1 $-1 + 2\sin^2\theta$	$\dfrac{2}{\sqrt{3}}$ $-\dfrac{1}{\sqrt{3}}$
$l_{R\alpha}$:	e_R	μ_R	τ_R	0	0	-1	-1	$2\sin^2\theta$	$\dfrac{1}{\sqrt{3}}$
Quarks $q_{L\alpha}$:	$\begin{pmatrix}\tilde{u}_L\\ d_L\end{pmatrix}$	$\begin{pmatrix}\tilde{c}_L\\ s_L\end{pmatrix}$	$\begin{pmatrix}\tilde{t}_L\\ b_L\end{pmatrix}$	$\dfrac{1}{2}$ $\dfrac{1}{2}$	$\dfrac{1}{2}$ $-\dfrac{1}{2}$	$\dfrac{1}{6}$ $\dfrac{1}{6}$	$\dfrac{2}{3}$ $-\dfrac{1}{3}$	$1 - \dfrac{4}{3}\sin^2\theta$ $-1 + \dfrac{2}{3}\sin^2\theta$	$\dfrac{4}{3\sqrt{3}}$ $-\dfrac{5}{3\sqrt{3}}$
$q_{R\alpha}$:	u_R	c_R	t_R	0	0	$\dfrac{2}{3}$	$\dfrac{2}{3}$	$-\dfrac{4}{3}\sin^2\theta$	$-\dfrac{2}{3\sqrt{3}}$
$q_{R\alpha}$:	d_R	s_R	b_R	0	0	$-\dfrac{1}{3}$	$-\dfrac{1}{3}$	$\dfrac{2}{3}\sin^2\theta$	$\dfrac{1}{3\sqrt{3}}$
Higgs	$\begin{pmatrix}\phi_+\\ \phi_0\end{pmatrix}$			$\dfrac{1}{2}$ $\dfrac{1}{2}$	$\dfrac{1}{2}$ $-\dfrac{1}{2}$	$\dfrac{1}{2}$ $\dfrac{1}{2}$	1 0	$1 - 2\sin^2\theta$ -1	$\dfrac{1}{\sqrt{3}}$ $-\dfrac{2}{\sqrt{3}}$

with the lepton contribution given by

$$\mathcal{L}_{\text{mass}}^{\text{lept}} = \sum_{\alpha,\beta=1}^{3} \bar{l}_{L\alpha} \frac{\phi}{\rho_0} m_{\alpha\beta} l_{R\beta} + \text{c.c.}, \tag{6.60}$$

where $m_{\alpha\beta}$ is an arbitrary constant complex mass matrix. In unitary gauge, in which ϕ is given by (6.32), we have

$$\mathcal{L}_{\text{mass}}^{\text{lept}} = \frac{\rho}{\rho_0} (\bar{e}_L \quad \bar{\mu}_L \quad \bar{\tau}_L) m \begin{pmatrix} e_R \\ \mu_R \\ \tau_R \end{pmatrix} + \text{c.c.} \tag{6.61}$$

Now, an arbitrary complex matrix can be brought to diagonal form with real non-negative diagonal elements. That is, there exists non-singular matrices A

Weinberg-Salam Model

and B, such that

$$AmB^{-1} = D, \qquad (6.62)$$

where D is diagonal, with non-negative diagonal elements. To prove this, note that $m^\dagger m$ is hermitian, with real non-negative eigenvalues, and the same is true of $mm^\dagger$. The eigenvalues of $m^\dagger m$ are the same as those of $mm^\dagger$. Therefore, there exist non-singular A and B such that

$$\begin{aligned} Amm^\dagger A^{-1} &= D^2, \\ Bm^\dagger m B^{-1} &= D^2. \end{aligned} \qquad (6.63)$$

The solutions to these equations are

$$\begin{aligned} m &= A^{-1}DB, \\ m^\dagger &= B^{-1}DA. \end{aligned} \qquad (6.64)$$

Thus, we make independent linear transformations on right- and left-handed particles:

$$\begin{pmatrix} e_R \\ \mu_R \\ \tau_R \end{pmatrix} \to B^{-1} \begin{pmatrix} e_R \\ \mu_R \\ \tau_R \end{pmatrix}, \qquad (6.65)$$

$$\begin{pmatrix} e_L \\ \mu_L \\ \tau_L \end{pmatrix} \to A^{-1} \begin{pmatrix} e_L \\ \mu_L \\ \tau_L \end{pmatrix}.$$

This makes the mass matrix diagonal:

$$\mathcal{L}^{\text{lept}}_{\text{mass}} = (\bar{e} \; \bar{\mu} \; \bar{\tau}) \begin{pmatrix} m_e & 0 & 0 \\ 0 & m_\mu & 0 \\ 0 & 0 & m_\tau \end{pmatrix} \begin{pmatrix} e \\ \mu \\ \tau \end{pmatrix} \frac{\rho}{\rho_0}, \qquad (6.66)$$

where m_e, m_μ, m_τ are arbitrary non-negative real parameters. Note that the transformations (6.65) mix leptons of different families. The same transformations, of course, must also be made in the kinetic part of $\mathcal{L}$, i.e., in the terms $(\bar{L}i\displaystyle{\not}\partial L + \bar{R}i\displaystyle{\not}\partial R)$ in (6.58). The effect is to transform the charge-changing current in the following way:

$$(\bar{e}_L \; \bar{\mu}_L \; \bar{\tau}_L)\gamma^\mu \begin{pmatrix} \nu_L \\ \nu'_L \\ \nu''_L \end{pmatrix} \to (\bar{e}_L \; \bar{\mu}_L \; \bar{\tau}_L)\gamma^\mu A \begin{pmatrix} \nu_L \\ \nu'_L \\ \nu''_L \end{pmatrix}. \qquad (6.67)$$

The non-charge-changing currents remain invariant, because they are of the form $\bar{L}\gamma^\mu L + \bar{R}\gamma^\mu R$.

The symbols e, μ, τ in (6.66) and in (6.67) denote mass eigenstates, because the mass matrix is diagonal in the corresponding basis. Therefore, the physical neutrinos are

$$\begin{pmatrix} \tilde{\nu} \\ \tilde{\nu}' \\ \tilde{\nu}'' \end{pmatrix}_L = A \begin{pmatrix} \nu \\ \nu' \\ \nu'' \end{pmatrix}_L. \qquad (6.68)$$

But this makes no difference, since the neutrinos are massless. Hence we can drop the tildes over the ν's. In the Weinberg-Salam model, there is no mixing of physical leptons of different families, when all neutrinos are assumed to be massless. For example, $\mu \to e + \gamma$ is forbidden.

We now turn to quark masses. The quarks denoted by q_R and $\tilde{q}_R$ in Table 6.2 are to be coupled to the Higgs doublet ϕ and the conjugation doublet $\tilde{\phi}$ respectively, to give a mass term invariant under weak hypercharge, as follows:

$$\mathcal{L}_{\text{mass}}^{\text{quark}} = \sum_{\alpha,\beta=1}^{3} \left(\bar{q}_{L\alpha} \frac{\phi}{\rho_0} M_{\alpha\beta} q_{R\beta} + \bar{q}_{L\alpha} \frac{\tilde{\phi}}{\rho_0} \tilde{M}_{\alpha\beta} \tilde{q}_{R\beta} \right) + \text{c.c.}, \quad (6.69)$$

where $\tilde{\phi}$ is defined in (6.15), and M and $\tilde{M}$ are arbitrary constant complex matrices. *A color sum over each flavor of quarks is understood.* In unitary gauge, this reads

$$\mathcal{L}_{\text{mass}}^{\text{quark}} = \frac{\rho}{\rho_0} \left[(\bar{d}_L \; \bar{s}_L \; \bar{b}_L) M \begin{pmatrix} d_R \\ s_R \\ b_R \end{pmatrix} + (\bar{u}_L \; \bar{c}_L \; \bar{t}_L) \tilde{M} \begin{pmatrix} u_R \\ c_R \\ t_R \end{pmatrix} \right] + \text{c.c.} \quad (6.70)$$

We diagonalize M and $\tilde{M}$ by making the transformations

$$\begin{pmatrix} d \\ s \\ b \end{pmatrix}_L \to A_L^{-1} \begin{pmatrix} d \\ s \\ b \end{pmatrix}_L, \quad \begin{pmatrix} u \\ c \\ t \end{pmatrix}_L \to B_L^{-1} \begin{pmatrix} u \\ c \\ t \end{pmatrix}_L,$$

$$\begin{pmatrix} d \\ s \\ b \end{pmatrix}_R \to A_R^{-1} \begin{pmatrix} d \\ s \\ b \end{pmatrix}_R, \quad \begin{pmatrix} u \\ c \\ t \end{pmatrix}_R \to B_R^{-1} \begin{pmatrix} u \\ c \\ t \end{pmatrix}_R, \quad (6.71)$$

where A_L, A_R, B_L, B_R are such that

$$A_L M A_R^{-1} = \begin{pmatrix} m_d & 0 & 0 \\ 0 & m_s & 0 \\ 0 & 0 & m_b \end{pmatrix},$$

$$B_L \tilde{M} B_R^{-1} = \begin{pmatrix} m_u & 0 & 0 \\ 0 & m_c & 0 \\ 0 & 0 & m_t \end{pmatrix}, \quad (6.72)$$

where the diagonal elements are arbitrary real non-negative parameters. With this, we have

$$\mathcal{L}_{\text{mass}}^{\text{quark}} = \frac{\rho}{\rho_0} \sum_q m_q \bar{q} q, \quad (6.73)$$

where q denotes the 4-component Dirac spinor for a quark, m_q its mass parameter, and the sum extends over all six flavors u, d, s, c, b, t (plus a color sum for each flavor). These quarks are now mass eigenstates.

The transformation (6.71) in the kinetic part of the Lagrangian density leaves the electromagnetic and neutral currents invariant, but gives a charge-changing

Weinberg-Salam Model

quark current of the form

$$J_\mu^{ch} \propto (\bar{d} \; \bar{s} \; \bar{b})_L \gamma_\mu C \begin{pmatrix} u \\ c \\ t \end{pmatrix}_L, \tag{6.74}$$

where the quark symbols refer to mass eigenstates, and C is the 3×3 matrix

$$C = A_L B_R^{-1}, \tag{6.75}$$

with the properties

$$C^\dagger C = 1,$$
$$|\det C|^2 = 1. \tag{6.76}$$

Since quarks are supposed to be confined by virtue of their strong interactions, the meaning of the quark mass is not obvious. We shall take this up when we discuss quantum chromodynamics, in chap. XII.

2 Cabibbo Angle

To understand the effect of the matrix C in (6.74), let us first ignore the b and t quarks, so that

$$J_\mu^{ch} \propto (\bar{d} \; \bar{s})_L \gamma_\mu C \begin{pmatrix} u \\ c \end{pmatrix}_L. \tag{6.77}$$

The most general form of the 2×2 matrix C is then

$$C = \begin{pmatrix} e^{i\theta} & 0 \\ 0 & 1 \end{pmatrix} \times [SU(2) \text{ matrix}]. \tag{6.78}$$

The first factor may be ignored, as it can be absorbed into a redefinition of the phases of u and c. The most general form of an $SU(2)$ matrix corresponds to a rotation through the Euler angles α, β, γ:

$$C = e^{i\gamma\tau_3/2} \, e^{i\beta\tau_2/2} \, e^{-i\alpha\tau_3/2}. \tag{6.79}$$

Again, the first and last factor can be absorbed into redefinitions of the phases of (d, s) and (u, c) respectively. Therefore, we may take $C = e^{-i\beta\tau_2/2}$. Putting $\theta = \beta/2$, we have

$$C = \begin{matrix} d \\ s \end{matrix} \begin{pmatrix} \cos\theta & \sin\theta \\ -\sin\theta & \cos\theta \end{pmatrix}, \tag{6.80}$$

where θ is an arbitrary angle to be determined by experiments. This is the Cabibbo angle, as discussed earlier in Chapter I, with the experimental value

$$\theta \cong \tfrac{1}{4}. \tag{6.81}$$

Since C is real, the Lagrangian density remains real, and is invariant under time-reversal. Through the CPT theorem, it is therefore invariant under CP.

3 Kobayashi-Maskawa Matrix

The most general form of C for the 3×3 case was first worked out by Kobayashi and Maskawa[6]. By (6.76), C can be reduced to an $SU(3)$ matrix, by

[6] M. Kobayashi and K. Maskawa, *Prog. Theo. Phys.* **49**, 652 (1975).

redefining the phases of the quark fields. Hence it is a sum of products of the matrices $e^{i\theta_1 \lambda_1}, \ldots, e^{i\theta_8 \lambda_8}$, where $\lambda_1, \ldots, \lambda_8$ are the Gell-Mann matrices, of which λ_3 and λ_8 are chosen to be diagonal. Therefore, there are 8 free parameters $\theta_1, \ldots, \theta_8$. We can write C in the form

$$C = e^{i\lambda_3 \theta_3} e^{i\lambda_8 \theta_8} U e^{i\lambda_3 \theta_3'} e^{i\lambda_8 \theta_8'}, \qquad (6.82)$$

where U is an $SU(3)$ matrix containing only 4 parameters, and consequently can be constructed from any 4 generators, excluding λ_3 and λ_8. For convenience we choose these to be $\lambda_1, \lambda_2, \lambda_5, \lambda_7$. We note that $\lambda_2, \lambda_5, \lambda_7$ generate a rotation group, because $[\lambda_2, \lambda_5] = i\lambda_7$. Therefore, U can be constructed from 3 orthogonal matrices with 3 arbitrary parameters, and one unitary matrix of determinant 1, with one arbitrary parameter:

$$C = R_2 \tilde{U} R_1 R_3 \qquad (6.83)$$

where

$$\tilde{U} = \begin{pmatrix} e^{-i\delta/3} & 0 & 0 \\ 0 & e^{-i\delta/3} & 0 \\ 0 & 0 & e^{2i\delta/3} \end{pmatrix} = e^{-i\delta/3} \begin{pmatrix} 1 & 0 & 0 \\ 0 & 1 & 0 \\ 0 & 0 & e^{i\delta} \end{pmatrix}, \qquad (6.84)$$

and

$$R_1 = \begin{pmatrix} c_1 & s_1 & 0 \\ -s_1 & c_1 & 0 \\ 0 & 0 & 1 \end{pmatrix}, \quad R_2 = \begin{pmatrix} 1 & 0 & 0 \\ 0 & c_2 & s_2 \\ 0 & -s_2 & c_2 \end{pmatrix}, \quad R_3 = \begin{pmatrix} 1 & 0 & 0 \\ 0 & c_3 & s_3 \\ 0 & -s_3 & c_3 \end{pmatrix}, \qquad (6.85)$$

$c_i \equiv \cos \theta_i, \quad s_i \equiv \sin \theta_i \quad (i = 1, 2, 3)$.

The overall phase $e^{-i\delta/3}$ in (6.84) may be absorbed into the quark fields. Therefore

$$C = \begin{pmatrix} 1 & 0 & 0 \\ 0 & c_2 & s_2 \\ 0 & -s_2 & c_2 \end{pmatrix} \begin{pmatrix} c_1 & s_1 & 0 \\ -s_1 & c_1 & 0 \\ 0 & 0 & e^{i\delta} \end{pmatrix} \begin{pmatrix} 1 & 0 & 0 \\ 0 & c_3 & s_3 \\ 0 & -s_3 & c_3 \end{pmatrix}$$

$$\begin{array}{c} \\ = \end{array} \begin{array}{c} \\ d \\ s \\ b \end{array} \begin{pmatrix} u & c & t \\ c_1 & s_1 c_3 & s_1 s_3 \\ -s_1 c_2 & c_1 c_2 c_3 - s_2 s_3 e^{i\delta} & c_1 c_2 c_3 + s_2 c_3 e^{i\delta} \\ s_1 s_2 & -c_1 s_2 c_3 - c_2 s_3 e^{i\delta} & -c_1 s_2 c_3 + c_2 c_3 e^{i\delta} \end{pmatrix}. \qquad (6.86)$$

In the literature, one may find (c_2, s_2) interchanged with $(c_3, \pm s_3)$, and $-\delta$ instead of δ. There are 3 arbitrary angles $\theta_1, \theta_2, \theta_3$, of which θ_1 is the Cabbibo angle. There is an arbitrary phase δ, which makes the Lagrangian density non-real. Therefore, a non-zero value of δ violates time-reversal invariance. By the CPT theorem, it violates CP invariance[c]. The only parameters determined

---

[c] It is difficult to put a bound on δ by the magnitude of the observed CP violation in K° decay, because δ appears only in flavor-mixed components involving the heavy quarks c and t. In K° decay, it can contribute only through the virtual effects of the heavy quarks. Thus, the observed smallness of CP violation does not necessarily imply that δ is small.

with precision from experimental data are:[7]
$$|c_1| = 0.9737 + 0.0025,$$
$$|s_1 c_2| = 0.219 \pm 0.003. \qquad (6.87)$$

The parameters θ_3 and δ are essentially unknown, except for bounds obtained from a combination of experimental data and theoretical hypotheses.[8]

We summarize the Weinberg-Salam model by displaying the complete Lagrangian density in Table 6.3, which is self-explanatory.

4 Solitons

There are no static topological solitons in the Weinberg-Salam model. The gauge group is $G = SU(2) \times U(1)$, with elements of the form
$$g \in G, \qquad g = e^{-i\boldsymbol{\omega} \cdot \mathbf{t}} e^{-i\nu t_0}.$$

Table 6.3 LAGRANGIAN DENSITY OF THE WEINBERG-SALAM MODEL

$\mathscr{L} = \mathscr{L}_V + \mathscr{L}_F + \mathscr{L}_H + \mathscr{L}_{int}$ [$\mathscr{L}_V$ given in (6.50) and (6.52)]

$\mathscr{L}_F = \bar{\psi}(i\partial\!\!\!/ - m)\psi$

$\mathscr{L}_H = \partial\eta \cdot \partial\eta - m_H^2 \eta^2$

$\mathscr{L}_{int} = \mathscr{L}_{VV} + \mathscr{L}_{VF} + \mathscr{L}_{HV} + \mathscr{L}_{HF} + \mathscr{L}_{HH}$

$\mathscr{L}_{VV} = \frac{1}{2}g[(\partial \times \mathbf{W}_1) \cdot (\mathbf{W}_2 \times \mathbf{W}_3) + (\partial \times \mathbf{W}_2) \cdot (\mathbf{W}_3 \times \mathbf{W}_1) + (\partial \times \mathbf{W}_3) \cdot (\mathbf{W}_1 \times \mathbf{W}_2)]$

$\quad -\frac{1}{4}g^2[(\mathbf{W}_1 \times \mathbf{W}_2)^2 + (\mathbf{W}_2 \times \mathbf{W}_3)^2 + (\mathbf{W}_3 \times \mathbf{W}_1)^2]$

$\mathscr{L}_{VF} = \frac{1}{2}g(W_1 \cdot J_1^{ch} + W_2 \cdot J_2^{ch}) + eA \cdot J^{em} + \frac{g}{\cos\theta_W} Z \cdot J^{neut}$

$\quad J_\mu^{ch} = \bar{l}_L \tau_i \gamma_\mu l_L + \bar{q}_L \tau_i \gamma_\mu C q_L$

$\quad\quad = \bar{\psi}\tau_i \gamma_\mu \frac{1 - \gamma_5}{2} C\psi$

$\quad J_\mu^{em} = \bar{\psi} Q \gamma_\mu \psi$

$\quad J_\mu^{neut} = \bar{\psi} Q' \gamma_\mu \psi$

$\mathscr{L}_{HV} = [m_W^2(W_1^2 + W_2^2) + m_Z^2 Z^2]\dfrac{\eta}{\rho_0}\left(1 + \dfrac{\eta}{2\rho_0}\right)$

$\mathscr{L}_{HF} = (\bar{\psi}M\psi)\dfrac{\eta}{\rho_0}$

$\mathscr{L}_{HH} = -\dfrac{1}{6}\rho_0 \lambda \eta^3 - \dfrac{1}{24}\lambda \eta^4$

---

[7] R. E. Shrock and L. L. Wang, *Phys. Rev. Lett.* **41**, 1692 (1978).
[8] R. E. Shrock, S. B. Treiman and L. L. Wang, *Phys. Rev. Lett.* **42**, 1589 (1979).

The little group is $H = [U(1)]_{em}$, with elements

$$h \in H, \quad h = e^{-i\theta Q} = e^{-i\theta(t_3 + t_0)}.$$

It is clear that H is not a normal subgroup of G. A coset with respect to H is, for fixed ω and ν,

$$gH = \{e^{-i\omega \cdot \mathbf{t}} e^{-i\nu t_0} e^{-i\theta Q} | -\infty < \theta < \infty\}.$$

Parametrize $e^{-i\omega \cdot \mathbf{t}}$ by Euler angles α, β, γ:

$$e^{-i\omega \cdot \mathbf{t}} = e^{-i\alpha t_3} e^{-i\beta t_2} e^{-i\gamma t_3}.$$

Using $t_0 = Q - t_3$, we have

$$e^{-i\omega \cdot \mathbf{t}} e^{-i\nu t_0} e^{-i\theta Q} = e^{-i\alpha t_3} e^{-i\beta t_2} e^{-i(\gamma - \theta)t_3} e^{-i(\nu - \theta)Q}.$$

Noting that $\nu - \theta$ and $\nu + \theta$ are independent parameters, we can write the set of cosets as

$$gH = \{e^{-i\omega' \cdot \mathbf{t}} e^{-i\theta' Q} | -\infty < \theta' < \infty\},$$

$$G/H = \{e^{-i\omega' \cdot \mathbf{t}}\}.$$

Since G/H has the topology of $SU(2)$, $\pi_2(G/H) = 0$.

Semi-classical non-topological solitons do exist. A class of these consists of a torus in space in which the Higgs field is expelled by one quantum of the Z field. These have been called "vorticons", and their masses estimated to be of the order of 3000 GeV/c^2.[9]

6.4 Comments

Experimental verification of the Weinberg-Salam model has so far been limited to energy domains far below the theoretical masses of the gauge bosons, where the interactions are indistinguishable from current-current interactions. In this domain, the structures of the charge and neutral currents have been verified, and the Weinberg angle has been measured. However, two important aspects of the theory have not been tested:

1. the existence of the W and the Z, with their predicted masses;
2. the existence of the Higgs boson, with its predicted couplings.

If the gauge bosons W and Z do not exist, or if they exist with the wrong masses, then the gauge concept would require re-examination. In particular, Bjorken[10], Hung and Sakurai[11] pointed out that there are alternatives to the gauge theory which reproduce all experimental results to date.

An intriguing possibility is that, in the Weinberg-Salam model, we have gauged the wrong symmetry; that the gauge bosons in this model are really bound states, on the same footing with quarks and leptons. The binding force would have to come from some still unknown source, perhaps a deeper gauged symmetry. We may make an analogy with the theory of strong interactions. A

---

[9] K. Huang and R. Tipton, *Phys. Rev.* **23**, 3050 (1981).
[10] J. D. Bjorken, *Phys. Rev.* **D19**, 335 (1979).
[11] P. Q. Hung and J. J. Sakurai, *Nucl. Phys.* **B143**, 81 (1978).

long time ago, before we knew anything about quarks, Sakurai[12] proposed that strong interactions arise from gauging isospin $SU(2)$ and baryon number $U(1)$, with the gauge bosons identified respectively with the ρ and the ω. The proposal ran into difficulty with giving mass to the particles without spoiling renormalizability, the Higgs mechanism being unknown then.

If Sakurai had known about the Higgs mechanism, he could have devised a renormalizable theory, albeit with a large number of arbitrary parameters. A huge industry of strong interaction phenomenology might have been spawned (for one cannot really calculate anything in theories with strong coupling). And, like the Regge phenomenology, it would fit almost anything, and come to a halt only when enough people believe in quarks. Is it possible that the W and Z are similar to the ρ and ω? A definite scenario for such a case has been suggested by Abbott and Farhi[13].

With regard to the Higgs mechanism, there is so far no experimental evidence to support an elementary Higgs field. The only similar concept in other physical models is the Ginzberg-Landau order parameter in the theory of superconductivity, which is a kind of mean field approximation to the condensate of Cooper pairs. It is possible that the Higgs field in the Weinberg-Salam model is of a similar nature. Replacing the Higgs field by something more fundamental, with fewer arbitrary parameters, would be aesthetically pleasing.

[12] J. J. Sakurai, *Ann. Phys.* **11**, 1 (1960).
[13] L. F. Abbot and E. Farhi, *Phys. Lett.* **B101**, 69 (1981); *Nucl. Phys.* **B189**, 547 (1981).

CHAPTER VII
METHOD OF PATH INTEGRALS

7.1 Non-Relativistic Quantum Mechanics

Feynman's method of path integrals[1] is a method of formulating a quantum theory. It is equivalent to the more familiar method of canonical quantization, but much more convenient to use for gauge theories. We introduce the method by first considering non-relativistic quantum mechanics with one degree of freedom. Generalizations will then become obvious.

In the method of canonical quantization, we work with Hilbert-space operators q_{op} and p_{op}, representing respectively the coordinate and the conjugate momentum, as defined by the commutation relation[a]

$$[p_{op}, q_{op}] = -i\hbar. \tag{7.1}$$

We denote the eigenstates of these operators by $|q\rangle$ and $|p\rangle$:

$$\begin{aligned} q_{op}|q'\rangle &= q'|q'\rangle, \\ p_{op}|p'\rangle &= p'|p'\rangle, \end{aligned} \tag{7.2}$$

and normalize them according to

$$\langle q''|q'\rangle = \delta(q'' - q'), \qquad \int_{-\infty}^{\infty} dq'|q'\rangle\langle q'| = 1,$$

$$\langle p''|p'\rangle = 2\pi\hbar\,\delta(p'' - p'), \qquad \int_{-\infty}^{\infty} \frac{dp_1}{2\pi\hbar}|p'\rangle\langle p'| = 1. \tag{7.3}$$

$$\langle p'|q'\rangle = e^{-ip'q'/\hbar}.$$

The dynamics of the system is completely specified by the transition amplitude governing the time evolution of the system:

$$\langle q'', t''|q', t'\rangle \equiv \left\langle q''\left|\exp\left[-\frac{i}{\hbar}(t'' - t')H_{op}\right]\right|q'\right\rangle, \tag{7.4}$$

where H_{op} is the time-independent Hamiltonian operator. We may regard $|q', t'\rangle$ as the eigenstate of the Heisenberg operator $q_{op}(t)$ with the eigenvalue q':

$$\begin{aligned} q_{op}(t) &\equiv e^{itH_{op}/\hbar} q_{op} e^{-itH_{op}/\hbar}, \\ q_{op}(t)|q', t'\rangle &= q'|q', t'\rangle. \end{aligned} \tag{7.5}$$

---

[a] In this chapter we do not set $\hbar = 1$.

[1] R. P. Feynman and A. R. Hibbs, *Quantum Mechanics and Path Integrals* (McGraw-Hill, New York, 1965).

Method of Path Integrals

The time evolution is then given by the equation

$$|q', t'\rangle = e^{it'H_{op}/\hbar}|q'\rangle. \tag{7.6}$$

The object of the method of path integrals is to express the transition amplitude entirely in terms of a classical Hamiltonian $H(p, q)$, without reference to operators and states in Hilbert space.

To proceed, we divide the time evolution interval $t'' - t'$ into N equal steps, and take the limit $N \to \infty$ later. Let

$$\begin{aligned}\Delta t &\equiv (t'' - t')/N, \\ \varepsilon &\equiv \Delta t/\hbar.\end{aligned} \tag{7.7}$$

We can write

$$\begin{aligned}\langle q'', t''|q', t'\rangle &= \langle q''|e^{-iH_{op}N\varepsilon}|q'\rangle = \langle q''|(1 - i\varepsilon H_{op})^N|q'\rangle \\ &= \int dq_1 \cdots dq_{N-1}\langle q''|(1 - i\varepsilon H_{op})|q_{N-1}\rangle \cdots \langle q_1|(1 - i\varepsilon H_{op})|q'\rangle.\end{aligned} \tag{7.8}$$

Next we rewrite a typical factor in the integrand above as

$$\langle q_2|(1 - i\varepsilon H_{op})|q_1\rangle = \int_{-\infty}^{\infty} \frac{dp_1}{2\pi\hbar} \langle q_2|p_1\rangle\langle p_1|(1 - i\varepsilon H_{op})|q_1\rangle, \tag{7.9}$$

and define the classical Hamiltonian $H(p, q)$ by

$$\langle p|H_{op}|q\rangle \equiv \langle p|q\rangle H(p, q). \tag{7.10}$$

This definition gives the usual connection between the classical and quantum mechanical Hamiltonian, provided $H(p, q)$ does not contain cross-products of p and q. Otherwise, H_{op} must be normal ordered, such that all p_{op}'s appear to the left of all q_{op}'s. Using (7.10), we obtain

$$\begin{aligned}\langle q_2|(1 - i\varepsilon H_{op})|q_1\rangle &= \int_{-\infty}^{\infty} \frac{dp_1}{2\pi\hbar} \langle q_2|p_1\rangle\langle p_1|q_1\rangle[1 - i\varepsilon H(p_1, q_1)] \\ &= \int_{-\infty}^{\infty} \frac{dp_1}{2\pi\hbar} e^{ip_1(q_2-q_1)/\hbar}[1 - i\varepsilon H(p_1, q_1)].\end{aligned} \tag{7.11}$$

Hence

$$\begin{aligned}\langle q'', t''|q', t'\rangle = \int \frac{dq_1\, dp_1}{2\pi\hbar} \cdots \int \frac{dq_{N-1}\, dp_{N-1}}{2\pi\hbar} \\ \exp\left[\frac{i}{\hbar}\sum_{n=0}^{N-1} p_n(q_{n+1} - q_n)\right]\prod_{n=1}^{N-1}[1 - i\varepsilon H(p_n, q_n)],\end{aligned} \tag{7.12}$$

with the conditions

$$q_0 \equiv q', \quad q_N \equiv q''. \tag{7.13}$$

Now comes the key step in the development: we note that in (7.12) the factor $(1 - i\varepsilon H)$ may be effectively replaced by $\exp(-i\varepsilon H)$. The reason is that we assume $N^{-1}\Sigma_n H(p_n, q_n)$ approaches a finite limit as $N \to \infty$: given N numbers $z_1, \ldots, z_n$ such that $X \equiv \text{Lim } N^{-1}\Sigma z_n$ exists, we have[b]

$$\underset{N\to\infty}{\text{Lim}} \prod_{n=1}^{N} \left(1 + \frac{z_n}{N}\right) = \underset{N\to\infty}{\text{Lim}} \prod_{n=1}^{N} e^{z_n/N} = e^X, \tag{7.14}$$

which can be proved by power series expansion in z_n. The advantage of replacing $(1 - i\varepsilon H)$ by $\exp(-i\varepsilon H)$ is that we can now express the amplitude (7.12) as integrals over unitary amplitudes:

$$\langle q'', t''|q', t'\rangle = \int \frac{dq_1\, dp_1}{2\pi\hbar} \cdots \int \frac{dq_{N-1}\, dp_{N-1}}{2\pi\hbar}$$
$$\exp\left\{\frac{i}{\hbar} \Delta t \sum_{n=1}^{N-1} \left[\frac{p_n(q_{n+1} - q_n)}{\Delta t} - H(p_n, q_n)\right]\right\} \tag{7.15}$$

To approach the limit $N \to \infty$ (or $\Delta t \to 0$), we adopt the view that the set of values $\{q_1, p_1, \ldots, q_{N-1}, p_{N-1}\}$ are successive values of certain functions $q(t)$, $p(t)$, which may be discontinuous functions. Accordingly, we use the notation

$$t_n = t' + n\,\Delta t,$$
$$q_n = q(t_n), \tag{7.16}$$
$$p_n = p(t_n),$$

and write

$$(q_{n+1} - q_n)/\Delta t \xrightarrow[\Delta t \to 0]{} \dot{q}(t_n),$$
$$\sum_{n=1}^{N-1} f(t_n)\,\Delta t \xrightarrow[\Delta t \to 0]{} \int_{t'}^{t''} dt\, f(t). \tag{7.17}$$

Thus, we can rewrite the transition amplitude in the form

$$\langle q'', t''|q', t'\rangle = \int (Dq)(Dp) \exp \frac{i}{\hbar} \int_{t'}^{t''} dt[p\dot{q} - H(p, q)],$$
$$q(t') = q', \quad q(t'') = q''. \tag{7.18}$$

This represents an integral over all paths $p(t)$ in momentum space, and all paths $q(t)$ in coordinate space, between the times t' and t'' with fixed values of the coordinates at the endpoints. The volume elements in path space are denoted by

$$(Dp) = \prod_{n=1}^{N-1} dq(t_n), \quad (Dp) = \prod_{n=1}^{N-1} \frac{dp(t_n)}{2\pi\hbar}. \tag{7.19}$$

---

[b] The fact that $(1 - i\varepsilon H)$ and $\exp(-i\varepsilon H)$ agree to first order in ε is not sufficient reason for replacing one by the other in (7.12). For example, it would be incorrect to replace $(1 - i\varepsilon H)$ by $(1 + i\varepsilon H)^{-1}$.

Method of Path Integrals

The generalization of (7.18) to more than one degree of freedom is immediate:

$$\langle q_1'', \ldots, q_n''; t'' | q_1', \ldots, q_n'; t' \rangle$$
$$= \int \prod_\alpha (Dq_\alpha)(Dp_\alpha) \exp \frac{i}{\hbar} \int_{t'}^{t''} dt \left[\sum_\alpha p_\alpha \dot{q}_\alpha - H(p, q) \right]. \quad (7.20)$$

Feynman's formula for the transition amplitude is derived from (7.18) by restricting the classical Hamiltonian to the special form

$$H(p, q) = \frac{p^2}{2m} + V(q). \quad (7.21)$$

One can then perform the momentum integrations explicitly to obtain

$$\int (Dp) \exp \frac{i}{\hbar} \int_{t'}^{t''} dt(p\dot{q} - H) = \left(\frac{m}{2\pi\hbar} \right)^{N/2} \exp \frac{i}{\hbar} \int_{t'}^{t''} dt\, L(q, \dot{q}), \quad (7.22)$$

where $L(q, \dot{q})$ is the classical Lagrangian:

$$L(q, \dot{q}) = \tfrac{1}{2} m \dot{q}^2 - V(q). \quad (7.23)$$

Substituting (7.22) into (7.18), we obtain

$$\langle q'', t'' | q', t' \rangle = \mathcal{N} \int (Dq) \exp \frac{i}{\hbar} \int_{t'}^{t''} dt\, L(q, \dot{q}), \quad (7.24)$$

which is the Feynman formula. Here, $\mathcal{N}$ is a normalization constant which is usually infinite in the limit $N \to \infty$, but irrelevant to physical results. This is because it cancels in matrix elements of the form $\langle q'', t'' | O | q', t' \rangle / \langle q'', t'' | q, t' \rangle$. For this reason, one need only define (Dq) up to a multiplicative constant (possibly infinite).

Under the assumption (7.21) one can show

$$\langle q'', t'' | T[q_{\mathrm{op}}(t_1) \cdots q_{\mathrm{op}}(t_n)] | q', t' \rangle$$
$$= \mathcal{N} \int (Dq)[q(t_1) \cdots q(t_n)] \exp \frac{i}{\hbar} \int_{t'}^{t''} dt\, L(q, \dot{q}). \quad (7.25)$$

We indicate the proof for $n = 2$, with $t_1 > t_2$:

$$\langle q'', t'' | q_{\mathrm{op}}(t_1) q_{\mathrm{op}}(t_2) | q', t' \rangle$$
$$= \int dq_1\, dq_2 \langle q'', t'' | q_{\mathrm{op}}(t_1) | q_1, t_1 \rangle \langle q_1, t_1 | q_{\mathrm{op}}(t_2) | q_2, t_2 \rangle \langle q_2, t_2 | q', t' \rangle$$
$$= \int dq_1\, dq_2\, q_1 q_2 \langle q'', t'' | q_1, t_1 \rangle \langle q_1, t_1 | q_2, t_2 \rangle \langle q_2, t_2 | q', t' \rangle.$$

From this point, we follow the steps beginning with (7.8) to obtain the final result.

The Feynman formulas (7.24) and (7.25) are in fact valid under conditions more general than (7.21). They hold for a classical Lagrangian of the general form

$$L(q, \dot{q}) = \frac{1}{2} \sum_{\alpha,\beta} \dot{q}_\alpha A_{\alpha\beta}(q) \dot{q}_\beta + \sum_\alpha b_\alpha(q) \dot{q}_\alpha - V(q), \tag{7.26}$$

whenever the matrix $A(q)$ is independent of q. (Here it is assumed that A is a real non-singular matrix). To show this, first define the canonical momenta:

$$p_\alpha \equiv \frac{\partial L}{\partial \dot{q}_\alpha} = \sum_\beta A_{\alpha\beta} \dot{q}_\beta + b_\alpha. \tag{7.27}$$

Using the matrix notation

$$\begin{aligned} p &= A\dot{q} + b, \\ \dot{q} &= A^{-1}(p - b), \end{aligned} \tag{7.28}$$

we can write the classical Hamiltonian in the form

$$H(p, q) = \tfrac{1}{2}(\tilde{p}, A^{-1}\tilde{p}) + V, \qquad \tilde{p} \equiv p - b. \tag{7.29}$$

From (7.18) we have

$$\langle q'', t'' | q', t' \rangle = \mathcal{N} \int \prod_\alpha dq_\alpha \, dp_\alpha \exp \frac{i}{\hbar} \int_{t'}^{t''} dt \left(\sum_\alpha p_\alpha \dot{q}_\alpha - H \right).$$

$$\sum_\alpha p_\alpha \dot{q}_\alpha - H = \sum_\alpha (\tilde{p}_\alpha + b_\alpha) \dot{q}_\alpha - \frac{1}{2} \sum_{\alpha,\beta} \tilde{p}_\alpha A_{\alpha\beta}^{-1} \tilde{p}_\beta - V$$

$$= -\frac{1}{2}(\tilde{p}, A^{-1}\tilde{p}) + (\tilde{p}, \dot{q}) + (b, \dot{q}) - V.$$

$$\int (Dp) \exp \frac{i}{\hbar} \int dt (p\dot{q} - H) = \prod_t \int (Dp) \exp \left[\frac{i}{\hbar} (\Delta t)(p\dot{q} - H) \right] \tag{7.30}$$

$$= \prod_t \int (D\tilde{p}) \exp \frac{i}{\hbar} (\Delta t) \left[-\frac{1}{2}(\tilde{p}, A^{-1}\tilde{p}) + (\tilde{p}, \dot{q}) + (b, \dot{q}) - V \right]$$

$$= \text{const.} \prod_t \left\{ \exp \frac{i}{\hbar} (\Delta t) \left[\frac{1}{2}(\dot{q}, A\dot{q}) + (b, \dot{q}) - V \right] \right\} (\det A)^{-1/2}$$

$$= \text{const.} \left[\exp \frac{i}{\hbar} \int_{t'}^{t''} dt \, L(q, \dot{q}) \right] \prod_t (\det A)^{-1/2}.$$

If $A(q)$ is independent of q, then the last factor is a constant, which may be absorbed into an overall normalization factor. ∎

If the matrix $A(q)$ in (7.26) does depend on q, then the Feynman formula requires modification. We rewrite

$$\prod_t (\det A)^{-1/2} = \exp \left[-\frac{1}{2} \sum_t \ln(\det A) \right]. \tag{7.31}$$

In the limit $\Delta t \to 0$,

$$\sum_t \to \delta(0) \int dt, \tag{7.32}$$

because a rectangle of unit area, with base dt, has height $\delta(0)$. Thus,

$$\prod_t (\det A)^{-1/2} \xrightarrow[\Delta t \to 0]{} \exp\left[-\frac{1}{2}\delta(0) \int_{t'}^{t''} dt \ln(\det A)\right]. \tag{7.33}$$

Therefore

$$\langle q'', t''|q', t'\rangle = \mathcal{N} \int (Dq) \exp \frac{i}{\hbar} \int_{t'}^{t''} dt \left[L(q, \dot{q}) - \frac{\hbar}{2i}\delta(0) \ln \det A(q)\right]. \tag{7.34}$$

An example of the necessity for the singular term with the factor $\delta(0)$ is encountered in massive vector boson theory.[2]

The Feynman formula (7.24) does not apply when the matrix $A_{\alpha\beta}$ in (7.26) is singular. This is the case when there is a coordinate whose time derivative does not appear in the Lagrangian. In such a case, one has to return to the Hamiltonian form (7.18).

7.2 Quantum Field Theory

The extension of the Feynman formula to a quantum field theory is straightforward; one merely allows the number of coordinates to become nondenumerably infinite. We shall illustrate this extension for the case of one boson field. Generalization to more than one boson field is straightforward. The case of fermion fields will be discussed separately later.

In canonical quantization, the coordinates are the Schrödinger field operators $\phi_{\text{op}}(\mathbf{x})$, where the spatial variable $\mathbf{x}$ serves as a continuous label for the coordinates. We denote eigenstates of $\phi_{\text{op}}(\mathbf{x})$ by $|\phi\rangle$:

$$\phi_{\text{op}}(\mathbf{x})|\phi\rangle = \phi(\mathbf{x})|\phi\rangle, \tag{7.35}$$

where $\phi(\mathbf{x})$ is a c-number function. The transition amplitude is defined by

$$\langle \phi_2, t_2|\phi_1, t_1\rangle \equiv \langle \phi_2|e^{-i(t_2-t_1)H_{\text{op}}/\hbar}|\phi_1\rangle, \tag{7.36}$$

where H_{op} is the time-independent Hamiltonian operator of the system. Let the classical Lagrangian density be

$$\mathcal{L}(x) \equiv \mathcal{L}(\phi(x), \partial^\mu \phi(x)), \tag{7.37}$$

which is assumed to be quadratic in $\partial_\mu \phi(x)$ with coefficients independent of $\phi(x)$. The Feynman formula reads:

$$\langle \phi_2, t_2|\phi_1, t_1\rangle = \mathcal{N} \int_{\phi_1}^{\phi_2} (D\phi) \exp \frac{i}{\hbar} \int_1^2 d^4x \, \mathcal{L}(x), \tag{7.38}$$

---
[2] T. D. Lee and C. N. Yang, *Phys. Rev.* **128**, 885 (1962), Appendix E.

where

$$\int_1^2 d^4x \equiv \int_{t_1}^{t_2} dx_0 \int_{\text{all space}} d^3x, \tag{7.39}$$

with the endpoint constraints

$$\phi(\mathbf{x}, t_2) = \phi_2(\mathbf{x}), \quad \phi(\mathbf{x}, t_1) = \phi_1(\mathbf{x}). \tag{7.40}$$

The functional integration $\int (D\phi)$ may be defined by first considering x to be a discrete variable, and then approaching the continuum limit in the final result of a physical calculation. Alternatively, we may enclose the system in a large but finite space-time volume, integrate independently over each of the discrete Fourier components of ϕ, and then approach the limit of infinite volume. We also have

$$\langle \phi_2, t_2 | T \phi_{\text{op}}(x_1) \cdots \phi_{\text{op}}(x_n) | \phi_1, t_1 \rangle$$
$$= \mathcal{N} \int_{\phi_1}^{\phi_2} (D\phi) \phi(x_1) \cdots \phi(x_n) \exp \frac{i}{\hbar} \int_1^2 d^4x \, \mathcal{L}(x). \tag{7.41}$$

The proof is similar to that for (7.25).

The transition amplitude may be continued analytically to complex times in a certain manner. To investigate this possibility, consider the eigenstates $|n\rangle$ of H_{op}, and assume that there is a unique vacuum state with zero energy:

$$\begin{aligned} H_{\text{op}}|n\rangle &= E_n|n\rangle, \quad (E_n \geq 0), \\ H_{\text{op}}|0\rangle &= 0, \quad \langle 0|0\rangle = 1. \end{aligned} \tag{7.42}$$

Then we can write

$$\begin{aligned} \langle \phi_2, t_2 | \phi_1, t_1 \rangle &= \left\langle \phi_2 \left| \exp\left[-\frac{i}{\hbar}(t_2 - t_1)H_{\text{op}}\right] \right| \phi_1 \right\rangle \\ &= \sum_n \langle \phi_2 | n \rangle \langle n | \phi_1 \rangle \exp\left[-\frac{i}{\hbar}(t_2 - t_1)E_n\right] \\ &= \int_0^\infty dE \, \rho_{21}(E) \exp\left[-\frac{i}{\hbar}(t_2 - t_1)E\right], \end{aligned} \tag{7.43}$$

where

$$\rho_{21}(E) = \sum_n \langle \phi_2 | n \rangle \langle n | \phi_1 \rangle \delta(E - E_n). \tag{7.44}$$

It is clear that the transition amplitude can be continued into the lower half plane of $t_2 - t_1$, for example:

$$\langle \phi_2, -i\tau | \phi_1, 0 \rangle = \int_0^\infty dE \, \rho_{21}(E) e^{-\tau E/\hbar} \quad (\tau > 0). \tag{7.45}$$

Method of Path Integrals

After calculating the right-hand side, the expression for real times can be recovered by analytically continuing τ to the positive imaginary axis.

Taking the limit $\tau \to \infty$, we obtain

$$\langle \phi_2, -i\tau | \phi_1, 0 \rangle \xrightarrow[\tau \to \infty]{} \langle \phi_2 | 0 \rangle \langle 0 | \phi_1 \rangle = \Psi_0[\phi_2]\Psi_0^*[\phi_1], \quad (7.46)$$

where $\Psi_0[\phi] \equiv \langle \phi | 0 \rangle$ is the ground state wave function of the system. Combining (7.46) with (7.38), we obtain

$$\Psi_0[\phi_2]\Psi_0^*[\phi_1] = \mathcal{N} \int_{\phi_1}^{\phi_2} (D\phi) \exp \frac{1}{\hbar} \int_{-\infty}^{\infty} d\tau \int d^3x\, \mathcal{L}(\mathbf{x}, -i\tau), \quad (7.47)$$

with the endpoint constraints

$$\phi(\mathbf{x}, t = i\infty) = \phi_2(\mathbf{x}), \qquad \phi(\mathbf{x}, t = -i\infty) = \phi_1(\mathbf{x}). \quad (7.48)$$

The Feynman formula (7.38) is not applicable as it stands, when there are fields whose conjugate momenta vanish identically. This is the case for gauge fields, which will be discussed separately in Chap. VIII. It suffices to mention at this point that the Feynman formula still holds, provided $\mathcal{L}(x)$ is supplemented by a "gauge fixing" term. The general methods discussed in the rest of this chapter still apply.

7.3 External Sources

The system can be coupled to an external source by adding the term $\phi(x)J(x)$ to the classical Lagrangian density, where $J(x)$ is an arbitrary function. As we shall see, this is a useful mathematical device.

In the presence of an external source, the transition amplitude is denoted by

$$\langle \phi_2, t_2 | \phi_1, t_1 \rangle_J \equiv \mathcal{N} \int_{\phi_1}^{\phi_2} (D\phi) \exp \frac{i}{\hbar} \int_1^2 d^4x [\mathcal{L}(x) + \phi(x)J(x)]. \quad (7.49)$$

Matrix elements of time-ordered products of fields can be obtained from the above by functional differentiations with respect to the external source:

$$\frac{\delta \langle \phi_2, t_2 | \phi_1, t_1 \rangle_J}{\delta J(y_1) \cdots \delta J(y_n)}$$

$$= \left(\frac{i}{\hbar}\right)^n \mathcal{N} \int_{\phi_1}^{\phi_2} (D\phi)\phi(y_1) \ldots \phi(y_n) \exp \frac{i}{\hbar} \int_1^2 d^4x [\mathcal{L}(x) + \phi(x)J(x)]. \quad (7.50)$$

The right-hand side reduces to that of (7.41) upon setting $J(x) \equiv 0$.

Suppose the Lagrangian density can be separated into an unperturbed term plus a perturbation term:

$$\mathcal{L}(x) = \mathcal{L}_0(x) + \mathcal{L}'(\phi(x)). \quad (7.51)$$

Note that $\mathcal{L}'$ is considered a function of $\phi(x)$, even though it may involve derivatives of $\phi(x)$. We can write

$$\langle \phi_2, t_2 | \phi_1, t_1 \rangle_J$$
$$= \left\{ \exp \frac{i}{\hbar} \int_1^2 d^4x \, \mathscr{L}'\left(\frac{\hbar}{i} \frac{\delta}{\delta J(x)}\right) \right\} \mathscr{N} \int_{\phi_1}^{\phi_2} (D\phi) \exp \frac{i}{\hbar} \int_1^2 d^4x [\mathscr{L}_0(x) + \phi(x) J(x)]. \tag{7.52}$$

A perturbation series can be obtained by expanding the first exponential factor as a power series in $\mathscr{L}'$. We shall illustrate this technique in Section 7.7 by deriving Feynman rules for ϕ^4 theory.

An important application of the method of path integrals is the calculation of the vacuum-vacuum amplitude in the presence of an external source. Knowing this amplitude determines the complete dynamics of the system. In other words, *all dynamical information can be deduced from the response of the vacuum state to an arbitrary external source.*

We first define the vacuum-vacuum amplitude, and exhibit its relation to the Green's functions of the system. In the presence of an external source, the Hamiltonian of the system is changed from H_{op} to

$$H^J_{op} = H_{op} - H'_{op}$$
$$H'_{op} = \int d^3x \, J(x) \phi_{op}(x), \tag{7.53}$$

where $\phi_{op}(x)$ is a Heisenberg operator, and $J(x)$ is a c-number function, arbitrary except for the condition that it is to be turned off in the infinite past and the infinite future:

$$J(x) \xrightarrow[|x_0| \to \infty]{} 0. \tag{7.54}$$

This insures that the system can be in the vacuum state $|0\rangle$ of the Hamiltonian H_{op} in the infinite past and infinite future. The response of the vacuum state is described by the amplitude

$$\langle 0^+ | 0^- \rangle_J \equiv \text{Probability amplitude that the system will be in state } |0\rangle \text{ at } x_0 = \infty,$$
$$\text{when it was known to be in state } |0\rangle \text{ at } x_0 = -\infty. \tag{7.55}$$

By unitarity, this amplitude can only be a phase factor:

$$\langle 0^+ | 0^- \rangle_J \equiv \exp \frac{i}{\hbar} W[J]. \tag{7.56}$$

Now, go to the interaction picture with respect to H'_{op}. At $x_0 = 0$, the state of the system is given by

$$|0^-\rangle_J = T\left[\exp \frac{i}{\hbar} \int_{-\infty}^0 dt \, H'_{op}(t)\right] |0\rangle, \tag{7.57}$$

where the time evolution of $H'_{op}(t)$ is governed by H_{op}. That is, $H'_{op}(t)$ is a Heisenberg operator of the system without external sources. On the other hand,

Method of Path Integrals

the state of the system at $x_0 = 0$ that will evolve into $|0\rangle$ at $x_0 = \infty$ is given by

$$|0^+\rangle_J = T\left[\exp\frac{i}{\hbar}\int_\infty^0 dt\, H'_{op}(t)\right]|0\rangle. \quad (7.58)$$

Therefore

$$\langle 0^+|0^-\rangle_J = \left\langle 0\left|T\left[\exp\frac{i}{\hbar}\int_{-\infty}^\infty dt\, H'_{op}(t)\right]\right|0\right\rangle$$

$$= \left\langle 0\left|T\left[\exp\frac{i}{\hbar}\int d^4x\, J(x)\phi_{op}(x)\right]\right|0\right\rangle. \quad (7.59)$$

We see immediately that

$$\left[\frac{\delta\langle 0^+|0^-\rangle_J}{\delta J(x_1)\cdots\delta J(x_n)}\right]_{J=0} = \left(\frac{i}{\hbar}\right)^n \mathcal{G}_n(x_1,\ldots,x_n), \quad (7.60)$$

where

$$\mathcal{G}_n(x_1,\ldots,x_n) \equiv \langle 0|T\phi_{op}(x_1)\cdots\phi_{op}(x_n)|0\rangle \quad (n \geq 1) \quad (7.61)$$

are the Green's functions which completely describe the dynamics of the system in the absence of external sources. From (7.60), we obtain the expansion

$$\langle 0^+|0^-\rangle_J = \sum_{n=0}^\infty \left(\frac{i}{\hbar}\right)^n \frac{1}{n!}\int d^4x_1\cdots d^4x_n \mathcal{G}_n(x_1,\ldots,x_n)J(x_1)\cdots J(x_n). \quad (7.62)$$

where $\mathcal{G}_0 \equiv 1$. This shows that $\langle 0^+|0^-\rangle_J$ is the generating functional for the Green's functions.

The phase $W[J]$ defined in (7.56) is the generating functional for *connected* Green's functions:

$$\frac{i}{\hbar}W[J] = \sum_{n=0}^\infty \left(\frac{i}{\hbar}\right)^n \frac{1}{n!}\int d^4x_1\cdots d^4x_n\, G_n(x_1,\ldots,x_n)J(x_1)\cdots J(x_n), \quad (7.63)$$

where $G_n(x_1,\ldots,x_n)$, $(n \geq 1)$, is the connected n-point Green's function, i.e., the sum of all *connected* Feynman graphs with n external lines terminating at $x_1,\ldots,x_n$. To show this, we note that $\mathcal{G}_n$, $(n \geq 1)$, is a sum of products of connected Green's functions G_m, $(m \geq 1)$, which may be displayed as follows. Let $\{\sigma_1,\ldots,\sigma_n\}$ be a partition of the integer n:

$$n = \sigma_1 + 2\sigma_2 + 3\sigma_3 + \cdots. \quad (7.64)$$

We can write

$$\mathcal{G}_n(x_1,\ldots,x_n) = \sum_{\substack{\{\sigma_l\} \\ \Sigma l\sigma_l = n}} \sum_P P[\underbrace{G_1(\cdot)\cdots G_1(\cdot)}_{\sigma_2\text{ factors}}][\underbrace{G_2(\cdot\cdot)\cdots G_2(\cdot\cdot)}_{\sigma_2\text{ factors}}]\cdots \quad (7.65)$$

The above parentheses contain a total of n dots which are to be put into one-to-one correspondence with $x_1,\ldots,x_n$ in some manner. The symbol P

denotes a distinct permutation of $x_1, \ldots, x_n$. The number of such permutations is

$$\frac{n!}{(\sigma_1! \, \sigma_2! \cdots)[(1!)^{\sigma_1}(2!)^{\sigma_2} \cdots]}. \tag{7.66}$$

When (7.65) is substituted into (7.62), every term in the sum Σ_P gives the same contribution, upon integration over $x_1, \ldots, x_n$. Hence

$$\langle 0^+|0^-\rangle_J = \sum_{n=0}^{\infty} \sum_{\substack{\{\sigma_l\} \\ \Sigma l \sigma_l = n}} \left(\frac{i}{\hbar}\right)^n \frac{[\int d^4x \, G_1(x)J(x)]^{\sigma_1}}{\sigma_1!} \frac{[\int d^4x \, d^4y \, G_2(x, y)J(x)J(y)]^{\sigma_2}}{\sigma_2! \, (2!)^{\sigma_2}} \cdots \tag{7.67}$$

The double sum in (7.67) is the same as the sum over each of the integers σ_l independently. Therefore

$$\langle 0^+|0^-\rangle_J = \sum_{\sigma_1=0}^{\infty} \frac{1}{\sigma_1!} \left[\frac{i}{\hbar} \int d^4x \, G_1(x)J(x)\right]^{\sigma_1} \cdot \sum_{\sigma_2=0}^{\infty} \frac{1}{\sigma_2!} \left[\left(\frac{i}{\hbar}\right)^2 \frac{1}{2!} \right.$$

$$\left. \times \int d^4x \, d^4y \, G_2(x, y)J(x)J(y)\right]^{\sigma_2} \cdots$$

$$= \exp \sum_{n=1}^{\infty} \left(\frac{i}{\hbar}\right)^n \frac{1}{n!} \int d^4x_1 \cdots d^4x_n \, G_n(x_1, \ldots, x_n)J(x_1) \cdots J(x_n). \blacksquare \tag{7.68}$$

We now show that the vacuum-vacuum amplitude can be expressed as a path integral. For simplicity let us turn on the source only for a finite but large time interval:

$$J(x) = 0 \quad \text{for } |x_0| > T, \quad (T \to \infty). \tag{7.69}$$

Consider the transition amplitude from a time t_1 before the source was turned on, to a time t_2 after the source has been turned off:

$$\langle \phi_2, t_2|\phi_1, t_1\rangle = \int (D\phi)(D\phi')\langle \phi_2, t_2|\phi, T\rangle\langle \phi, T|\phi', -T\rangle_J \langle \phi', -T|\phi_1, t_1\rangle$$

$$(t_1 < -T, \, t_2 > T). \tag{7.70}$$

We can calculate the source-free amplitudes immediately. For example,

$$\langle \phi_2, t_2|\phi, T\rangle = \left\langle \phi_2 \left| \exp\left[-\frac{i}{\hbar}(t_2 - T)H_{\text{op}}\right] \right| \phi \right\rangle$$

$$= \sum_n \langle \phi_2|n\rangle\langle n|\phi\rangle \exp\left[-\frac{i}{\hbar}(t_2 - T)E_n\right] \tag{7.71}$$

$$\xrightarrow[\substack{t_2 \to -i\infty \\ T \to i\infty}]{} \langle \phi_2|0\rangle\langle 0|\phi\rangle.$$

Method of Path Integrals

Thus, continuing the times:

$$t_1 \to -i\infty, \quad t_2 \to i\infty, \quad T \to i\infty, \tag{7.72}$$

we obtain

$$\frac{\langle \phi_2, t_2 | \phi_1, t_1 \rangle_J}{\langle \phi_2 | 0 \rangle \langle 0 | \phi_1 \rangle} \to \int (D\phi)(D\phi') \langle \phi_2, T | \phi_1, -T \rangle_J \tag{7.73}$$

$$= \left\langle 0 \left| \exp\left(-\frac{i}{\hbar} T H_{\text{op}}^J\right) \right| 0 \right\rangle = \langle 0^+ | 0^- \rangle_J,$$

or,

$$\langle 0^+ | 0^- \rangle_J = \lim_{\substack{t_2 \to -i\infty \\ t_1 \to i\infty}} \frac{\langle \phi_2, t_2 | \phi_1, t_1 \rangle_J}{\langle \phi_2 | 0 \rangle \langle 0 | \phi_1 \rangle}. \tag{7.74}$$

Using (7.49) we can express the right-hand side as a path integral over fields defined in Euclidean 4-space. Note that the endpoint constraints are irrelevant because the right-hand side of (7.74) is independent of $\phi_2(x)$ and $\phi_1(x)$.

7.4 Euclidean 4-Space

Euclidean 4-space is obtained from Minkowski space by clockwise rotation of the real axis in the complex x_0 plane into the negative imaginary axis, as indicated in Fig. 7.1. A point in Euclidean 4-space is denoted by x_E, and is related to the corresponding point $x = (x_0, \mathbf{x})$ in Minkowski space by

$$\begin{aligned}
x_E &= (\mathbf{x}, x_4), \\
x_4 &= ix_0 \quad \text{(real)}, \\
d^4x &= -i\,d^4x_E, \\
x_E^2 &= x_1^2 + x_2^2 + x_3^2 + x_4^2 = -x^2.
\end{aligned} \tag{7.75}$$

The corresponding Euclidean momentum space is defined so that $k_4 x_4 = k^0 x^0$. This convention is chosen so that in the propagation of a plane wave, a positive sense of x_4 corresponds to a positive sense of x^0. Accordingly, we rotate the k^0 axis counter-clockwise into the positive imaginary axis, as indicated in Fig. 7.1. We have

$$\begin{aligned}
k_E &= (\mathbf{k}, k_4), \\
k_4 &= -ik_0 \quad \text{(real)}, \\
d^4k &= i\,d^4k_E, \\
k_E^2 &= k_1^2 + k_2^2 + k_3^2 + k_4^2 = -k^2.
\end{aligned} \tag{7.76}$$

Note that $k \cdot x = k^0 x^0 - \mathbf{k} \cdot \mathbf{x}$ transforms into $k_4 x_4 - \mathbf{k} \cdot \mathbf{x}$, but in taking the Fourier transform of $f(k^2)$, we can replace $k \cdot x$ by $k_E \cdot x_E = k_4 x_4 + \mathbf{k} \cdot \mathbf{x}$.

Continuation to Euclidean space means that we consider the dynamical evolution of the system in imaginary time. That is, we must in principle solve the

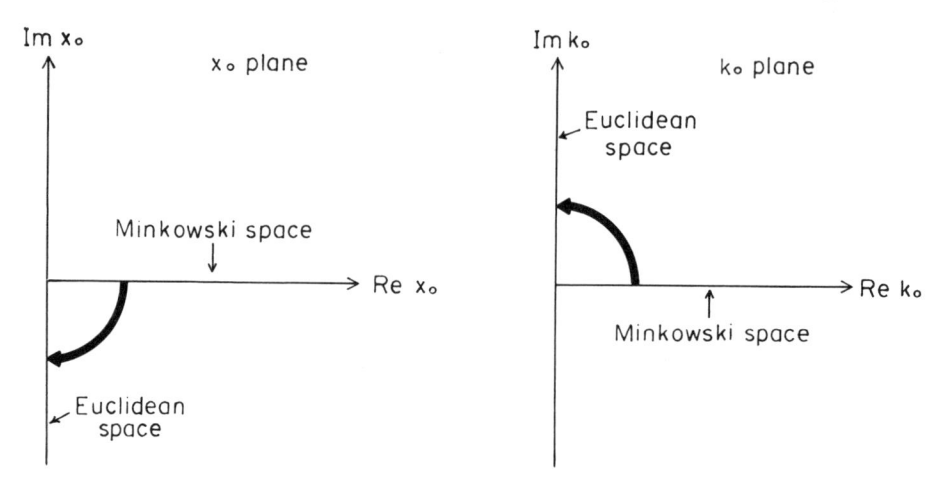

Fig. 7.1 How to continue Minkowski space into Euclidean space

equations of motion in which the time x^0 is replaced by $-ix_4$, where x_4 is a real parameter. The Lorentz invariance of the Lagrangian density is replaced by invariance with respect to O_4 rotations in Euclidean space. The equations of motion will determine how the fields are to be continued into Euclidean space.

A real scalar field $\phi(x)$ defined in Minkowski space is replaced by a real scalar field $\phi(x_E)$ invariant under O_4.

A massive vector field $A^\mu(x)$ with real components is replaced by a Euclidean vector field $A^\mu(x_E)$ with real components, according to the rule

$$A^k(x) \to A^k(x_E) \quad (k = 1, 2, 3),$$
$$A^0(x) \to iA_4(x_E). \tag{7.77}$$

Note that A^0 continues to A_4 with sign opposite to that of x^0 because it should transform like $\partial/\partial x^0$. The subsidiary condition $\partial_\mu A^\mu = 0$ is replaced by

$$\mathbf{\nabla} \cdot \mathbf{A} + \frac{\partial A_4}{\partial x_4} = 0.$$

For Euclidean vectors, there is no distinction between upper and lower indices.

For a gauge field, the answer depends on how we choose to handle the unphysical degrees of freedom associated with gauge invariance. These have no physical significance, and covariance arguments are not compelling. In a covariant gauge, (7.77) applies. In a non-covariant gauge, we could still use (7.77) as a formal device although it is not necessary.

The functional integral for the vacuum-vacuum amplitude in Euclidean form is as follows:

$$\langle 0^+|0^-\rangle_J \equiv \exp\frac{i}{\hbar} W[J] = \mathcal{N} \int (D\phi) \exp\left(-\frac{1}{\hbar}\{S_E[\phi] - (J, \phi)_E\}\right), \tag{7.78}$$

where

$$S_E[\phi] = \int d^4x_E \mathcal{L}(x_E) = -iS[\phi].$$
$$(J, \phi)_E = \int d^4x_E\, J(x_E)\phi(x_E). \tag{7.79}$$

In this form, the functional integral is formally the partition function of a classical system in 4 dimensions, at temperature $\hbar$.

The Euclidean formulation is very practical when one wants to investigate which paths make important contributions to the transition amplitude. The path integral in Minkowski space assigns a phase factor to each path, the classical path being the one with stationary phase. The Euclidean formulation assigns a Boltzmann-like factor to each path, with the classical path corresponding to the one with least Euclidean action. The neighboring paths give contributions that are damped out, instead of oscillating even more wildly.

In the Feynman graph expansion of the vacuum-vacuum amplitude, the Euclidean prescription merely supplies the correct $i\varepsilon$ in the propagators [see (7.99) and (7.100)].

7.5 Calculation of Path Integrals

Path integrals of the Gaussian type can be calculated by an extension of the formula for the ordinary Gaussian integral. Suppose $Q(x)$ is a quadratic form in one variable:

$$Q(x) = \frac{1}{2}ax^2 - bx = -\frac{b^2}{2a} + \frac{1}{2}a(x - x_0)^2, \tag{7.80}$$

where

$$x_0 = b/a. \tag{7.81}$$

It is well-known that

$$\int_{-\infty}^{\infty} dx\, e^{-Q(x)} = \left(\frac{2\pi}{a}\right)^{1/2} \exp\left(\frac{b^2}{2a}\right). \tag{7.82}$$

This result can be immediately generalized to a quadratic form $Q(u)$ involving n variables $\{u_1, u_2, \ldots, u_n\}$:

$$Q(u) = \tfrac{1}{2}(u, Au) - (b, u), \tag{7.83}$$

where b is a constant n-vector, A is a symmetric non-singular $n \times n$ matrix, and

$$(b, u) \equiv \sum_{i=1}^{n} b_i u_i. \tag{7.84}$$

We can also write

$$Q(u) = -\tfrac{1}{2}(b, A^{-1}b) + \tfrac{1}{2}((u - u_0), A(u - u_0)),$$
$$u_0 = A^{-1}b. \tag{7.85}$$

Then,

$$\int (Du)\, e^{-Q(u)} = (\det A)^{-1/2} \exp \tfrac{1}{2}(b, A^{-1}b),$$
$$(Du) \equiv (2\pi)^{-n/2}\, du_1 \cdots du_2, \qquad (7.86)$$
$$\det A = \prod_{i=1}^{n} a_i.$$

For a real field $\phi(x)$, define a quadratic form $Q[\phi]$ by

$$Q[\phi] = \tfrac{1}{2}(\phi, A\phi) - (b, \phi), \qquad (7.87)$$

where A is non-singular and self-adjoint, and

$$(\phi_1, \phi_2) = \int d^4x_E\, \phi_1(x_E)\phi_2(x_E). \qquad (7.88)$$

We can rewrite $Q[\phi]$ as

$$Q[\phi] = -\tfrac{1}{2}(b, A^{-1}b) + \tfrac{1}{2}((\phi - \phi_0), A(\phi - \phi_0)), \qquad (7.89)$$
$$\phi_0 = A^{-1}b.$$

Then,

$$\int (D\phi)\, e^{-Q[\phi]} = \mathcal{N}(\det A)^{-1/2} \exp \tfrac{1}{2}(b, A^{-1}b). \qquad (7.90)$$

The volume element $(D\phi)$ may be defined up to multiplicative constant as $\Pi_x\, d\phi(x)$, the ambiguous constant (possibly infinite) being absorbed into the normalization constant $\mathcal{N}$.

7.6 The Feynman Propagator

We calculate the generating functional $W[J]$ for a free scalar field, to illustrate the fact that the Feynman propagator is the inverse of the "kinetic operator" in the Lagrangian density.

The Lagrangian density is

$$\mathcal{L}_0(x) = \tfrac{1}{2}\partial_\mu \phi(x)\partial^\mu \phi(x) - \tfrac{1}{2}m^2\phi^2(x). \qquad (7.91)$$

The classical action is

$$S_0[\phi] = -\frac{1}{2}\int d^4x\, \phi(x)(\Box^2 + m^2)\phi(x)$$
$$= \frac{i}{2}\int d^4x_E\, \phi(x_E)(-\Box_E^2 + m^2)\phi(x_E). \qquad (7.92)$$

Method of Path Integrals

The kinetic operator is $(-\Box_E^2 + m^2)$. Using (7.78), we have

$$\exp \frac{i}{\hbar} W_0[J] = \mathcal{N} \int (D\phi) \exp \frac{i}{\hbar} \left\{ S_0[\phi] - i \int d^4x_E \, \phi(x_E) J(x_E) \right\}$$

$$= \mathcal{N} \int (D\phi) e^{-Q[\phi, J]}, \tag{7.93}$$

where

$$Q[\phi, J] = \frac{1}{2}(\phi, A\phi) - (b, \phi), \tag{7.94}$$

with

$$A = \frac{1}{\hbar}(-\Box_E^2 + m^2), \tag{7.95}$$

$$b = \frac{1}{\hbar} J(x_E).$$

According to (7.90), therefore,

$$\exp \frac{i}{\hbar} W_0[J] = \mathcal{N} (\det A)^{-1/2} \exp \tfrac{1}{2}(b, A^{-1}b), \tag{7.96}$$

or,

$$\frac{i}{\hbar} W_0[J] = \frac{1}{2\hbar}(J, (-\Box_E^2 + m^2)^{-1}J) + \ln[\mathcal{N}(\det A)^{-1/2}]. \tag{7.97}$$

The operator $(-\Box_E^2 + m^2)^{-1}$ can be studied best in momentum space:

$$(-\Box_E^2 + m^2)^{-1} f(x_E) = (-\Box_E^2 + m^2)^{-1} \int \frac{d^4k_E}{(2\pi)^4} e^{-ik_E \cdot x_E} \tilde{f}(k_E).$$

$$= \int \frac{d^4k_E}{(2\pi)^4} \frac{e^{-ik_E \cdot x_E}}{k_E^2 + m^2} \tilde{f}(k_E)$$

$$= i \int d^4y_E \, \Delta_F(x_E - y_E) f(y_E), \tag{7.98}$$

where Δ_F is the Feynman propagator, defined by

$$\Delta_F(x_E) \equiv -i \int \frac{d^4k_E}{(2\pi)^4} \frac{e^{-ik_E \cdot x_E}}{k_E^2 + m^2}. \tag{7.99}$$

When x_E is rotated into Minkowski space, this takes the familiar form

$$\Delta_F(x) = \int \frac{d^4k}{(2\pi)^4} \frac{e^{-ik \cdot x}}{k^2 - m^2 + i\varepsilon} \quad (\varepsilon \to 0^+). \tag{7.100}$$

We see that the Euclidean-space formulation amounts to specifying the contour of the k_0 integration in the usual manner.

The eigenvalues of $(-\Box_E^2 + m^2)$ are $(k_E^2 + m^2)$. Hence

$$\det A = \prod_{k_E} \hbar^{-1}(k_E^2 + m^2), \qquad (7.101)$$

which is a divergent quantity. However, since it is independent of $J(x)$, we can cancel it in (7.97) by choosing $\mathcal{N}$ appropriately. Thus we obtain

$$\begin{aligned} W_0[J] &= \frac{1}{2} \int d^4x_E \, d^4y_E \, J(x_E) \, \Delta_F(x_E - y_E) J(y_E) \\ &= -\frac{1}{2} \int d^4x \, d^4y \, J(x) \, \Delta_F(x - y) J(y). \end{aligned} \qquad (7.102)$$

Comparison with (7.63) shows that the only non-vanishing connected Green's function for the free field theory is the Feynman propagator:

$$G_2(x, y) = i\hbar \, \Delta_F(x - y). \qquad (7.103)$$

7.7 Feynman Graphs

We illustrate Feynman graphs using ϕ^4 theory as an example. The Lagrangian density in the presence of an external source is

$$\mathcal{L}(x) = \mathcal{L}_0(x) + \mathcal{L}'(\phi(x)) + \phi(x)J(x), \qquad (7.104)$$

where

$$\begin{aligned} \mathcal{L}_0(x) &= \frac{1}{2} \partial_\mu \phi(x) \, \partial^\mu \phi(x) - \frac{1}{2} m^2 \phi^2(x), \\ \mathcal{L}'(\phi(x)) &= \frac{\lambda}{4!} \phi^4(x). \end{aligned} \qquad (7.105)$$

Using (7.74) and (7.52), we can write

$$\exp \frac{i}{\hbar} W[J] = \left\{ \exp \frac{i}{\hbar} \int d^4x \, \mathcal{L}'\left(\frac{\hbar}{i} \frac{\delta}{\delta J(x)}\right) \right\} \cdot \exp \frac{i}{\hbar} W_0[J], \qquad (7.106)$$

where $W_0[J]$ is given by (7.102). The usual Feynman graph expansion for the Green's functions is obtained by expanding the above in powers of $\mathcal{L}'$.

Without using the specific form of $\mathcal{L}'$, we can see that the classical limit (defined formally as the limit of the theory as $\hbar \to 0$), is obtained by retaining only the "tree graphs". Consider any connected Feynman graph, and let

$$\begin{aligned} V &= \text{No. of vertices}, \\ I &= \text{No. of internal lines}, \\ E &= \text{No. of external lines}. \end{aligned} \qquad (7.107)$$

Each internal line carries an internal 4-momentum that is integrated over. Each vertex imposes one condition of 4-momentum conservation. There is a

Method of Path Integrals

condition of total 4-momentum conservation for the external lines. Thus, the number of independent internal 4-momenta is

$$l = I - V + 1, \tag{7.108}$$

which is the number of closed loops in the graph. The graph is proportional to a certain power of $\hbar$, which comes from internal lines and vertices. Powers of $\hbar$ coming from external lines are ignored, because the latter are to be replaced by wave functions in a S-matrix element.

According to (7.106), a vertex (which is associated with $\mathscr{L}'$) comes with a factor $\hbar^{-1}$. Each internal line corresponds to a factor $\hbar\Delta_F$, which arises when we remove the two factors $J/\hbar$ in the quantity $\hbar^{-1}(J, \Delta_F J)$ in $W_0[J]$, through differentiation by $\delta/\delta(J/\hbar)$. The contribution of an amputated connected graph to $(i/\hbar)W[J]$ is therefore proportional to $\hbar^{I-V}$, and the contribution to $W[J]$ is proportional to

$$\hbar^{I-V+1} = \hbar^l. \tag{7.109}$$

Since $W[J]$ is the generating functional for connected Green's functions, an l-loop graph of an amputated connected Green's function is proportional to $\hbar^l$. The tree graphs are those with $l = 0$, and are therefore the only surviving ones in the classical limit. Quantum corrections may be classified by the power of $\hbar$, and hence by the number of loops.

We now derive the Feynman rules, which depend on the explicit form of $\mathscr{L}'$. Eq. (7.106) can be written more explicitly as

$$\exp\frac{i}{\hbar} W[J] = \left\{\exp\int d^4x \frac{i\lambda\hbar^3}{4!}\left[\frac{\delta}{\delta J(x)}\right]^4\right\} \cdot \left\{\exp\int d^4x\, d^4y\, J(x) \frac{\Delta_F(x-y)}{2i\hbar} J(y)\right\}. \tag{7.110}$$

Note that

$$\frac{\delta J(x)}{\delta J(z)} = \delta^4(x-z),$$

$$\left[\frac{\delta}{\delta J(z)}\right]^4 \prod_{i=1}^{4} J(x_i) = 4! \prod_{i=1}^{4} \delta^4(z - x_i). \tag{7.111}$$

To work out (7.110), it is convenient to rewrite it using the following graphical short-hand:

Line: $x\!-\!y \equiv \Delta_F(x-y)/i\hbar$

Line End: $\underset{x}{\circ} \equiv J(x)$

Terminal: $\underset{z}{(\mathbf{X})} \equiv (\lambda\hbar^3/4!)[\delta/\delta J(z)]^4$

Vertex: $\bullet \equiv \underset{3\ \ 4}{\overset{1\ \ 2}{\times}} = i\lambda\hbar^3 \prod_{i=1}^{4} \delta^4(z - x_i).$

As an example, we write

$$\underset{x}{\bigcirc}\!\!-\!\!\underset{y}{\bigcirc} \equiv J(x)\frac{\Delta_F(x-y)}{i\hbar}J(y). \tag{7.113}$$

With this notation, we have

$$\exp\frac{i}{\hbar}W[J] = \left\{\exp\int d^4z(\mathbf{X})\right\} \cdot \left\{\exp\int d^4x\, d^4y\, \frac{1}{2}(\underset{x}{\bigcirc}\!\!-\!\!\underset{y}{\bigcirc})\right\}$$

$$= \sum_{m=0}^{\infty}\sum_{n=0}^{\infty}\frac{1}{m!\,n!}\int (dx)(dy)(dz)\left[\underset{z_1}{(\mathbf{X})}\cdots\underset{z_m}{(\mathbf{X})}\right]\left[\frac{1}{2}(\underset{x_1}{\bigcirc}\!\!-\!\!\underset{y_1}{\bigcirc})\cdots\frac{1}{2}(\underset{x_n}{\bigcirc}\!\!-\!\!\underset{y_n}{\bigcirc})\right], \tag{7.114}$$

where $\int (dx)\,(dy)\,(dz)$ denote integrations over all the x's, y's and z's in the integrand. A non-vanishing term in the sum above has the following properties:

1. A terminal must seek out 4 line ends, and join them together to form a vertex.
2. A line end need not be joined to a terminal. [If it is not joined to a terminal, it is the free end of an external line and corresponds to a factor $J(x)$].
3. If both ends of a line are joined to the same terminal, the result is an infinite constant, which can be absorbed by mass renormalization. We can thus ignore this type of connection.

According to the above, each terminal must seek out 4 *different* lines and join them to form a vertex. A Feynman graph is a drawing showing the connectivity of lines and terminals corresponding to a non-vanishing term in (7.114). Since the coordinates attached to terminals and line ends are integrated over, two distinct terms in (7.114) may have the same connectivity, and hence, share the same Feynman graph. Thus, (7.114) is a weighted sum of Feynman graphs and $(i/\hbar)W[J]$ is a weighted sum of connected Feynman graphs.

A terminal can pick out one or the other end of a line, producing two distinct terms in (7.114) that are equal in numerical value. Hence the factor $\frac{1}{2}$ in $\frac{1}{2}(\bigcirc\!\!-\!\!\bigcirc)$ can be omitted, provided we do not distinguish the two ends of a line, for the purpose of drawing Feynman graphs. To incorporate this rule, we write symbolically

$$\exp\frac{i}{\hbar}W[J] = \sum_{m=0}^{\infty}\sum_{n=0}^{\infty}\frac{1}{m!\,n!}\int\left[\underset{1}{(\mathbf{X})}\cdots\underset{m}{(\mathbf{X})}\right]\left[\underset{1}{(\bigcirc\!\!-\!\!\bigcirc)}\cdots\underset{n}{(\bigcirc\!\!-\!\!\bigcirc)}\right]. \tag{7.115}$$

It is understood that attaching one end of a line to a terminal is not counted as distinct from attaching the other end. The integration is performed after all vertices have been formed, and extends over the coordinates of all vertices and all free ends of external lines.

From now on, we only consider connected graphs. Let N be the number of terms in (7.115) that have the same pattern of connectivity corresponding to a connected graph, having m vertices and n lines (internal or external). Its weight is given by $N/(m!\,n!)$. [By convention the factor $1/(m!\,n!)$ in (7.115) is not included in the definition of a graph]. We define the *symmetry number*[3] s of a

---
[3] T. T. Wu, *Phys. Rev.* **125**, 1436 (1962).

connected graph as

$$\frac{1}{s} \equiv \frac{N}{m!\, n!} \qquad (7.116)$$

Then

$$\frac{i}{\hbar} W[J] = \sum_G \frac{G}{s}, \qquad (7.117)$$

where the sum extends over all connected graphs G.

Consider a connected non-vacuum graph (i.e., a graph with at least two external lines), with m vertices and n lines (external or internal). Every permutation of the m vertices corresponds to a distinct term in (7.115), because a vertex can be uniquely identified by its ordered position along some line that goes from one external end to another. Renumbering the vertices along this line amounts to a change in the assignment of the terminals in (7.115) to the vertices, and hence gives a new term. Therefore, N contains a factor $m!$. The interchange of two lines corresponds to a distinct term in (7.115), *unless* the two lines are internal lines that share the same vertices at both ends. Such internal lines are called "equivalent". The number of distinct permutations of lines is therefore $n!/(k_1!\, k_2! \ldots)$, where $k_1, k_2, \ldots$ are the numbers of internal lines in equivalent sets. Therefore, $N = n!\, m!/(k_1!\, k_2! \ldots)$, and the symmetry number of a connected non-vacuum graph is

$$s = \prod_i (k_i!). \qquad (7.118)$$

The symmetry number of a vacuum graph does not follow any simple general rule, and has to be worked out for each individual case[c]. Some examples of symmetry numbers are given in Fig. 7.2.

7.8 Boson Loops and Fermion Loops

Consider a free complex scalar field coupled to an external source $\Omega(x)$ that creates particle-antiparticle pairs:

$$\mathcal{L}(x) = \partial_\mu \phi^* \partial^\mu \phi - m^2 \phi^* \phi + \phi^* \phi \Omega$$
$$= \tfrac{1}{2}[\partial_\mu \phi_1 \partial^\mu \phi_1 + (\Omega - m^2)\phi_1^2] + \tfrac{1}{2}[\partial_\mu \phi_2 \partial^\mu \phi_2 + (\Omega - m^2)\phi_2^2], \quad (7.119)$$

where

$$\phi = (\phi_1 + i\phi_2)/\sqrt{2},$$
$$\phi^* = (\phi_1 - i\phi_2)/\sqrt{2}. \qquad (7.120)$$

---

[c] Consider, for example, the term

$$\frac{1}{2!}\left[(\mathbf{X})_1 (\mathbf{X})_2\right] \frac{1}{4!}\left[(\mathrm{O}\!\!-\!\!\mathrm{O})_1 \cdots (\mathrm{O}\!\!-\!\!\mathrm{O})_4\right].$$

There is only one way of connecting the lines to produce a vacuum graph, i.e., the fifth graph in Fig. 7.2. The symmetry number is therefore $2!\, 4! = 48$.

142   *Quarks, Leptons and Gauge Fields*

Any connected Feynman graph of this theory consists of a single loop, on which any number of vertices can be placed.

We can calculate the generating functional $W[\Omega]$ in closed form as follows: the classical action is

$$S[\phi_1, \phi_2, \Omega] = -i \int d^4x_E \, \mathcal{L}(x_E) = i\hbar[(\phi_1, A\phi_1) + (\phi_2, A\phi_2)], \quad (7.121)$$

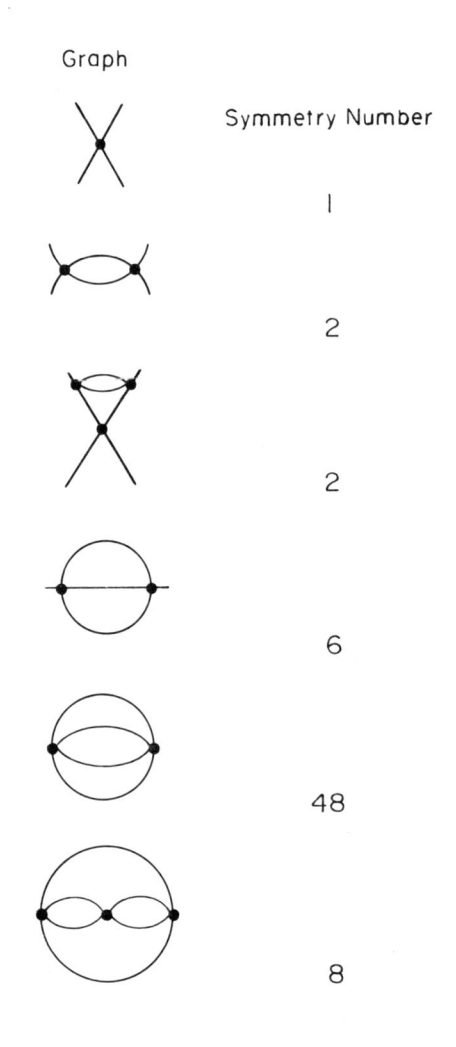

Graph	Symmetry Number
✕	1
✕✕	2
✕ (with bubble)	2
⊖	6
⊜	48
⊗	8

Fig. 7.2   Examples of symmetry number for Feynman graphs in ϕ^4 theory

Method of Path Integrals

where
$$A = \hbar^{-1}(-\Box_E^2 + m^2 - \Omega). \tag{7.122}$$

Hence
$$\exp \frac{i}{\hbar} W[\Omega] = \mathcal{N} \int (D\phi_1)(D\phi_2) \exp \frac{i}{\hbar} S[\phi_1, \phi_2, \Omega]$$
$$= \mathcal{N} \left[\int (D\phi_1) e^{-(\phi_1, A\phi_1)} \right]^2 = \mathcal{N} \det A^{-1}. \tag{7.123}$$

The Feynman graph expansion is obtained by expanding the above in powers of Ω. To do this, let us introduce the notation
$$i\Delta_F = (-\Box_E^2 + m^2)^{-1},$$
$$\langle x|\Delta_F|y\rangle = \Delta_F(x - y),$$
$$\langle x|\Omega|y\rangle = \delta^4(x - y)\Omega(x), \tag{7.124}$$
$$\text{Tr } f = \int d^4x_E \langle x_E|f|x_E\rangle.$$

We can then write
$$A = \hbar^{-1}[(i\Delta_F)^{-1} - \Omega]$$
$$\det A = \det(i\hbar \Delta_F)^{-1} \cdot \det(1 - i\Delta_F\Omega). \tag{7.125}$$

Choosing $\mathcal{N} = \det(i\hbar \Delta_F)$, we have
$$\frac{i}{\hbar} W[\Omega] = -\ln \det(1 - i\Delta_F\Omega) = -\text{Tr} \ln(1 - i\Delta_F\Omega). \tag{7.126}$$

Expanding this in powers of Ω gives
$$\frac{i}{\hbar} W[\Omega] = \sum_{n=1}^{\infty} \frac{1}{n} \text{Tr}(i\Delta_F\Omega)^n$$
$$= \sum_{n=1}^{\infty} \frac{1}{n} \int d^4y_1 \cdots d^4y_n [i\Delta_F(y_1 - y_2) i\Delta_F(y_2 - y_3) \cdots i\Delta_F(y_n - y_1)] \cdot \tag{7.127}$$
$$\cdot \Omega(y_1) \cdots \Omega(y_n),$$

where all the y's are Euclidean coordinates. This shows that there is only one nth order connected graph, consisting of one closed loop with n vertices on it, and the symmetry number is n.

If the field $\phi(x)$ obeyed Fermi statistics instead of Bose statistics, then every closed loop would be associated with an extra factor -1, but the Feynman rules are otherwise unchanged. Since in this example all connected graphs are one-loop graphs, they would simply change sign, and so would $W[\Omega]$. To reproduce the Feynman rules, we must redefine the meaning of the path integral in a manner such that (7.123) turns into its reciprocal, i.e.,

$$\int (D\phi_1)(D\phi_2) \exp \frac{i}{\hbar} S[\phi_1; \phi_2, \Omega] = \det A. \tag{7.128}$$

In this particular example, the use of Fermi statistics is of course unphysical (making ϕ a "ghost field," which is used only as a formal device in the quantization of gauge fields). The main point of the example is to motivate a way for the treatment of a physical fermion field theory by the method of path integrals, which we shall discuss in the next section.

7.9 Fermion Fields

In canonical quantization, fermion fields $\psi_j(x)$, $\psi_k^\dagger(x)$ are defined by anticommutation rules:

$$\{\psi_j(x), \psi_k^*(y)\}_{x_0 = y_0} = \hbar \delta^3(\mathbf{x} - \mathbf{y}),$$
$$\{\psi_j(x), \psi_k(y)\}_{x_0 = y_0} = 0.$$
(7.129)

This suggests that in the formal classical limit $\hbar \to 0$, $\psi_j(x)$ and $\psi_k^\dagger(x)$ should be represented not by numbers, but by anticommuting c-numbers. To extend the method of path integrals to the fermion case, we would have to define functional integrals over anticommuting c-numbers.

For this purpose, we adopt the point of view that a quantum field theory is defined by its Feynman rules. Thus, fermion fields differ from boson fields *only* in that a closed fermion loop in a Feynman graph is associated with an extra factor -1. We shall simply write down a path integral for the generating functional that reproduces this rule.

Consider a free spinor field coupled to an external source $\Omega(x)$ that creates fermion-antifermion pairs:

$$\mathcal{L}(x) = \bar{\psi}(i\gamma \cdot \partial - m)\psi + \bar{\psi}\psi\Omega,$$
(7.130)

with the action

$$\int d^4x \, \mathcal{L}(x) = -i \int d^4x_E \, \mathcal{L}(x_E) = i\hbar[(\bar{\psi}, S_F^{-1}\psi) + (\bar{\psi}, \Omega\psi)]$$
(7.131)

where

$$(f, g) \equiv \int d^4x_E f(x_E) g(x_E),$$
$$S_F \equiv (i\gamma \cdot \partial - m)^{-1}.$$
(7.132)

All connected Feynman graphs are one-loop graphs, just as in the example in the last section. The Feynman rules would be correctly given if we could define the generating functional $W[\Omega]$ through a path integral formula such that

$$\exp \frac{i}{\hbar} W[\Omega] \equiv \mathcal{N} \int (D\psi)(D\bar{\psi}) \exp\left[\frac{i}{\hbar} (\bar{\psi}, (S_F^{-1} + \Omega)\psi)\right]$$
$$= \mathcal{N}' \det(S_F^{-1} + \Omega).$$
(7.133)

To see how this can be done, first consider two anticommuting c-numbers η_1 and η_2, defined by

$$\{\eta_1, \eta_2\} = 0,$$
$$\eta_1^2 = \eta_2^2 = 0.$$
(7.134)

Method of Path Integrals

Any function of η_1 and η_2 is of the form

$$f(\eta_1, \eta_2) = C_0 + C_1\eta_1 + C_2\eta_2 + C_3\eta_1\eta_2. \qquad (7.135)$$

where C_i is a complex number. The operation of "integration" is defined by the rules[4]

$$\int d\eta_1 = 0, \qquad \int d\eta_2 = 0,$$
$$\int d\eta_1 \eta_1 = 1, \qquad \int d\eta_2 \eta_2 = 1, \qquad (7.136)$$

where $d\eta_1$ and $d\eta_2$ are supposed to anticommute with each other, and with η_1 and η_2. We also define

$$\int d\eta_1 \eta_2 \equiv -\eta_2 \int d\eta_1, \qquad (7.137)$$

from which follows

$$\int d\eta\, f(\eta - \eta_0) = \int d\eta\, f(\eta). \qquad (7.138)$$

Integrating (7.135) by these rules, we find

$$\int d\eta_1\, f(\eta_1, \eta_2) = C_1 + C_3 \eta_2,$$
$$\int d\eta_2 \int d\eta_1\, f(\eta_1, \eta_2) = C_3. \qquad (7.139)$$

In particular, for any number A,

$$\int d\eta_2 \int d\eta_1\, e^{A\eta_1\eta_2} = \int d\eta_2 \int d\eta_1 (1 + A\eta_1\eta_2) = A. \qquad (7.140)$$

We can generalize the definitions above to any number of anticommuting c-numbers $\{\eta_i\}$:

$$\{\eta_i, \eta_j\} = 0, \qquad \eta_j^2 = 0,$$
$$\int d\eta_i = 0, \qquad \int d\eta_i \eta_i = 1 \quad \text{(no sum)}. \qquad (7.141)$$

All $d\eta_i$ anticommute among themselves, and with all η_j. Let us divide $\{\eta_i\}$ into two disjoint sets $\{\eta_\alpha\}$ and $\{\eta_\alpha{}^*\}$. (The asterisk merely serves to distinguish between the two sets, and does not denote complex conjugation). Consider the quadratic form

$$X = \sum_{\alpha,\beta} \eta_\alpha{}^* A_{\alpha\beta} \eta_\beta \equiv \eta^* A \eta, \qquad (7.142)$$

---
[4] F. A. Berezin, *Method of Second Quantization* (Academic Press, New York, 1966).

where A is a symmetric matrix whose elements are numbers. Through a linear transformation, this can be reduced to diagonal form:

$$X = \sum_{\alpha=1}^{n} A_\alpha \eta_\alpha^* \eta_\alpha, \tag{7.143}$$

where A_α are the eigenvalues of the matrix A. Because of (7.141),

$$e^X = 1 + X + \frac{1}{2!} X^2 + \cdots + \frac{1}{n!} X^n, \tag{7.144}$$

$$\int (D\eta)(D\eta^*) e^X = \int (D\eta)(D\eta^*) \frac{X^n}{n!},$$

where

$$(D\eta) = \prod_{\alpha=1}^{n} d\eta_\alpha, \qquad (D\eta^*) = \prod_{\alpha=1}^{n} d\eta_\alpha^*. \tag{7.145}$$

The signs of these quantities depend on the order in which the factors are arranged. To obtain one term in the expansion of X^n, we must choose one term from each of the n factors X and multiply them together. Since $\eta_\alpha^{*2} = \eta_\alpha^2 = 0$, all the terms we choose must be different from one another. Hence, the only possible result is $\Pi_\alpha A_\alpha \eta_\alpha^* \eta_\alpha$. Since there are $n!$ ways of choosing, we have

$$X = n! \prod_\alpha A_\alpha \eta_\alpha^* \eta_\alpha. \tag{7.146}$$

Hence (up to a sign),

$$\int (D\eta)(D\eta^*) e^{\eta^* A \eta} = \prod_{\alpha=1}^{n} A_\alpha = \det A. \tag{7.147}$$

We may look upon this formula as a novel way of representing a determinant.
More generally, let

$$Q(\eta, \eta^*) = \eta^* A \eta - b^* \eta^* - b\eta, \tag{7.148}$$

where A is a non-singular symmetric matrix whose elements are numbers, and b^* and b are n-vectors whose components are anticommuting c-numbers. We rewrite

$$Q(\eta, \eta^*) = (\eta^* - \eta_0^*) A (\eta - \eta_0) - b^* A^{-1} b,$$
$$\eta_0 = A^{-1} b, \tag{7.149}$$
$$\eta_0^* = b^* A^{-1}.$$

Then, by (7.138) and (7.147), we have (up to a sign)

$$\int (D\eta)(D\eta^*) e^{Q(\eta, \eta^*)} = e^{-b^* A^{-1} b} \det A. \tag{7.150}$$

It is now obvious that to obtain (7.133), all we have to do is to represent the spinor fields $\psi(x)$ and $\bar{\psi}(x)$ by anticommuting c-numbers, regarding x as a discrete label first, and then pass to the continuum limit in the final result.

Method of Path Integrals

For a free spinor field coupled to external sources that create particles singly, we take

$$\mathcal{L}(x) = \bar{\psi}(i\gamma \cdot \partial - m)\psi + J\psi + J^*\bar{\psi} \tag{7.151}$$

where $J(x)$ and $J^*(x)$ are 4-component anticommuting c-numbers. The generating functional $W[J, J^*]$, which generates connected Feynman graphs with external lines, is then given by

$$\exp\frac{i}{\hbar} W[J, J^*] = \mathcal{N} \int (D\psi)(D\bar{\psi}) \exp\left\{\frac{i}{\hbar} [(\bar{\psi}, S_F^{-1}\psi) + (J, \psi) + (J^*, \bar{\psi})]\right\}$$

$$= \mathcal{N}' \, e^{\hbar^{-1}(J^*, S_F J)} \det S_F^{-1}. \tag{7.152}$$

To use the final result, it matters little whether J and J^* are anticommuting c-numbers or just ordinary numbers, for they occur paired. Indeed, the result may be derived easily by more conventional means; the use of anticommuting c-numbers here may seem to be a quaint way of doing it. The virtue of the method, however, lies in the fact that it enables us to treat fermion fields and boson fields on equal footing by the method of path integrals. This is useful when we consider fermion and boson fields that interact with each other.

CHAPTER VIII
QUANTIZATION OF GAUGE FIELDS

8.1 Canonical Quantization

The quantization of gauge fields presents special problems associated with the freedom of gauge transformations. To see how they arise, we first review the canonical quantization of the free Maxwell field, whose classical Lagrangian density is given by

$$\mathcal{L}(x) = -\tfrac{1}{4}F^{\mu\nu}F_{\mu\nu} = \tfrac{1}{2}(\mathbf{B}\cdot\mathbf{B} - \mathbf{E}\cdot\mathbf{E}),$$

$$\mathbf{B} \equiv \mathbf{\nabla}\times\mathbf{A}, \qquad (8.1)$$

$$\mathbf{E} \equiv -\frac{\partial\mathbf{A}}{\partial t} - \mathbf{\nabla}A^0.$$

The independent variables are the field variables $A^\mu(x)$, regarded as coordinates, and their time derivatives $\dot{A}^\mu(x)$. Changing A^μ by a gauge transformation $A^\mu \to A^\mu + \partial^\mu\omega$ leaves $\mathcal{L}(x)$ invariant.

According to the rules of canonical quantization, we must first cast the theory in Hamiltonian form, and then impose canonical commutation relations between the coordinates and their corresponding conjugate momenta. Noting that $\mathcal{L}(x)$ is independent of $\dot{A}^0(x)$, we see that the momentum conjugate to $A^0(x)$ is identically zero. Therefore $A^0(x)$ is not an independent coordinate, and may be eliminated through the classical constraint equation

$$\nabla^2 A^0 + \frac{\partial}{\partial t}\mathbf{\nabla}\cdot\mathbf{A} = 0.$$

The simplest choice is to set $A^0 = 0$ (thereby committing ourselves to a particular Lorentz frame). Classically, this requires $\partial(\mathbf{\nabla}\cdot\mathbf{A})/\partial t = 0$, which will be replaced by a subsidiary condition restricting the Hilbert space in the quantized theory. After setting $A^0 = 0$, $\mathcal{L}(x)$ is invariant under a more restrictive residual gauge transformation $\mathbf{A}\to\mathbf{A} - \mathbf{\nabla}\omega$, where ω is a time-independent function. The classical Hamiltonian, which is also invariant under the residual gauge transformation, is given by

$$H = \frac{1}{2}\int d^3x[\mathbf{E}\cdot\mathbf{E} + (\mathbf{\nabla}\times\mathbf{A})^2]. \qquad (8.2)$$

The independent variables are the fields $\mathbf{A}(x)$, regarded as coordinates, and their conjugate momenta $-\mathbf{E}(x)$. The theory is quantized by imposing the equal-time commutation relations

$$[E^j(x), A^k(y)]_{x^0=y^0} = i\delta_{jk}\delta^3(\mathbf{x} - \mathbf{y}). \tag{8.3}$$

However, not all three components of **A** represent dynamical degrees of freedom, owing to the residual gauge invariance. To complete the definition of the theory, we must explicitly eliminate the unphysical degrees of freedom.

The residual gauge invariance is expressed through the fact

$$[H, \boldsymbol{\nabla} \cdot \mathbf{E}] = 0. \tag{8.4}$$

In other words, $\boldsymbol{\nabla} \cdot \mathbf{E}$ is a constant of the motion. Consequently, the longitudinal part $\mathbf{E}_\parallel$ is not a dynamical variable, but is completely determined by initial data and boundary conditions. The longitudinal part $\mathbf{A}_\parallel$ is also completely determined, because

$$\frac{\partial \mathbf{A}}{\partial t} = -\mathbf{E}, \tag{8.5}$$

which follows directly from $\partial \mathbf{A}/\partial t = i[H, \mathbf{A}]$. Therefore, only the transverse parts $\mathbf{E}_\perp$ and $\mathbf{A}_\perp$ are truly dynamical variables. Their commutation relations are the transverse projections of (8.3):

$$[E_\perp{}^j(x), A_\perp{}^k(y)]_{x^0=y^0} = i\delta^{\mathrm{Tr}}_{jk}(\mathbf{x} - \mathbf{y}), \tag{8.6}$$

where

$$\delta^{\mathrm{Tr}}_{jk}(\mathbf{x} - \mathbf{y}) = \int \frac{d^3p}{(2\pi)^3} e^{i\mathbf{p}\cdot(\mathbf{x}-\mathbf{y})} \left[\delta_{jk} - \frac{p^j p^k}{|\mathbf{p}|^2}\right]. \tag{8.7}$$

The Hamiltonian can be rewritten in the form

$$H = \frac{1}{2} \int d^3x [\mathbf{E}_\perp \cdot \mathbf{E}_\perp + (\boldsymbol{\nabla} \times \mathbf{A}_\perp)^2] + \frac{1}{2} \int d^3x\, \mathbf{E}_\parallel \cdot \mathbf{E}_\parallel. \tag{8.8}$$

The last term is a c-number constant, which has no physical significance.

The residual gauge transformation $\mathbf{A} \to \mathbf{A} - \boldsymbol{\nabla}\omega$ only affects $\mathbf{A}_\parallel$, and therefore merely changes the reference point of the total energy. Thus, we may choose $\mathbf{A}_\parallel = 0$ for simplicity. All states of the system will then satisfy the subsidiary condition $\boldsymbol{\nabla} \cdot \mathbf{E} = 0$, which represents a restriction on the Hilbert space of the system.

To see this in more formal detail, let us represent a state in Hilbert space by the coordinate representative $\Psi[A]$. Then the Schrödinger operator $\mathbf{E}(\mathbf{x})$ is represented by

$$E^k(\mathbf{x}) = i\frac{\delta}{\delta A^k(\mathbf{x})}. \tag{8.9}$$

By virtue of (8.4), H and $\boldsymbol{\nabla} \cdot \mathbf{E}$ can be simultaneously diagonalized. Therefore, an eigenfunction $\Psi[A]$ of H can be chosen to be an eigenfunction of $\boldsymbol{\nabla} \cdot \mathbf{E}$:

$$\boldsymbol{\nabla} \cdot \mathbf{E}(\mathbf{x})\Psi[A] = \rho(\mathbf{x})\Psi[A], \tag{8.10}$$

where $\rho(\mathbf{x})$ is an arbitrary function.

Under the residual gauge transformation, $\Psi[\mathbf{A}]$ changes to $\Psi[\mathbf{A} - \boldsymbol{\nabla}\omega]$, which we can express in the form[a]

$$\Psi[\mathbf{A} - \boldsymbol{\nabla}\omega] = \left\{\exp \int d^3x[-\partial_i\omega(\mathbf{x})]\frac{\delta}{\delta A^i(\mathbf{x})}\right\} \Psi[\mathbf{A}]$$

$$= \exp\left(i\int d^3x(\boldsymbol{\nabla}\omega)\cdot\mathbf{E}\right)\Psi[\mathbf{A}] = \exp\left(-i\int d^3x\,\omega\boldsymbol{\nabla}\cdot\mathbf{E}\right)\Psi[\mathbf{A}]. \quad (8.11)$$

Now, any $\mathbf{A}(\mathbf{x})$ can be written as $\mathbf{A}(\mathbf{x}) = \mathbf{A}_\perp(\mathbf{x}) - \boldsymbol{\nabla}\omega(\mathbf{x})$. That is, the longitudinal part of $\mathbf{A}$ is pure-gauge. In the absence of external magnetic flux, all pure-gauge fields can be transformed to zero continuously. Thus, by (8.10) and (8.11), we can always write, for any $\mathbf{A}$,

$$\Psi[\mathbf{A}] = e^{-i\chi}\Psi[\mathbf{A}_\perp], \quad (8.12)$$

where

$$\chi = \int d^3x\,\omega\rho = \int d^3x\,\omega\nabla^2\left(\frac{1}{\nabla^2}\rho\right) = \int d^3x(\nabla^2\omega)\frac{1}{\nabla^2}\rho$$

$$= -\int d^3x(\boldsymbol{\nabla}\cdot\mathbf{A})\frac{1}{\nabla^2}\rho = \int d^3x\,d^3y\,\frac{[\boldsymbol{\nabla}\cdot\mathbf{A}(\mathbf{x})]\rho(\mathbf{y})}{4\pi|\mathbf{x}-\mathbf{y}|}. \quad (8.13)$$

The wave function $\Psi[\mathbf{A}_\perp]$ is an eigenfunction of the new Hamiltonian

$$e^{i\chi}H\,e^{-i\chi} = \frac{1}{2}\int d^3x\left[E_\perp^2 + (\boldsymbol{\nabla}\times\mathbf{A}_\perp)^2\right] + \frac{1}{2}\int d^3x\,d^3y\,\frac{\rho(\mathbf{x})\rho(\mathbf{y})}{4\pi|\mathbf{x}-\mathbf{y}|}, \quad (8.14)$$

which differs from H only by a constant, the energy of an arbitrary static charge distribution. In the physical Hilbert space spanned by $\Psi[\mathbf{A}_\perp]$, (8.10) becomes the subsidiary condition

$$\boldsymbol{\nabla}\cdot\mathbf{E}(\mathbf{x})\Psi[\mathbf{A}_\perp] = 0, \quad (8.15)$$

which is the quantum version of Gauss' law. One can obtain the same results as above by imposing the Coulomb gauge $\boldsymbol{\nabla}\cdot\mathbf{A} = 0$ on the classical theory before quantization. Our discussion brings out more clearly what is meant by choosing a gauge in the quantized theory.

Demonstration of Lorentz covariance is complicated in Coulomb gauge, and simpler in a covariant gauge, such as the Lorentz gauge $\partial_\mu A^\mu = 0$. However, to change the gauge, one would have to go back to the classical theory and start again. In general, a different gauge choice leads to a different definition of the Hilbert space of the system. For this reason, canonical quantization is clumsy to use, if one has to switch gauge in a discussion.

For non-Abelian gauge theories, canonical quantization[1] is far more compli-

---

[a] We drop surface integrals, rather carelessly. The real justification is that in Maxwell theory there is no topological charge. (See the treatment of Yang-Mills fields in Sec. 8.6.).

[1] J. Schwinger, *Phys. Rev.*, **125**, 1043 (1962); **127**, 324 (1962).

Quantization of Gauge Fields

cated. For example, the analogue of Gauss' law is the non-linear constraint

$$\nabla \cdot \mathbf{E}_a + gC_{abc}\mathbf{A}_b \cdot \mathbf{E}_c = 0, \tag{8.16}$$

whose solution does not naturally separate $\mathbf{E}_a$ and $\mathbf{A}_a$ into transverse and longitudinal parts. Thus, the Coulomb gauge does not appear to be natural here. Efforts to implement the Coulomb gauge in the quantized theory has led to ambiguities[2], whose origin may be traced to the existence of the topological charge[3]. Of all the linear gauge conditions commonly used, none seem "natural." Attempts to find a "natural" gauge based on exact solutions to (8.16) leads to intractible formulas[4]. The general problem is how to handle constrained Hamiltonian systems, and Dirac has proposed a general formalism to do this[5,6]. The formalism has been applied to non-Abelian gauge theories[7], but the difficulty in finding a practical calculational method remains. So far, the best method seems to be the Fadeev-Popov version of the Feynman path integral, which we shall discuss later.

For non-Abelian gauge fields, the last step in (8.11) fails, again owing to the existence of the topological charge. This leads to the "θ-worlds," to be discussed later.

8.2 Path Integral Method in Hamiltonian Form

The path integral in Hamiltonian form, given in the simplest instance by (7.18), was derived directly from canonical quantization and contains nothing new in principle. However, it enables us to eliminate the unphysical degrees of freedom in a much simpler fashion. We illustrate this with free Maxwell theory.

The classical Hamiltonian is

$$H = \int d^3x \{\tfrac{1}{2}[\mathbf{E} \cdot \mathbf{E} + (\nabla \times \mathbf{A})^2] + \mathbf{E} \cdot \nabla A^0\}. \tag{8.17}$$

The independent coordinates are $\mathbf{A}(x)$ and $A^0(x)$, with $-\mathbf{E}(x)$ conjugate to $\mathbf{A}(x)$, while the momentum conjugate to $A^0(x)$ is absent. Thus $A^0(x)$ is not a dynamical variable. Let $|A, t\rangle$ be defined by

$$\mathbf{A}_{op}(\mathbf{x}, t)|\mathbf{A}, t\rangle = \mathbf{A}(\mathbf{x})|\mathbf{A}, t\rangle, \tag{8.18}$$

which is the analogue of (7.5). The transition amplitude is then given by

$$\langle A_2, t_2 | A_1, t_1 \rangle = \mathcal{N} \int (DA^0)(D\mathbf{A})(D\mathbf{E})$$

$$\cdot \exp \frac{1}{i} \int_1^2 d^4x \left[\frac{1}{2} E^2 + \frac{1}{2}(\nabla \times \mathbf{A})^2 + \mathbf{E} \cdot \left(\frac{\partial \mathbf{A}}{\partial t} + \nabla A^0 \right) \right], \tag{8.19}$$

---

[2] V. N. Gribov, *Nucl. Phys.* **B139**, 1 (1978).
[3] R. Jackiw, I. Muzinich, and C. Rebbi, *Phys. Rev.*, **D17**, 1576 (1978).
[4] J. Goldstone and R. Jackiw, *Phys. Letters*, **74B**, 81 (1978).
[5] P.A.M. Dirac, *Cand. J. Phys.* **2**, 129 (1950); **3**, 1 (1951).
[6] A. Hanson, T. Regge, and C. Teitelboim, *Constrained Hamiltonian Systems* (Academia Nazionale dei Lincei, Rome, 1976).
[7] N. Christ, A. Guth, and E. Weinberg, *Nucl. Phys.* **B114**, 61 (1976).

where A^0, $\mathbf{A}$, $\mathbf{E}$, are functions of $\mathbf{x}$ and t, and the integral $\int_1^2 d^4x$ extends over all space between the times t_1 and t_2, with $\mathbf{A}(\mathbf{x}, t)$ constrained by

$$\mathbf{A}(\mathbf{x}, t_1) = \mathbf{A}_1(\mathbf{x}),$$
$$\mathbf{A}(\mathbf{x}, t_2) = \mathbf{A}_2(\mathbf{x}). \tag{8.20}$$

We can immediately carry out the path integration over A^0:

$$\int (DA^0) \exp i \int d^4x (\nabla \cdot \mathbf{E}) A^0 \propto \delta[\nabla \cdot \mathbf{E}]. \tag{8.21}$$

This requires the longitudinal part of $\mathbf{E}$ to vanish. Thus, $\int (D\mathbf{E})$ may be replaced by $\int (D\mathbf{E}_\perp)$, and $(\mathbf{E} \cdot \partial \mathbf{A}/\partial t)$ by $(\mathbf{E}_\perp \cdot \partial \mathbf{A}_\perp/\partial t)$, (since $\int d^3x\, \mathbf{E}_\perp \cdot \partial \mathbf{A}_\parallel/\partial t = 0$):

$$\langle \mathbf{A}_2, t_2 | \mathbf{A}_1, t_1 \rangle = \mathcal{N} \int (D\mathbf{A}_\parallel)(D\mathbf{A}_\perp)(D\mathbf{E}_\perp)$$

$$\cdot \exp \frac{1}{i} \int_1^2 d^4x \left[\frac{1}{2} E_\perp^2 + \frac{1}{2} (\nabla \times \mathbf{A}_\perp)^2 + \mathbf{E} \cdot \frac{\partial \mathbf{A}_\perp}{\partial t} \right]$$

$$= \mathcal{N} \int (D\mathbf{A}_\parallel)(D\mathbf{A}_\perp) \exp \frac{i}{2} \int_1^2 d^4x \left[\left(\frac{\partial \mathbf{A}_\perp}{\partial t}\right)^2 - (\nabla \times \mathbf{A}_\perp)^2 \right]. \tag{8.22}$$

Since the integrand is independent of $\mathbf{A}_\parallel$, $\int (D\mathbf{A}_\parallel)$ may be absorbed into $\mathcal{N}$. This means that the longitudinal parts of $\mathbf{A}_2$ and $\mathbf{A}_1$ have no physical significance, for they only affect the normalization of the transition amplitude. This reflects the gauge invariance of the system, as $\mathbf{A}_\parallel$ can be changed at will through a gauge transformation. Thus, we have recovered all the results of canonical quantization in Coulomb gauge in a much more streamlined fashioned.

We note that (8.22) is in Feynman form:

$$\langle \mathbf{A}_{\perp 2}, t_2 | \mathbf{A}_{\perp 1}, t_1 \rangle = \mathcal{N} \int (D\mathbf{A}_\perp) \exp i \int_1^2 d^4x\, \mathcal{L}_\perp(x),$$

$$\mathcal{L}_\perp(x) = \frac{1}{2} \left[\left(\frac{\partial \mathbf{A}_\perp}{\partial t}\right)^2 - (\nabla \times \mathbf{A}_\perp)^2 \right]. \tag{8.23}$$

This result is obtain by eliminating the unphysical degrees of freedom in Coulomb gauge. Unphysical fields can be more generally defined as those that can be eliminated through a gauge transformation. Since physical results should be gauge invariant, we expect the Feynman form to be valid when any suitable gauge condition is used to eliminate the unphysical fields.

8.3 Feynman Path Integral: Fadeev-Popov Method

The previous example suggests that the Feynman path integral is valid provided we fix the gauge. Fadeev and Popov[8] showed how this could be done.

---

[8] L. Fadeev and V. N. Popov, *Phys. Letters*, **25B**, 29 (1967).

Quantization of Gauge Fields

The failure of the unmodified Feynman formula and its possible remedy can be seen most clearly in Maxwell theory as follows. The action is given by

$$S[A] = -\frac{1}{4}\int d^4x \, F^{\mu\nu}F_{\mu\nu} = -\frac{1}{2}\int d^4x (\partial^\mu A^\nu - \partial^\nu A^\mu)\partial_\mu A_\nu$$

$$= \frac{1}{2}\int d^4x \, A_\mu(g^{\mu\nu}\Box^2 - \partial^\mu\partial^\nu)A_\nu$$

$$= -\frac{1}{2}\int \frac{d^4k}{(2\pi)^4} \tilde{A}_\mu(g^{\mu\nu}k^2 - k^\mu k^\nu)\tilde{A}_\nu, \qquad (8.24)$$

where $\tilde{A}^\mu$ is the Fourier transform of A^μ.[b] The kinetic operator $(g^{\mu\nu}k^2 - k^\mu k^\nu)$ has no inverse, because it has eigenfunctions k^ν, with eigenvalue zero. The propagator therefore does not exist. One remedy is to restrict $\tilde{A}^\mu$ with the condition $k_\mu \tilde{A}^\mu = 0$ (Lorentz gauge), so that the eigenvectors k^ν are no longer physically relevant.

To see the problem from another point of view, note that $S[A]$ is gauge invariant, and therefore independent of any fields that can be gauge-transformed away. If we include such fields in the Feynman path integral, we would obtain an infinite volume factor. This can be absorbed into the normalization constant only if these fields are clearly identified, and hence gauge fixing is called for.

From yet another point of view, the necessity for gauge fixing comes from our knowledge that the Green's functions of the theory are gauge-dependent. Hence, the generating functional $W[J]$ is not uniquely defined unless the gauge is specified.

A gauge transformation

$$U(x) = e^{-i\omega_a(x)L_a}$$

is parametrized by the functions $\omega_a(x)$, denoted collectively by ω. The gauge fields $A^\mu(x) \equiv A_a^\mu(x)L_a$ change according to

$$A_\mu \to A_\mu^\omega,$$

$$A_\mu^\omega = UA_\mu U^{-1} + \frac{i}{g}U\partial_\mu U^{-1}. \qquad (8.25)$$

Fixing the gauge means that we impose some condition on A_a^μ, called a gauge condition. Some commonly used gauge conditions are

$$\begin{array}{ll}\text{Lorentz gauge:} & \partial_\mu A_a^\mu = 0, \\ \text{Coulomb gauge:} & \nabla \cdot \mathbf{A}_a = 0, \\ \text{Axial gauge:} & A_a^3 = 0, \\ \text{Temporal gauge:} & A_a^0 = 0.\end{array} \qquad (8.26)$$

[b] The two $\tilde{A}_\mu$'s should have respective arguments k and $-k$; but $\tilde{A}_\mu(k) = \tilde{A}_\mu(-k)$ since $\tilde{A}_\mu(x)$ is real.

A general gauge condition may be represented in the form

$$\hat{f}_a A = 0 \quad (a = 1, \ldots, N), \tag{8.27}$$

where $\hat{f}_a$ is a mapping of the function space of $A_a^\mu(x)$ into itself. The admissible A's are those that are mapped into zero by each $\hat{f}_a$ ($a = 1, \ldots, N$).

As long as the gauge condition can eliminate all unphysical degrees of freedom, there will be no difficulty in passing from the Hamiltonian form of the path integral to the Feynman form. In that case, the transition amplitude in Feynman form is given by

$$\langle A_2, t_2 | A_1, t_1 \rangle = \mathcal{N} \int_{A_1}^{A_2} (DA)\, e^{iS_{21}[A]} \mathcal{J}[A] \prod_{a=1}^{N} \delta[\hat{f}_a A], \tag{8.28}$$

where

$$S_{21}[A] \equiv \int_{1}^{2} d^4x\, \mathcal{L}(x),$$

and $\mathcal{J}[A]$ is a Jacobian defined by

$$\int (DA)\mathcal{J}[A] \prod_{a=1}^{N} \delta[\hat{f}_a A] = 1. \tag{8.29}$$

In practice, one can tell whether a particular gauge condition works by seeing whether the kinetic operator has an inverse. In labelling the transition amplitude $\langle A_2, t_2 | A_1, t_1 \rangle$, we need not specify explicitly which components of A are the dynamical ones, as gauge-fixing does the job automatically; the unphysical ones affect only the normalization of the transition amplitude.

Fadeev and Popov showed a convenient way of calculating the Jacobian $\mathcal{J}[A]$. A side benefit of their method is that one can relax the gauge condition. Instead of demanding $\hat{f}_a A = 0$, one may allow $\hat{f}_a A$ to take on any functional form, but assign a definite weight to each functional form. This results in greater flexibility of the method. We now follow their development.

We assume that $\hat{f}_a$ have the property that, given any A, there always exists a gauge transformation ω such that

$$\hat{f}_a A^\omega = 0. \tag{8.30}$$

Thus, given A, $\{\hat{f}_a\}$ determines a gauge transformation $U(x)$. As x varies, $U(x)$ traces out an orbit in the gauge-group manifold. The volume of this orbit is denoted by $1/\Delta_f[A]$:

$$\Delta_f[A] \int (D\omega)\, \delta[\hat{f} A^\omega] = 1, \tag{8.31}$$

where $(D\omega) = \Pi_x \Pi_a\, d\omega_a(x)$, and

$$\delta[\hat{f} A] \equiv \prod_{a=1}^{N} \delta[\hat{f}_a A]. \tag{8.32}$$

Clearly $\Delta_f[A]$ is gauge invariant:

$$\Delta_f[A] = \Delta_f[A^\omega]. \tag{8.33}$$

Quantization of Gauge Fields

By integrating both sides of (8.31) over A, one can manipulate it into the form

$$[\int(D\omega)/\int(DA)] \int (DA) \, \Delta_f[A] \delta[\hat{f}A] = 1, \tag{8.34}$$

which shows that $\Delta_f[A]$ is proportional to the Jacobian $\mathcal{J}[A]$.

To calculate $\Delta_f[A]$, let $f_a(x)$ denote the numerical value of the function $\hat{f}_a A$ at x:

$$f_a(x) \equiv \hat{f}_a A(x). \tag{8.35}$$

Then

$$\begin{aligned}
\Delta_f^{-1}[A] &= \prod_x \prod_a \int_{-\infty}^{\infty} d\omega_a(x) \, \delta(f_a(x)) \\
&= \prod_x \prod_a \int_{-\infty}^{\infty} df_a(x) \, \delta(f_a(x)) \, \frac{\partial(\omega_1(x), \ldots, \omega_N(x))}{\partial(f_1(x), \ldots, f_N(x))} \\
&= \prod_x \det \left\| \frac{\partial \omega_a(x)}{\partial f_b(x)} \right\|_{f=0} \equiv \det\left(\frac{\delta \omega}{\delta f}\right)_{f=0}.
\end{aligned} \tag{8.36}$$

The last step defines the functional determinant of the continuous matrix $\delta\omega_a(x)/\delta f_b(y)$, with rows labelled by (a, x) and columns by (b, y).

If A satisfies the gauge condition, then the condition $f = 0$ in (8.36) may be replaced by $\omega = 0$, and we can write

$$\Delta_f[A] = \det\left(\frac{\delta f}{\delta \omega}\right)_{\omega=0} \quad \text{(for } A \text{ satisfying } \hat{f}_a A = 0\text{)}. \tag{8.37}$$

This in fact determines $\Delta_f[A']$ for arbitrary A' because, by assumption, A' can be gauge transformed to A, and by (8.33), $\Delta_f[A'] = \Delta_f[A]$. To calculate the functional determinant in (8.37), it is only necessary to make infinitesimal gauge transformations from A, which is a great convenience.

The identity (8.31) can be generalized to

$$\Delta_f[A] \int (D\omega) \prod_a \delta(\hat{f}_a A^\omega - g_a) = 1, \tag{8.38}$$

where $\{g_a\}$ is a set of arbitrary functions. This merely redefines $\hat{f}_a$ to make the gauge condition read $\hat{f}_a A = g_a$. Since g_a are independent of A, $\Delta_f[A]$ is unaffected. Now multiply both sides by an arbitrary functional $G[g] \equiv G[g_1, \ldots, g_N]$, and integrate over all g_a. The result is a more general identity:

$$\Delta_f[A] \frac{\int (D\omega) G[\hat{f}A]}{\int (Dg) G[g]} = 1. \tag{8.39}$$

Taking advantage of this, we can derive a more general representation for the transition amplitude than (8.28). The idea is to insert (8.39) into the unmodified Feynman path integral, and then try to factor out the infinite volume factor corresponding to integration over the unphysical fields.

We start by writing

$$\langle A_2, t_2 | A_1, t_1 \rangle = \frac{\mathcal{N}}{\int (Dg) G[g]} \int_{A_1}^{A_2} (DA)\, e^{iS_{21}[A]} \Delta_f[A] \int (D\omega) G[\hat{f}A^\omega]. \quad (8.40)$$

Now interchange the order of $\int (DA)$ and $\int (D\omega)$. Since (DA), $S[A]$ and $\Delta_f[A]$ are all gauge invariant, they can be replaced respectively by (DA^ω), $S[A^\omega]$ and $\Delta_f[A^\omega]$. The new integration variable A^ω can be renamed A. Thus we obtain

$$\langle A_2, t_2 | A_1, t_1 \rangle = \frac{\mathcal{N} \int (D\omega)}{\int (Dg) G[g]} \int_{A_1}^{A_2} (DA)\, e^{iS_{21}[A]} G[\hat{f}A]\, \Delta_f[A]. \quad (8.41)$$

Note that we have factored out $\int (D\omega)$, which is just the volume coming from integrating over fields that can be gauged away. Absorbing this constant together with $\int (Dg) G[g]$ into $\mathcal{N}$, we finally have

$$\langle A_2, t_2 | A_1, t_1 \rangle = \mathcal{N} \int_{A_1}^{A_2} (DA)\, e^{iS_{21}[A]} G[\hat{f}A]\, \Delta_f[A]. \quad (8.42)$$

The generating functional $W[J]$ for Green's functions in the gauge $\hat{f}_a A = 0$ is given by

$$e^{iW[J]} = \mathcal{N} \int (DA)\, e^{iS[A] + i(J, A)} G[\hat{f}A]\, \Delta_f[A], \quad (8.43)$$

where

$$S[A] = \int d^4x\, \mathcal{L}(x),$$

$$(J, A) = \int d^4x\, J_{a\mu}(x) A_a^\mu(x). \quad (8.44)$$

The integration $\int d^4x$ in $S[A]$ and (J, A) is defined by passage to Euclidean space, then continued back to Minkowski space. Since we know that this merely supplies the correct $i\varepsilon$ prescription in the propagators, we shall not always indicate it explicitly, but it should be understood as being done.

8.4 Free Maxwell Field

The action is

$$S[A] = -\frac{1}{2} \int \frac{d^4k}{(2\pi)^4} \widetilde{A}_\mu (g^{\mu\nu} k^2 - k^\mu k^\nu) \widetilde{A}_\nu. \quad (8.45)$$

It is convenient to decompose $\widetilde{A}^\mu$ into transverse and longitudinal parts (in the 4-dimensional sense):

$$\widetilde{A}^\mu = \widetilde{A}_T^\mu + \widetilde{A}_L^\mu, \quad (k \cdot \widetilde{A}_T = 0). \quad (8.46)$$

Quantization of Gauge Fields

The transverse and longitudinal projection operators are

$$P_T^{\mu\nu}(k) = g^{\mu\nu} - \frac{k^\mu k^\nu}{k^2},$$

$$P_L^{\mu\nu}(k) = \frac{k^\mu k^\nu}{k^2}.$$
(8.47)

As matrices in μ, ν with metric $g^{\mu\nu}$, they have the properties

$$\begin{gathered}P_T^2 = P_T, \qquad P_L^2 = P_L,\\ P_T + P_L = 1, \qquad P_T \cdot P_L = 0,\\ \tilde{A}_T = P_T \tilde{A}, \qquad \tilde{A}_L = P_L \tilde{A}.\end{gathered}$$
(8.48)

We can then write, in matrix notation,

$$S[A] = -\tfrac{1}{2}(\tilde{A}, \tilde{K}\tilde{A}),$$
(8.49)

where

$$\tilde{K}^{\mu\nu} = k^2 g^{\mu\nu} - k^\mu k^\nu = k^2 P_T(k).$$
(8.50)

1 Lorentz Gauge

Choose $\hat{f}$ such that

$$\begin{aligned}\hat{f}A(x) &\equiv \partial_\mu A^\mu(x),\\ \hat{f}\tilde{A}(x) &= ik \cdot \tilde{A}(k),\end{aligned}$$
(8.51)

where $\tilde{A}^\mu$ is the Fourier transform of A^μ. The weight assigned to $\hat{f}A$ is left open pending the choice of the functional G. First we calculate $\Delta_f[A]$. Let $\tilde{A}$ be such that $k \cdot \tilde{A} = 0$. Under an infinitesimal gauge transformation, $\tilde{A}$ and $\tilde{f} \equiv \hat{f}A$ respectively change by

$$\begin{aligned}\delta \tilde{A}^\mu &= ik^\mu \delta\tilde{\omega},\\ \delta\tilde{f} &= -k^2 \delta\tilde{\omega}.\end{aligned}$$
(8.52)

Hence

$$\left(\frac{\delta \tilde{f}}{\delta \tilde{\omega}}\right)_{\omega=0} = -k^2,$$

$$\Delta_f[A] = \det\left(\frac{\delta \tilde{f}}{\delta \tilde{\omega}}\right)_{\omega=0} = -\prod_k k^2.$$
(8.53)

We see that $\Delta_f[A]$ is divergent, but independent of A. Hence it can be absorbed into $\mathcal{N}$. Therefore

$$e^{iW[J]} = \mathcal{N} \int (DA)\, e^{iS[A]+i(J,A)} G[\hat{f}A].$$
(8.54)

We still have the freedom to choose G.

First consider the choice $G[f] = \delta[f]$. This corresponds to the transverse Lorentz gauge, or Landau gauge. We have

$$\delta[\hat{f}A] = \prod_k \delta(k \cdot \tilde{A}) = \prod_k \delta(k \cdot \tilde{A}_L). \tag{8.55}$$

This requires $k \cdot \tilde{A}_L = 0$, which can be satisfied only by $\tilde{A}_L \equiv 0$. Hence

$$e^{iW[J]} = \mathcal{N} \int (DA_T) \, e^{iS[A_T] + i(J, A_T)}. \tag{8.56}$$

In fact, $S[A_T] = S[A]$. However, the restriction to the space of $\tilde{A}_T$ renders $\tilde{K}$ non-singular, for the only eigenfunctions of $\tilde{K}$ with zero eigenvalue are longitudinal fields. It can be readily verified that

$$\tilde{K}^{-1} = \frac{1}{k^2} P_T(k), \tag{8.57}$$

in the sense

$$\tilde{K}\tilde{K}^{-1} = P_T(k). \tag{8.58}$$

Thus we can perform the path integral in (8.55), obtaining

$$\begin{aligned} e^{iW[J]} &= \mathcal{N} \int (DA_T) \exp\left(-\frac{i}{2}(A_T, \tilde{K}A_T) + i(\tilde{J}, \tilde{A}_T)\right) \\ &= \mathcal{N} (\det \tilde{K})^{-1/2} \exp\left(\frac{i}{2}(\tilde{J}_T, \tilde{K}^{-1}\tilde{J}_T)\right). \end{aligned} \tag{8.59}$$

Since $\tilde{K}^{-1}$ is proportional to P_T, $\tilde{J}_T$ may be replaced by J. Choosing the constant in front to be unity, we obtain the well-known result

$$W[J] = \frac{1}{2}(J, \tilde{K}^{-1}J) = \frac{1}{2}\int d^4x \, J_\mu(x) D^{\mu\nu}(x-y) J_\nu(y), \tag{8.60}$$

$$D^{\mu\nu}(x) = \int \frac{d^4k}{(2\pi)^4} \frac{e^{-ik \cdot x}}{k^2 + i\varepsilon} \left(g^{\mu\nu} - \frac{k^\mu k^\nu}{k^2 + i\varepsilon}\right).$$

The $i\varepsilon$ comes from the definition of $(\tilde{J}, \tilde{K}^{-1}\tilde{J})$ as an Euclidean integral.

Next, consider the choice

$$G[\hat{f}A] = \exp\frac{i}{2\lambda} \int d^4x (\hat{f}A)^2 = \exp\frac{i}{2\lambda} \int d^4x (\partial_\mu A^\mu)^2, \tag{8.61}$$

where λ is a real parameter. Common choices for λ are

$$\begin{aligned} \lambda &= 1 \quad \text{(Feynman gauge)}, \\ \lambda &= 0 \quad \text{(Landau gauge)}. \end{aligned} \tag{8.62}$$

The latter reduces to the previous case with $G[f] = \delta[f]$. We now have

$$e^{iW[J]} = \mathcal{N} \int (DA) \exp i[-\tfrac{1}{2}(A, KA) + (J, A)], \tag{8.63}$$

Quantization of Gauge Fields

where

$$K^{\mu\nu} = -\left[g^{\mu\nu}\Box^2 - \left(1 - \frac{1}{\lambda}\right)\partial^\mu\partial^\nu\right], \tag{8.64}$$

with Fourier transform

$$\begin{aligned}\tilde{K}^{\mu\nu} &= k^2\left[g^{\mu\nu}k^2 - \left(1 - \frac{1}{\lambda}\right)\frac{k^\mu k^\nu}{k^2}\right] \\ &= k^2\left[P_T^{\mu\nu}(k) + \frac{1}{\lambda}P_L^{\mu\nu}(k)\right].\end{aligned} \tag{8.65}$$

It is easily verified that

$$\begin{aligned}(\tilde{K}^{-1})^{\mu\nu} &= \frac{1}{k^2}[P_T^{\mu\nu}(k) + \lambda P_L^{\mu\nu}(k)] \\ &= \frac{1}{k^2}\left[g^{\mu\nu} - (1-\lambda)\frac{k^\mu k^\nu}{k^2}\right].\end{aligned} \tag{8.66}$$

Thus

$$\begin{aligned}W[J] &= \frac{1}{2}(J, \tilde{K}^{-1}J) = \frac{1}{2}\int d^4x\, J_\mu(x)D^{\mu\nu}(x-y)J_\nu(x), \\ D^{\mu\nu}(x) &= \int \frac{d^4k}{(2\pi)^4}\frac{e^{-ik\cdot x}}{k^2 + i\varepsilon}\left[g^{\mu\nu} - (1-\lambda)\frac{k^\mu k^\nu}{k^2 + i\varepsilon}\right].\end{aligned} \tag{8.67}$$

As is well-known, the term proportional to $k^\mu k^\nu$ does not contribute in Feynman graphs when the Maxwell field is coupled to a conserved current. Thus Lorentz gauges with different λ's are physically equivalent.

The fact that $D^{\mu\nu}$ transforms under a Lorentz transformation as a tensor does not in itself make the theory Lorentz invariant, for $D^{\mu\nu}$ is not a physical quantity. To show Lorentz invariance, one must show that the S matrix is invariant. In the free field case, the matrix is $\varepsilon_\mu D^{\mu\nu}\varepsilon_\nu$, where ε^μ is the polarization vector of a free photon. Lorentz invariance depends on the fact that ε^μ is a 4-vector that can be brought to the form $\varepsilon^\mu = (0, \boldsymbol{\varepsilon}_\perp)$ in some Lorentz frame. Thus, the manifest invariance of (8.67) is merely a formal nicety; the components $\mathbf{J}_\parallel$ and $\mathbf{J}_0$ actually play no role in extracting physical results.

2 Coulomb Gauge

We choose $\hat{f}$ and G as follows:

$$\begin{aligned}\hat{f}A(x) &\equiv \boldsymbol{\nabla}\cdot\mathbf{A}(x), \\ G[\hat{f}A] &\equiv \exp\frac{i}{2\lambda}\int d^4x[\boldsymbol{\nabla}\cdot\mathbf{A}(x)]^2.\end{aligned} \tag{8.68}$$

Then
$$\Delta_f[A] = -i\prod_k |\mathbf{k}|^2, \tag{8.69}$$

which is independent of A, and hence may be ignored. Thus
$$e^{iW[J]} = \mathcal{N} \int (DA) \exp i\left\{S[A] + (J, A) + \frac{1}{2\lambda}\int d^4x (\nabla \cdot \mathbf{A})^2\right\}. \tag{8.70}$$

To present this in mock-invariant form, we write
$$\int d^4x (\nabla \cdot \mathbf{A})^2 = -\int d^4x A_i \partial^i \partial^j A_j = \int \frac{d^4k}{(2\pi)^4} \widetilde{A}_i k^i k^j \widetilde{A}_j$$
$$= \int \frac{d^4k}{(2\pi)^4} \widetilde{A}_\mu s^\mu s^\nu \widetilde{A}_\nu, \tag{8.71}$$

where
$$s^\mu \equiv (0, \mathbf{k}). \tag{8.72}$$

Then
$$e^{iW[J]} = \mathcal{N} \int (DA) \exp i[-\tfrac{1}{2}(\widetilde{A}, \widetilde{K}_c \widetilde{A}) + (\widetilde{J}, \widetilde{A})]$$
$$= \exp \frac{i}{2}(\widetilde{J}, \widetilde{K}_c^{-1}\widetilde{J}), \tag{8.73}$$

where
$$\widetilde{K}_c{}^{\mu\nu} = k^2 g^{\mu\nu} - k^\mu k^\nu + \frac{1}{\lambda} s^\mu s^\nu. \tag{8.74}$$

To find the inverse of $\widetilde{K}_c$, we write
$$(\widetilde{K}_c^{-1})^{\mu\nu} = C_1 g^{\mu\nu} + C_2 k^\mu k^\nu + C_3 s^\mu s^\nu + C_4 k^\mu s^\nu + C_5 s^\mu k^\nu, \tag{8.75}$$
and determine the coefficients C_i by requiring $(\widetilde{K}\widetilde{K}^{-1})^{\mu\nu} = g^{\mu\nu}$. The result is
$$\widetilde{D}_c{}^{\mu\nu} \equiv (\widetilde{K}_c^{-1})^{\mu\nu} = \frac{1}{k^2 + i\varepsilon}\left[g^{\mu\nu} + \left(1 + \frac{\lambda k^2}{s^2}\right)\frac{k^\mu k^\nu}{s^2} - \frac{1}{s^2}(k^\mu s^\nu + s^\mu k^\nu)\right]$$
$$= \frac{1}{k^2 + i\varepsilon}\left[g^{\mu\nu} - \frac{1}{|\mathbf{k}|^2}(k^\mu k^\nu - k^\mu s^\nu - s^\nu k^\mu)\right] + \frac{\lambda k^\mu k^\nu}{|\mathbf{k}|^4}, \tag{8.76}$$

which is the photon propagator in Coulomb gauge. More explicitly,
$$\widetilde{D}_c{}^{ij} = -\frac{1}{k^2 + i\varepsilon}\left(\delta_{ij} - \frac{k_i k_j}{|\mathbf{k}|^2}\right) + \frac{\lambda k_i k_j}{|\mathbf{k}|^4},$$
$$\widetilde{D}_c{}^{i0} = \widetilde{D}_c{}^{0i} = \frac{\lambda k^0 k^i}{|\mathbf{k}|^4}, \tag{8.77}$$
$$\widetilde{D}_c{}^{00} = -\frac{1}{|\mathbf{k}|^2} + \frac{\lambda k_0^2}{|\mathbf{k}|^4}.$$

Quantization of Gauge Fields

The final result is

$$W[J] = \tfrac{1}{2}(\tilde{J}, \tilde{D}_c \tilde{J}). \tag{8.78}$$

For the case $\lambda = 0$, which is equivalent to $G[f] = \delta[f]$,

$$W[J] = \frac{1}{2} \int \frac{d^4k}{(2\pi)^4} \tilde{J}^i(k) \frac{1}{k^2 + i\varepsilon} \left(-\delta_{ij} + \frac{k_i k_j}{|\mathbf{k}|^2}\right) \tilde{J}^j(k) - \int \frac{d^4k}{(2\pi)^4} \frac{\tilde{J}_0^2(k)}{|\mathbf{k}|^2}, \tag{8.79}$$

or

$$W[J] = \frac{1}{2} \int d^4x \, J^i(x) D_c^{ij}(x-y) J^j(y) - \int dx_0 \, d^3x \, d^3y \, \frac{J_0(x) J_0(y)}{4\pi |\mathbf{x} - \mathbf{y}|}, \tag{8.80}$$

where D_c is the Fourier transform of $\tilde{D}_c$. The last term contains only $\mathbf{J}_\perp$, because $\tilde{D}_c(k)$ is proportional to $P_\perp(\mathbf{k})$. The last term is the analogue of the last term in (8.14). We see that the system is described in the most succinct physical terms: when one turns on an external current J^μ, only $\mathbf{J}_\perp$ disturbs the system. $\mathbf{J}_\parallel$ does not couple to the system at all, and J_0 merely shifts the reference point of the total energy.

3 Temporal and Axial Gauges

The temporal gauge $A^0 = 0$, and the axial gauge $A^3 = 0$, can be treated as one gauge, as $S[A]$ is really defined as an integral over Euclidean space, and the integrand is rotationally invariant. The distinction between the two gauges comes when we transform back to Minkowski space, in which physical photons are defined.

Denoting Euclidean quantities by the convention

$$\begin{aligned} p_\mu &= (\mathbf{k}, -ik_0), \\ A_\mu &= (\mathbf{A}, iA_0), \end{aligned} \tag{8.81}$$

we have

$$\begin{aligned} S[A] &= -\frac{1}{2} \int \frac{d^4k}{(2\pi)^4} \left[k^2 A^\mu A_\mu - (k^\mu A_\mu)^2 \right] \\ &= -\frac{i}{2} \int \frac{d^4p}{(2\pi)^4} A_\mu (p^2 \delta_{\mu\nu} - p_\mu p_\nu) A_\nu, \end{aligned} \tag{8.82}$$

where the tildes denoting Fourier transforms have been omitted for simplicity. We set a component of A_μ to zero, say,

$$A_4 = 0. \tag{8.83}$$

The rest of the components are denoted by A_i ($i = 1, 2, 3$). Then

$$e^{iW[J]} = \mathcal{N} \int (\mathbf{D}\mathbf{A}) \exp[\tfrac{1}{2}(A, KA) + (J, A)], \tag{8.84}$$

where

$$(A, B) \equiv \int \frac{d^4p}{(2\pi)^4} A_i(p) B_i(p), \qquad (8.85)$$

$$K_{ij} \equiv p^2 \delta_{ij} - p_i p_j.$$

Calculating the path integral leads to

$$W[J] = -\frac{i}{2}(J, K^{-1}J),$$

$$K_{ij}^{-1} = \frac{1}{p^2}\left(\delta_{ij} + \frac{p_i p_j}{p_4^2}\right). \qquad (8.86)$$

Note that $p^2 = |\mathbf{p}|^2 + p_4^2$. The condition $A_4 = 0$ does not fix the gauge completely. This is reflected in the fact that K^{-1} diverges when $p_4 = 0$. However, this has no physical consequences, as we shall see.

Now we have to transform (8.86) back to Minkowski space, so that we can calculate the S-Matrix by replacing normal components of $\mathbf{J}$ (with respect to $\mathbf{k}$) by photon wave functions.

If we choose to regard p_4 as the imaginary energy, we would let

$$p_4 \to -ik_0 - i\varepsilon, \quad (\varepsilon \to 0^+),$$

and obtain $W[J]$ in the temporal gauge $A^0 = 0$:

$$W[J] = \frac{1}{2}\int \frac{d^4k}{(2\pi)^4} \frac{1}{k^2 + i\varepsilon} J^i \left(\delta_{ij} - \frac{k_i k_j}{k_0^2 + i\varepsilon}\right) J^j \qquad (8.87)$$

Only the components of $\mathbf{J}_\perp(\mathbf{k})$, defined by $\mathbf{J}_\perp(\mathbf{k}) \cdot \mathbf{k} = 0$, are physically relevant.

If we choose to regard a component other than p_4 as the imaginary energy, say p_3, we would take

$$p_3 \to -ik_0 - i\varepsilon, \quad (\varepsilon \to 0^+),$$
$$p_1 = k_1, \quad p_2 = k_2, \quad p_4 = k_3,$$

and obtain $W[J]$ in the axial gauge $A^3 = 0$:

$$W[J] = \frac{1}{2}\int \frac{d^4k}{(2\pi)^4} \frac{1}{k^2 + i\varepsilon} \left[\mathbf{J}_\perp^2 + \mathbf{J}_\parallel^2 - J_0^2 + \frac{(\mathbf{J}_\parallel \cdot \mathbf{k} - J_0 k_0)^2}{k_3^2}\right]. \qquad (8.88)$$

Again, only the dependence on $\mathbf{J}_\perp$ is physically relevant.

8.5 Pure Yang-Mills Fields

Yang-Mills fields are more complicated than the Maxwell field, because of the non-linear relation

$$F_a^{\mu\nu} = \partial^\mu A_a^\nu - \partial^\nu A_a^\mu - g C_{abc} A_b^\mu A_c^\nu. \qquad (8.89)$$

Quantization of Gauge Fields

This makes the Lagrangian density more complicated:

$$\mathcal{L}_0(x) = -\tfrac{1}{4}F_a{}^{\mu\nu}F_{a\mu\nu} = \mathcal{L}_0(x) + \mathcal{L}_1(x), \tag{8.90}$$

where

$$\begin{aligned}\mathcal{L}_0(x) &= -\tfrac{1}{2}(\partial^\mu A_a{}^\nu - \partial^\nu A_a{}^\mu)\partial_\mu A_{a\nu}, \\ \mathcal{L}_1(x) &= -g C_{abc} A_a{}^\nu A_b{}^\mu \partial_\mu A_{c\nu} \\ &\quad - \tfrac{1}{4}g^2 C_{abc}C_{ab'c'}A_b{}^\mu A_{b'\mu} A_c{}^\nu A_{c'\nu}.\end{aligned} \tag{8.91}$$

The term $\mathcal{L}_0(x)$ contributes to the action a term of the same form as that of the Maxwell theory:

$$S_0[A] \equiv \int d^4x\, \mathcal{L}_0(x) = \frac{1}{2}\int d^4x\, A_{a\mu}(g^{\mu\nu}\Box^2 - \partial^\mu\partial^\nu)A_{a\nu}. \tag{8.92}$$

Unfortunately, $\mathcal{L}_1(x)$ makes the Feynman path integral non-Gaussian, and not calculable in closed form.

One way to proceed is to make use of (7.52) and write

$$\begin{aligned}e^{iW[J]} &= \mathcal{N}\int (DA)\, e^{iS[A]+i(J,A)} G[\hat{f}A]\,\Delta_f[A] \\ &= \left[\exp i\int d^4x\, \mathcal{L}_1\!\left(\frac{1}{i}\frac{\delta}{\delta J(x)}\right)\right] e^{iW_0[J]},\end{aligned} \tag{8.93}$$

where

$$e^{iW_0[J]} = \mathcal{N}\int (DA)\, e^{iS_0[A]+i(J,A)} G[\hat{f}A]\,\Delta_f[A]. \tag{8.94}$$

Gauge fixing is done in $W_0[J]$, from which $W[J]$ may be obtained by perturbation expansion in powers of g. This procedure is far from satisfactory for two reasons. First, we lose sight of the basic symmetry of the problem, because neither $\mathcal{L}_0$ nor $\mathcal{L}_1$ are separately gauge invariant. Secondly, non-perturbative effects, such as instantons and possibly quark confinement, would be difficult to discuss in this approach. Nevertheless, (8.93) is at least useful for perturbation theory, when one has reason to believe that to be valid.

In some problems for which standard perturbation theory is definitely unworkable, a better approach might be to expand $S[A]$ about the classical path, thus basing a new perturbation theory on the semi-classical approximation[9]. But we shall not go into that.

Even though $S_0[A]$ has the same form as in Maxwell theory, the non-linear nature of the gauge transformation can make $\Delta_f[A]$, and therefore $W_0[J]$, quite different. In general, this gives rise to fictitious "ghost fields", as we shall see. In the rest of this section, we illustrate the calculation of $W_0[J]$, but forego the derivation of Feynman rules.

---

[9] C. G. Callan, R. Dashen, and D. J. Gross, *Phys. Rev.*, D**17**, 2717 (1978).

1 Axial Gauge

Choose $\hat{f}_a A = A_a{}^3$. Under an infinitesimal gauge transformation ω about fields satisfying $\hat{f}_a A = 0$, we have

$$\hat{f}_a A^\omega = \frac{1}{g} \partial^3 \omega_a. \tag{8.95}$$

Hence, $\delta f/\delta \omega = g^{-1} \partial^3$, and $\Delta_f[A]$ are independent of A. Consequently $\Delta_f[A]$ can be absorbed into the normalization constant. For this reason, the canonical quantization of Yang-Mills fields is simplest in axial gauge[10], and the equivalence between the Feynman path integral and canonical quantization can be demonstrated explicitly in a straightforward fashion[11].

2 Lorentz Gauge: Fadeev-Popov Ghosts

Choose $\hat{f}_a A = \partial_\mu A_a{}^\mu$. For A such that $\partial_\mu A_a{}^\mu = 0$, an infinitesimal gauge transformation on $A_a{}^\mu$ gives

$$\begin{aligned}\hat{f}_a A_a{}^\omega &= \partial_\mu \left(A_a{}^\mu + \frac{1}{g} \partial^\mu \omega_a + C_{abc} \omega_b A_c{}^\mu \right) \\ &= \frac{1}{g} \Box^2 \omega_a + C_{abc} (\partial_\mu \omega_b) A_c{}^\mu.\end{aligned} \tag{8.96}$$

Hence

$$\begin{aligned}\frac{\delta f_a}{\delta \omega_b} &= \frac{1}{g} \delta_{ab} \Box^2 + C_{abc} A_c{}^\mu \partial_\mu, \\ \Delta f[A] &= \det\left(\frac{1}{g} \delta_{ab} \Box^2 + C_{abc} A_c{}^\mu \partial_\mu \right).\end{aligned} \tag{8.97}$$

Since $\Delta_f[A]$ depends on A, it cannot be absorbed. A convenient way of proceeding is to rewrite the determinant in terms of a path integral over anti-commuting c-number fields, as we did in Sec. 7.9:

$$\begin{aligned}\Delta_f[A] &= \int (D\eta^*)(D\eta) \exp\left[i \int d^4x\, \mathcal{L}_{\text{ghost}}(x) \right], \\ \mathcal{L}_{\text{ghost}}(x) &= \eta_a^*(x)[\delta_{ab} \Box^2 + g C_{abc} A_c{}^\mu(x) \partial_\mu] \eta_b(x),\end{aligned} \tag{8.98}$$

where η^* and η are independant scalar fields obeying Fermi statistics, with anti-commutation and integration rules given in (7.141). They are called Fadeev-Popov ghosts.

Choosing

$$G[f] = \exp\left[\frac{i}{2\lambda} \int d^4x\, f^2 \right] = \exp\left[\frac{i}{2\lambda} \int d^4x (\partial_\mu A_a{}^\mu)^2 \right], \tag{8.99}$$

---

[10] R. Arnowitt and S. Fickler, *Phys. Rev.*, **127**, 1821 (1962), Sec. 5.
[11] S. Coleman, *Secret Symmetry*, in *Law of Hadronic Matter*, Ed. A. Zichichi (Academic Press, New York, 1975), p. 193.

Quantization of Gauge Fields

we obtain

$$e^{iW[J]} = \int (D\eta^*)(D\eta)(DA) \exp\left[i\int d^4x(\mathscr{L}_{\text{eff}} + J_{a\mu}A_a^{\mu})\right], \quad (8.100)$$

where

$$\mathscr{L}_{\text{eff}} = \frac{1}{2} A_{a\mu}\left[g^{\mu\nu}\Box^2 - \left(1 - \frac{1}{\lambda}\right)\partial^\mu\partial^\nu\right]A_{a\nu} + \frac{1}{2}\eta_a^*\Box^2\eta_b + \mathscr{L}',$$

$$\mathscr{L}' = \mathscr{L}_1 + \frac{1}{2}gC_{abc}(\eta_a^*\partial_\mu\eta_b)A_a^{\mu}, \quad (8.101)$$

with $\mathscr{L}_1$ given in (8.91). Thus, the system is described as gauge fields with extra couplings to massless spinless ghost fields with derivative couplings. Since there is no physical need to introduce sources for the ghost fields, they only occur in closed loops in Feynman graphs.

8.6 The θ-World and the Instanton

1 Discovering the θ-World

To begin, let us set $A_a^0(x) = 0$. In the representation in which the Schrödinger operator $A_a(x)$ is diagonal, we denote a wave function of the system by $\Psi[A]$, and represent the conjugate momentum $-E_a(x)$ by

$$E_a^k(x) = i\frac{\delta}{\delta A_a^k(x)}. \quad (8.102)$$

The Hamiltonian is given by

$$H = \frac{1}{2}\int d^3x(\mathbf{E}_a\cdot\mathbf{E}_a + \mathbf{B}_a\cdot\mathbf{B}_a), \quad (8.103)$$

where

$$\mathbf{B}_a \equiv \nabla\times\mathbf{A}_a + \frac{1}{2}gC_{abc}\mathbf{A}_b\times\mathbf{A}_c. \quad (8.104)$$

Using $\partial \mathbf{A}_a/\partial t = i[H, \mathbf{A}_a]$, we find the operator equality

$$\frac{\partial \mathbf{A}_a}{\partial t} = -\mathbf{E}_a. \quad (8.105)$$

Classically, there is a residual gauge transformation $\mathbf{A} \to \mathbf{A}^\omega$, $(\mathbf{A} \equiv \mathbf{A}_a L_a)$, with

$$\mathbf{A}^\omega = U\mathbf{A}U^{-1} + \frac{i}{g}U\nabla U^{-1},$$

$$U(\mathbf{x}) = \exp[-i\omega_a(\mathbf{x})L_a], \quad (8.106)$$

where $\omega_a(\mathbf{x})$ is time-independent. For infinitesimal ω_a, the change in $\mathbf{A}_a$ is given by

$$\delta\mathbf{A}_a = -\frac{1}{g}\boldsymbol{\nabla}\omega_a + C_{abc}\omega_b\mathbf{A}_c. \tag{8.107}$$

Quantum-mechanically, the residual gauge invariance is expressed through the fact that

$$[\mathbf{D}\cdot\mathbf{E}_a, H] = 0, \tag{8.108}$$
$$\mathbf{D}\cdot\mathbf{E}_a \equiv \boldsymbol{\nabla}\cdot\mathbf{E}_a + gC_{abc}\mathbf{A}_b\cdot\mathbf{E}_c.$$

Therefore, an eigenfunction $\Psi[\mathbf{A}]$ of H can be chosen to be a simultaneous eigenfunction of $\mathbf{D}\cdot\mathbf{E}_a$. Without further ado, we choose the eigenvalue of $\mathbf{D}\cdot\mathbf{E}_a$ to be zero, thus imposing on the Hilbert space of the system the subsidiary condition

$$\mathbf{D}\cdot\mathbf{E}_a(\mathbf{x})\Psi[\mathbf{A}] = 0. \tag{8.109}$$

This is the generalization of Gauss' law, and is equivalent to imposing Coulomb gauge:

$$\mathbf{D}\cdot\mathbf{A}_a = \boldsymbol{\nabla}\cdot\mathbf{A}_a = 0. \tag{8.110}$$

Consider continuous gauge transformations $U(\mathbf{x})$ that approach a constant at spatial infinity[c]. The constant can always be reduced to 1, through multiplication by the inverse of the constant, which is always a group element. Therefore, consider continuous gauge transformations obeying the condition

$$U(\mathbf{x}) \xrightarrow[|\mathbf{x}|\to\infty]{} 1. \tag{8.111}$$

We can look upon $U(\mathbf{x})$ as a continuous mapping of ordinary 3-space into the gauge group G. The above condition identifies spatial infinity as one point for this purpose, thus making the topology of space that of a 3-dimensional sphere S^3. Hence, we are led to consider continuous mappings $S^3 \to G$. A theorem due to Bott[12] states:

Any continuous mapping of S^3 into a simple Lie group G can be continuously deformed into a mapping into an $SU(2)$ subgroup of G.

Therefore, for a Yang-Mills theory with simple gauge group G, it is only necessary to consider $S^3 \to SU(2)$. Since the manifold of $SU(2)$ has the topology of S^3 [See (4.22)], we consider continuous mappings $S^3 \to S^3$. As mentioned in Sec. 5.2, these mappings fall into homotopy classes characterized by winding numbers n, the number of times the spatial S^3 is covered by the group manifold

---

[c] We only consider $U \to$ Const, because this is sufficient for our purpose. We can even rule out the possibility $U \nrightarrow$ Const. by placing the system in a large but finite box, and impose any definite boundary conditions on $\mathbf{A}$. It is a common article of faith to assume that boundary conditions at large distances have no effect on local phenomena.

[12] R. Bott, *Bull. Soc. Math. France*, **84**, 251 (1956).

Quantization of Gauge Fields

S^3. As a representative of the nth class, we choose

$$U_n(\mathbf{x}) = [v(\mathbf{x})]^n \quad (n = 0, \pm 1, \pm 2, \ldots),$$

$$v(\mathbf{x}) = -\exp\frac{-i\pi \mathbf{x} \cdot \boldsymbol{\tau}}{(x^2 + \rho^2)^{1/2}} \quad (x^2 \equiv |\mathbf{x}|^2), \tag{8.112}$$

where ρ is an arbitrary number, and $\tau_a/2$ ($a = 1, 2, 3$), are the generators of a $SU(2)$ subgroup of G. A general member of class 0 is any gauge transformation that can be continuously deformed into the identity. A general member of class 1 is $v(\mathbf{x})$ times a member of class 0. A general member of class $n + m$ is a product of a member of class n with one of class m.

Suppose $\mathbf{A}$ undergoes an infinitesimal gauge transformation. Then $\Psi[\mathbf{A}]$ changes according to

$$\Psi[\mathbf{A} + \delta\mathbf{A}] - \Psi[\mathbf{A}] = -\frac{i}{g}\int d^3x(-\nabla\omega_a + gC_{abc}\omega_b A_c) \cdot \mathbf{E}_a \Psi[\mathbf{A}]$$

$$= \frac{i}{g}\left[\int d\mathbf{S} \cdot \mathbf{E}_a \omega_a - \int d^3x\, \omega_a \mathbf{D} \cdot \mathbf{E}_a\right] \Psi[\mathbf{A}]. \tag{8.113}$$

For class 0 gauge transformations, the first term vanishes since ω_a approach zero at spatial infinity. Thus, $\mathbf{D} \cdot \mathbf{E}_a$ is the generator of class 0 gauge transformations, under which $\Psi[\mathbf{A}]$ is invariant by (8.108). Under a gauge of transformation of class $n \neq 0$, $\Psi[\mathbf{A}]$ is not necessarily invariant; but since the Hamiltonian is locally gauge invariant, all energy eigenfunctions can be chosen so that they change at most by a constant phase, and this phase must be the same for all eigenfunctions:

$$\Psi[\mathbf{A}_n] = e^{in\theta}\Psi[\mathbf{A}] \quad (n = 0, \pm 1, \pm 2, \ldots), \tag{8.114}$$

where $\mathbf{A}_n$ is the transform of $\mathbf{A}$ by a class n gauge transformation, a typical one being

$$\mathbf{A}_n = v^n \mathbf{A} v^{-n} + \frac{i}{g} v^n \nabla v^{-n}. \tag{8.115}$$

We shall denote the vacuum-state wave function by $\Psi_\theta[\mathbf{A}]$. The vacuum state characterized by θ is called the "θ-vacuum".

The Hilbert space of the system is divided into sectors labelled by the continuous parameter θ, each containing states built on the θ-vacuum. These "θ-worlds" are isolated from one another by superselection rules, i.e., they are bridgeable only by non-gauge-invariant interactions. In each θ-world the vacuum state is unique.

If $\theta \neq 0$, then (8.114) requires the unique vacuum state to be complex, thus violating time-reversal invariance. By the CPT theorem, this implies CP violation[d]. From (8.112), we see that under spatial reflection, $v(\mathbf{x}) \to [v(\mathbf{x})]^{-1}$,

---

[d] CP violation in the θ-world does directly translate into CP violation in the real world, because the situation changes dramatically when the gauge fields are coupled to fermions (See Sec. 12.6).

and hence $n \to -n$. This leads to the conclusion that Ψ_θ is not an eigenstate of parity, if $\theta \neq 0$. The different θ-worlds are therefore physically inequivalent.

Let us call a gauge transformation of class 0 a "small" gauge transformation, and that of class ± 1 a "large" gauge transformation. Thus, Ψ_θ is invariant under a small gauge transformation, but changes by a phase factor $e^{\pm i\theta}$ under a large gauge transformation. To get a rough idea of what Ψ_θ might look like, let $\chi_n[A]$ be a functional of A that is "peaked" about $\mathbf{A}_n$ in the following sense: a small gauge transformation leaves χ_n invariant, while a large gauge transformation takes χ_n into χ_{n+1}. The χ_n's are like wave functions that peak about a particular potential minimum of a periodic potential in quantum mechanics. An intuitive representation of Ψ_θ might be

$$\Psi_\theta[A] \cong \sum_{n=-\infty}^{\infty} e^{in\theta} \chi_n[A], \tag{8.116}$$

for this realizes the property (8.114) in a concrete way. If we were to do a variational calculation, this might be a good guess for a trial wave function; but there is probably no χ_n that can make (8.116) an exact representation. This intuitive picture does suggest that quantum-mechanical tunnelling is responsible for the structure of Ψ_θ, and that the θ-vacua have different energies. This is indeed the case. The tunnelling mechanism is the instanton, as we now explain.

2 Instanton as Tunnelling Solution

Let us enclose the system in a large box of volume V. The transition amplitude over Euclidean time T is

$$\langle \mathbf{A}''| e^{-HT} |\mathbf{A}'\rangle = \mathcal{N} \int_{\mathbf{A}'}^{\mathbf{A}''} (D\mathbf{A})\, e^{-S_E[\mathbf{A}]}, \tag{8.117}$$

where we have left as understood the gauge-fixing factor $G[f]\, \Delta_f[A]$, with $\hat{f} A_a = \nabla \cdot \mathbf{A}_a$. The Euclidean action is

$$S_E[\mathbf{A}] = -\int_{-T/2}^{T/2} dx_4 \int_V d^3x\, \mathcal{L}(x_E). \tag{8.118}$$

The path integral extends over all paths $A_a(\mathbf{x}, x_4)$ such that

$$\begin{aligned} A_a(\mathbf{x}, T/2) &= A_a''(\mathbf{x}),\\ A_a(\mathbf{x}, -T/2) &= A_a'(\mathbf{x}). \end{aligned} \tag{8.119}$$

Let us choose $\mathbf{A}'$ to be some $\mathbf{A}_0$, and $\mathbf{A}''$ to be some $\mathbf{A}_1$. In fact, choose them to be class 0 and class 1 pure gauges respectively:

$$\begin{aligned} \mathbf{A}'(\mathbf{x}) &= 0,\\ \mathbf{A}''(\mathbf{x}) &= \frac{i}{g} v(\mathbf{x}) \nabla v^{-1}(\mathbf{x}), \end{aligned} \tag{8.120}$$

where $v(\mathbf{x})$ is given in (8.112). For $T \to \infty$, a path connecting $\mathbf{A}'$ to $\mathbf{A}''$ is the instanton. To see this, recall the instanton solutions (5.24) and (5.28), which can be written more explicitly as

$$\mathbf{A}_{\text{Ins}}(\mathbf{x}, x_4) = \frac{1}{g} \frac{i[\boldsymbol{\tau}(\boldsymbol{\tau} \cdot \mathbf{x}) - \mathbf{x}] + x_4}{x^2 + x_4^2 + \rho^2}, \quad (x^2 \equiv |\mathbf{x}|^2),$$

$$A_{\text{Ins}}^4(\mathbf{x}, x_4) = -\frac{1}{g} \frac{\boldsymbol{\tau} \cdot \mathbf{x}}{x^2 + x_4^2 + \rho^2},$$

(8.121)

where ρ is an arbitrary parameter. To conform to our gauge choice here, we must transform the above so that $A_{\text{Ins}}^4 = 0$. This may be achieved through the gauge transformation

$$U(\mathbf{x}, x_4) = -\exp\left[\frac{-i\pi \mathbf{x} \cdot \boldsymbol{\tau}}{(x^2 + \rho^2)^{1/2}} \phi(x^2, x_4)\right],$$

$$\phi(x^2, x_4) = \frac{1}{2} + \frac{1}{\pi} \tan^{-1} \frac{x_4}{(x^2 + \rho^2)^{1/2}}.$$

(8.122)

Note that

$$\phi(x^2, x_4) \xrightarrow[x_4 \to \infty]{} 1$$

$$\xrightarrow[x_4 \to -\infty]{} 0.$$

(8.123)

Hence

$$U(\mathbf{x}, x_4) \xrightarrow[x_4 \to \infty]{} v(\mathbf{x})$$

$$\xrightarrow[x_4 \to -\infty]{} 1.$$

(8.124)

Therefore, in the $A_4 = 0$ gauge, we have

$$\mathbf{A}_{\text{Ins}}(\mathbf{x}, x_4) = U(\mathbf{x}, x_4) \left\{ \frac{1}{g} \frac{i[\boldsymbol{\tau}(\boldsymbol{\tau} \cdot \mathbf{x}) - \mathbf{x}] + x_4}{x^2 + x_4^2 + \rho^2} \right\} U^{-1}(\mathbf{x}, x_4)$$

$$+ \frac{i}{g} U(\mathbf{x}, x_4) \boldsymbol{\nabla} U^{-1}(\mathbf{x}, x_4).$$

(8.125)

The first term vanishes for $|x_4| \to \infty$. Hence

$$\mathbf{A}_{\text{Ins}}(\mathbf{x}, x_4) \xrightarrow[x_4 \to \infty]{} \frac{i}{g} v(\mathbf{x}) \boldsymbol{\nabla} v^{-1}(\mathbf{x})$$

$$\xrightarrow[x_4 \to -\infty]{} 0.$$

(8.126)

This shows that the instanton interpolates, in Euclidean time, between two field configurations differing by a large gauge transformation. In this sense it is a "tunnelling solution"[13,14].

---

[13] R. Jackiw and C. Rebbi, *Phys. Rev. Lett.*, **37**, 172 (1976); C. G. Callan, R. F. Dashen and D. J. Gross, *Phys. Lett.*, **63B**, 334 (1976).
[14] K. M. Bitar and S.-J. Chang, *Phys. Rev.*, **D17**, 486 (1978), give a description of tunnelling in Minkowski space.

The energy E_θ of the θ-vacuum is given by

$$\begin{aligned}e^{-E_\theta T} &= \langle\theta|\,e^{-HT}|\theta\rangle \\ &= \mathcal{N}\int (d\mathbf{A}'')(d\mathbf{A}')\int_{\mathbf{A}'}^{\mathbf{A}''}(D\mathbf{A})\Psi_\theta^*[\mathbf{A}'']\,e^{-S_E[A]}\Psi_\theta[\mathbf{A}'],\end{aligned} \tag{8.127}$$

where $\int (d\mathbf{A}')$ denotes integration over time-independent functions $\mathbf{A}'(\mathbf{x})$. E_θ can be calculated approximately by assuming that the path integral (8.127) is dominated by a dilute "instanton gas"[15].

3 The θ-Action

We can make a transformation in Hilbert space to take away the phase factor $e^{in\theta}$ in (8.114), thus making all wave functions completely gauge invariant, even under large gauge transformations. This is similar to what we did in Maxwell theory to make $\nabla \cdot E = 0$. The difference is that in the Maxwell case, the transformed Hamiltonian is the same as the original one, except in being shifted by a constant. However, here the Hamiltonian will change in a substantial way. The easiest way to make the transformation is to do it with path integrals. We shall then obtain a transformed action, which will be called the "θ-action."

To develop some useful tools for this purpose, recall the definition of the topological charge from (5.5) and (5.7):

$$q[A] = \frac{g^2}{16\pi^2}\int d^4x_E\,\mathrm{Tr}(\widetilde{F}_{\mu\nu}F_{\mu\nu}) = \frac{g^2}{4\pi^2}\int d^4x_E\,\partial_\mu X_\mu(A),$$

$$X_\mu(A) = \varepsilon^{\mu\alpha\beta\gamma}\mathrm{Tr}\left[\frac{1}{2}A_\alpha\partial_\beta A_\gamma + \frac{i}{g}A_\alpha A_\beta A_\gamma\right]. \tag{8.128}$$

We remind the reader that $\mathbf{A} = A_a\tau_a/2$, and no distinction is made between upper and lower indices. For example, A_k are the components of $\mathbf{A}$, and ∂_k are the components of ∇. We recall from Sec. 5.2 that $q[A]$ is invariant under continuous gauge transformations, and that

$$q[A_{\mathrm{Ins}}] = 1. \tag{8.129}$$

Throughout the following discussion, we adopt some convenient spatial boundary conditions to make all spatial surface integrals vanish, e.g., periodic boundary conditions. That this procedure does not overlook important physical effects is an assumption we have to make (cf. footnote c). Accordingly, we write

$$\begin{aligned}q[A] &= \frac{g^2}{4\pi^2}\int d^4x_E\left[\frac{\partial X_4(A)}{\partial x_4} + \nabla\cdot X(A)\right] \\ &= \frac{g^2}{4\pi^2}\left[\int d^3x\,X_4(A)\right]_{x_4=-\infty}^{x_4=\infty}.\end{aligned} \tag{8.130}$$

---

[15] S. Coleman, in *The Whys of Subnuclear Physics*, Ed. A. Zichichi (Plenum, New York, 1980).

Quantization of Gauge Fields

Now consider the Schrödinger operator

$$N[\mathbf{A}] \equiv \frac{g^2}{4\pi^2} \int d^3x\, X_4(\mathbf{A}). \tag{8.131}$$

A straightforward calculation shows that $N[\mathbf{A}]$ is invariant under an infinitesimal gauge transformation with parameter ω:

$$N[\mathbf{A} + \delta\mathbf{A}] = -\frac{g^2}{8\pi^2}\int d^3x\, \nabla \cdot \text{Tr}[\mathbf{A} \times \delta\mathbf{A} + i\omega\mathbf{A} \times \mathbf{A}] = 0. \tag{8.132}$$

Hence, $N[\mathbf{A}]$ is invariant under small gauge transformations:

$$N[\mathbf{A}_0] = N[\mathbf{A}]. \tag{8.133}$$

Under a large gauge transformation, $N[\mathbf{A}]$ changes by ± 1. This can be shown as follows. Any large gauge transformation can be made by first making a small one, under which $N[\mathbf{A}]$ is invariant, followed by $v(\mathbf{x})$. Hence by (8.126)

$$\int_{-\infty}^{\infty} dx_4\, \frac{\partial N[\mathbf{A}_{\text{Ins}}]}{\partial x_4} = N[\mathbf{A}_1] - N[\mathbf{A}_0]. \tag{8.134}$$

On the other hand, (8.129) and (8.128) give

$$\int_{-\infty}^{\infty} dx_4\, \frac{\partial N[\mathbf{A}_{\text{Ins}}]}{\partial x_4} = 1. \tag{8.135}$$

Therefore, observing (8.133), we have

$$N[\mathbf{A}_1] = N[\mathbf{A}] + 1. \tag{8.136}$$

We can now proceed with our main task. Define new wave functions $\Phi[\mathbf{A}]$ by

$$\Phi[\mathbf{A}] \equiv e^{-i\theta N[\mathbf{A}]}\Psi[\mathbf{A}]. \tag{8.137}$$

It is clear that $\Phi[\mathbf{A}]$ is completely gauge invariant, even under large gauge transformations. The transformed θ-vacuum state is denoted by Φ_θ. The formula (8.127) giving E_θ can be rewritten in terms of $\Phi_\theta[\mathbf{A}]$ as

$$e^{-E_\theta T} = \mathcal{N} \int (d\mathbf{A}')(d\mathbf{A}'') \int_{\mathbf{A}'}^{\mathbf{A}''} (D\mathbf{A})\Phi_\theta^*[\mathbf{A}''] \, e^{-S_E[\mathbf{A},\,\theta]} \Phi_\theta[\mathbf{A}'], \tag{8.138}$$

where the new action, called the "θ-action," is given by

$$S_E[\mathbf{A}, \theta] = S_E[\mathbf{A}] + i\theta q[\mathbf{A}]. \tag{8.139}$$

The proof is (by now) straightforward. In Minkowski space, the θ-action is given by[e]

$$S[\mathbf{A}, \theta] = S[\mathbf{A}] - \theta q[\mathbf{A}]. \tag{8.140}$$

We note that the additional term does not change the equations of motion,

---

[e] When transforming $q[A]$ from Euclidean to Minkowski space, note that $\tilde{F}_{\mu\nu}F_{\mu\nu} \to \tilde{F}^{\mu\nu}F_{\mu\nu}$ (because $F_{\mu\nu}F_{\mu\nu} = -4\mathbf{B}\cdot\mathbf{E} = 4\mathbf{B}\cdot\partial\mathbf{A}/\partial x_4$), and the factor $-i$ cancels the i from $d^4x_E = id^4x$.

because it is a total 4-divergence. In Maxwell theory, such a term is irrelevant because $q[\mathbf{A}] \equiv 0$.

Let the "partition function" of the θ-world be denoted by Z_θ:

$$Z_\theta = \mathcal{N} \int (D\mathbf{A}) \, e^{-S_E[\mathbf{A},\, \theta]} \qquad (8.141)$$

The average topological charge of the θ-world is given by

$$\langle q \rangle = \frac{1}{i} \frac{\partial}{\partial \theta} \ln Z_\theta. \qquad (8.142)$$

We may look upon θ as the Lagrange multiplier that fixes the θ-world average of the topological charge. An analogy may be made with the chemical potential in statistical mechanics, which is the Lagrange multiplier that fixes the ensemble average of the baryonic charge.

CHAPTER IX

RENORMALIZATION

9.1 Charge Renormalization

To provide background for the topics we shall discuss in this chapter, we begin with a calculation—charge renormalization to lowest order in quantum electrodynamics.

The full photon propagator to second order in the unrenormalized charge e_0, is represented by the Feynman graphs[1] in Fig. 9.1, which give the expression (in Feynman gauge)

$$iD'_{\mu\nu}(k) = \frac{g_{\mu\nu}}{ik^2} + \frac{g_{\mu\alpha}}{ik^2} i\Pi^{\alpha\beta}(k) \frac{g_{\beta\nu}}{ik^2}, \qquad (9.1)$$

where $\Pi^{\mu\nu}(k)$ is the vacuum polarization tensor. Gauge invariance requires $k_\mu \Pi^{\mu\nu}(k) = 0$. Thus $\Pi^{\mu\nu}(k)$ must have the form

$$\Pi^{\mu\nu}(k) = e_0^2(g^{\mu\nu}k^2 - k^\mu k^\nu)\Pi(k^2). \qquad (9.2)$$

From the lowest-order Feynman graph, one obtains

$$\Pi^{\mu\nu}(k) = ie_0^2 \int \frac{d^4p}{(2\pi)^4} \text{Tr}\left(\gamma^\mu \frac{1}{\not{p} - m} \gamma^\nu \frac{1}{\not{p} - \not{k} - m}\right), \qquad (9.3)$$

where m is the physical electron mass. This expression is quadratically divergent, and is not of the form (9.2). The recipe to remedy the situation, as will be discussed in Sec. 9.3, is to replace $\Pi^{\mu\nu}(k)$ by $\Pi^{\mu\nu}(k) - \Pi^{\mu\nu}(0)$, which satisfies (9.2). This procedure also reduces the quadratic divergence to a logarithmic one, and gives

$$\Pi(k^2) = -\frac{1}{12\pi^2} \ln \frac{\Lambda^2}{cm^2} + C(k^2), \qquad (9.4)$$

where Λ is a cutoff momentum, c is a numerical constant that depends on the

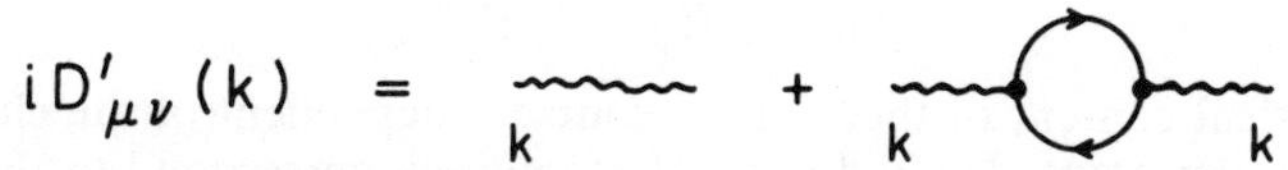

Fig. 9.1 Photon propagator to second order

---

[1] For the Feynman rules, see J. D. Bjorken and S. D. Drell, *Relativistic Quantum Fields* (McGraw-Hill, New York, 1965), Appendix B.

method of cutting off, and $C(k^2)$ is convergent:[2]

$$C(k^2) = \frac{1}{2\pi^2} \int_0^1 dx\, x(1-x) \ln\left[1 - x(1-x)\frac{k^2}{m^2}\right]. \tag{9.5}$$

This function is real for k^2 below the threshold of pair production, i.e., $k^2 < 4m^2$.

With (9.2), we have

$$D'_{\mu\nu}(k) = -\left(g_{\mu\nu} - \frac{k_\mu k_\nu}{k^2}\right)\frac{d'(k^2)}{k^2} + \begin{pmatrix}\text{gauge-dependent}\\ \text{terms}\end{pmatrix}, \tag{9.6}$$

$$d'(k^2) = 1 + e_0^2 \Pi(k^2).$$

Charge renormalization consists of subtracting off the logarithmic divergence in $d'(k^2)$, and absorbing it into a redefinition of the charge.

Define a finite function $\widetilde{\Pi}$ by

$$\widetilde{\Pi}(k^2, \mu^2) \equiv \Pi(k^2) - \Pi(\mu^2) = C(k^2) - C(\mu^2), \tag{9.7}$$

where μ^2 is an arbitrary number (the renormalization point). Then,

$$d'(k^2) = [1 + e_0^2 \Pi(\mu^2)] + e_0^2 \widetilde{\Pi}(k^2, \mu^2). \tag{9.8}$$

Let

$$Z(\mu^2) \equiv 1 + e_0^2 \Pi(\mu^2). \tag{9.9}$$

To second-order accuracy, we can rewrite

$$d'(k^2) = Z(\mu^2)[1 + e_0^2 Z(\mu^2)\widetilde{\Pi}(k^2, \mu^2)]. \tag{9.10}$$

Charge renormalization is carried out by identifying the physical charge as

$$e^2(\mu^2) = e_0^2 Z(\mu^2). \tag{9.11}$$

Using the notation $\alpha = e^2/4\pi$, we rewrite this as

$$\alpha(\mu^2) = \alpha_0 Z(\mu^2). \tag{9.12}$$

The physically relevant quantity $\alpha_0 d'(k^2)$ can now be written as

$$\alpha_0 d'(k^2) = \alpha(\mu^2)[1 + 4\pi\alpha(\mu^2)\widetilde{\Pi}(k^2, \mu^2)]. \tag{9.13}$$

The standard renormalized propagator $d_c(k^2)$ is obtained by choosing $\mu = 0$:

$$d_c(k^2) = 1 + 4\pi\alpha \widetilde{\Pi}(k^2, 0), \tag{9.14}$$

where α is the experimentally measured fine-structure constant:

$$\alpha \equiv \alpha(0) \cong 1/137. \tag{9.15}$$

The physical content of theory is of course independent of the choice of the renormalization point. The different $\alpha(\mu^2)$ merely correspond to different but

---

[2] J. M. Jauch and F. Rohrlich, *The Theory of Photons and Electrons* (Addison-Wesley, Reading, Mass. 1935), p. 194, eq. (4–65); N. N. Bogolubov and D. V. Shickov, *Introduction to the Theory of Quantized Fields* (Interscience, New York, 1959), p. 296, Eq. (24.35).

Renormalization

equivalent definitions of the coupling constant. However, the function $\alpha(\mu^2)$, known as the "running coupling constant", contains important physical information. To see this, note that the right side of (9.13) is actually independent of μ, and that $\widetilde{\Pi}(k^2, k^2) = 0$ by (9.7). Thus, by choosing successively $\mu^2 = k^2$ and $\mu^2 = 0$, we obtain

$$\alpha(k^2) = \alpha d_c(k^2). \tag{9.16}$$

Thus, the effects of vacuum polarization may be viewed in two ways. We might say that it modifies the propagator of a virtual photon, or alternatively, that it makes the effective fine-structure constant momentum-dependent. It is important to keep in mind that $\alpha(k^2)$ and $\alpha d_c(k^2)$ are interchangeable concepts.

Our earlier calculations give, to lowest order in α,

$$\frac{\alpha(k^2)}{\alpha} = 1 + \frac{2\alpha}{\pi} \int_0^1 dx\, x(1-x) \ln\left[1 - x(1-x)\frac{k^2}{m^2}\right]. \tag{9.17}$$

For large $|k^2/m^2|$, we can write

$$\frac{1}{\alpha(k^2)} = \frac{1}{\alpha} - \frac{1}{3\pi} \ln\left|\frac{k^2}{m^2}\right| \quad (|k^2/m^2| \gg 1). \tag{9.18}$$

Since perturbation theory certainly becomes invalid when $\alpha \ln|k^2/m^2| \sim 1$, (9.18) should be used only for

$$1 \ll |k^2/m^2| \ll e^{137}. \tag{9.19}$$

The electrostatic potential energy between two unit test charges (in units of the electronic charge) is given by

$$\begin{aligned}V(r) &= e^2 \int \frac{d^3k}{(2\pi)^3} e^{i\mathbf{k}\cdot\mathbf{r}} \left[\frac{d_c(k^2)}{-k^2}\right]_{k_0=0} \\ &= \int \frac{d^3k}{(2\pi)^3} \frac{e^{i\mathbf{k}\cdot\mathbf{r}}}{\mathbf{k}\cdot\mathbf{k}} 4\pi\alpha(-\mathbf{k}^2).\end{aligned} \tag{9.20}$$

We can also write

$$V(r) = \int d^3r' \frac{\rho(r')}{|\mathbf{r} - \mathbf{r}'|}, \tag{9.21}$$

where

$$\rho(r) = \int \frac{d^3k}{(2\pi)^3} e^{i\mathbf{k}\cdot\mathbf{r}} 4\pi\alpha(-\mathbf{k}^2). \tag{9.22}$$

Thus, $4\pi\alpha(-\mathbf{k}^2)/e$ is the Fourier transform of the charge density surrounding a bare charge e_0 placed in the vacuum—the "charge form factor" associated with vacuum polarization.

Let us describe qualitatively the charge cloud induced by vacuum polarization. Far from the bare charge, one sees the renormalized charge, because

$$V(r) \xrightarrow[r\to\infty]{} \frac{\alpha}{r}, \tag{9.23}$$

which follows from the fact that $\alpha(0) = \alpha$. At distances comparable to the electron Compton wavelength $1/m$, one begins to see an effective charge *larger* than the renormalized charge. This can be deduced from the fact that $\alpha(-\mathbf{k}^2)$ *increases* as $\mathbf{k}^2$ becomes large. Thus, the effect of vacuum polarization is to screen the bare charge. As one penetrates deeper into the charge cloud, the screening becomes less effective, and one sees more of the bare charge. Our understanding stops at $r \sim e^{-137}/m$, because perturbation theory fails for shorter distances. From perturbation theory, we cannot tell whether the bare charge is infinite (as suggested by perturbation theory), or that it is really finite. The question probably has a mathematical answer within quantum electrodynamics; but it is not physically relevant, for long before we reach $r \sim e^{-137}/m$, other interactions not taken into account in quantum electrodynamics would surely become important.

In quantum chromodynamics the analog of $\alpha(-\mathbf{k}^2)$ is a *decreasing* function of $\mathbf{k}^2$. Thus the effective coupling vanishes in the limit $\mathbf{k}^2 \to \infty$—a phenomenon known as "asymptotic freedom". In this case, instead of screening a bare charge, vacuum polarization "anti-screens".

We shall expand on renormalization in the next section. Readers not concerned with the technical details may skip to Sec. 9.3.

9.2 Renormalization in Quantum Electrodynamics

We give a highly compressed account of the renormalization scheme to all orders in quantum electrodynamics[3]. The purpose is to show how the second order calculation in the last section can be extended to higher orders, and to illustrate the intricacies and subtleties involved in any serious attempt to renormalize a field theory.

1 Divergences in Feynman graphs

For any Feynman graph, let

n = No. of vertices,
E_i = No. of internal electron lines,
E_e = No. of external electron lines,
P_i = No. of internal photon lines,
P_e = No. of external photon lines,

with the relations

$$E_i = n - \tfrac{1}{2}E_e,$$
$$P_i = \tfrac{1}{2}(n - P_e). \tag{9.24}$$

The number of independent internal 4-momenta in the graph is given by

---

[3] The method described here is that of Dyson and Ward: F. J. Dyson, *Phys. Rev.*, **75**, 486, 1736 (1949); J. C. Ward, *Proc. Phys. Soc.* (London), A **64**, 54 (1951).

Renormalization

$N = E_i + P_i - (n - 1)$, which, by (9.24), may be reexpressed as

$$N = \tfrac{1}{2}(n - E_e - P_e) + 1. \tag{9.25}$$

A graph may be represented schematically in the form

$$\text{Graph} \sim \int \frac{d^{4N}k}{(k^2)^{P_i} k^{E_i}}. \tag{9.26}$$

The integral above is generally divergent, and must be made finite by introducing a cut-off momentum Λ, which eventually tends to infinity. Renormalization is a procedure through which we subtract appropriate terms from the integral to render it finite in the limit $\Lambda \to \infty$, and absorb the subtracted terms into mass and charge renormalization.

We define a *primitively divergent graph* as a graph that is divergent, but becomes convergent when any internal line is cut (*i.e.*, when any integration variable is held fixed). Any divergent graph can be reduced to a primitively divergent one by cutting a sufficient number of internal lines. This is obvious because the graph becomes convergent when all internal lines are cut.

The superficial degree of divergence d of a primitively divergent graph may be obtained through naive power counting:[a]

$$d = 4N - 2P_i - E_i. \tag{9.27}$$

The actual degree of divergence may be smaller than d. Using (9.24) and (9.25), we can rewrite

$$d = 4 - P_e - E_e. \tag{9.28}$$

Note that d is independent of n, and decreases as the number of external lines is increased. This is what makes the theory renormalizable.

There are only a finite number of types of primitively divergent graphs, and they can be classified according to P_e and E_e, as shown in Table 9.1. More detailed considerations, as indicated in Table 9.1, reduce these to only 3 types:

— proper electron self-energy (SE),
— proper photon self-energy,
— proper vertex.

The term "proper" is synonymous with "one-particle irreducible". It denotes a connected graph that cannot be made disconnected by cutting only one internal line.

It is sufficient to consider only connected non-vacuum graphs. For any such graph, we define its *skeleton graph* as the graph obtained by removing all SE and vertex insertions inside the original graph. The skeleton graph may be convergent or divergent. If divergent, it must be primitively divergent.

To prove this last statement, assume the contrary. Then, by cutting a sufficient number of internal lines, the graph can be reduced to (possibly disconnected)

---

[a] The degree of divergence of a non-primitively divergent integral cannot be obtained by power counting. For example, the integral $\int dk\, dp\, k^{-1} p^{-2}$ is logarithmically divergent; but naive power counting would give $d = -1$.

components, of which one is primitively divergent. The latter must be either an SE or vertex graph, as indicated in Table 9.1. But these have been removed by definition. (Contradiction).

For an arbitrary connected non-vacuum graph, we deal with its possible divergence as follows:

a. Obtain the skeleton by removing all insertions.

b. If the skeleton is convergent, no subtraction is needed. The graph is then renormalized by re-inserting renormalized SE and vertex insertions.

c. If the skeleton is divergent, it must be either an SE or vertex graph. The problem is thus reduced to the renormalization of SE and vertex graphs.

We see from the above that, to renormalize a general graph of order n, it suffices to renormalize SE and vertex graphs to order less than or equal to n. Thus, the renormalization can be carried out order by order in perturbation theory.

A complication arises in the renormalization of SE graphs, in that the way of removing insertions is not always unique, due to the possibility of "overlapping

Table 9.1 PRIMITIVELY DIVERGENT GRAPHS

P_e	E_e	d	Example	Remarks
0	0	4		Vacuum graph. (May be ignored)
0	2	1		Electron SE. Superficially lin. div. Actually Log. div.
1	2	0		Vertex. log. div
2	0	2		Photon SE. Superficially quad. div. Actually log. div. by gauge invar.
3	0	1		Cancelled by graph with electron arrow reversed, by Furry's theorem. (Ignore)
4	0	0		Sum of 4! graphs corresponding to permutation of ext. mom. is convergent. Actually even more convergent than superficially indicated, by gauge invariance

Renormalization

divergences", as illustrated by the examples in Fig. 9.2. The problem is that, although the skeleton graph is unique, the way to restore the insertions is not unique, and one runs the risk of double counting. As we shall see, the Ward-Takahashi identity circumvents the problem in the case of electron SE graphs. A similar method can be devised for photon SE graphs.

There is an alternative renormalization scheme, the BPH method[4], that dispenses with the notion of skeleton graphs; but we shall not discuss it here.

We adopt the following notation:

$S'(p)$ = Full electron propagator of momentum p,

$D'_{\mu\nu}(k)$ = Full photon propagator of momentum k,

$\Gamma_\mu(p, p')$ = Full vertex with external lines omitted, in which the electron enters with momentum p and exits with momentum p'.

2 Vertex

We put

$$\Gamma_\mu(p, p') = \gamma_\mu + \Lambda_\mu(p, p'). \tag{9.29}$$

The graphs for Λ_μ are shown in Fig. 9.3 (a), and the skeleton expansion is shown in Fig. 9.3 (b). The problem of overlapping divergences does not arise in this case. The sum of all skeleton graphs, denoted by Λ_μ^*, is logarithmically divergent.

To obtain Λ_μ from Λ_μ^*, we replace all free propagators and bare vertices in the graphs of Λ_μ^* by the corresponding full propagators and full vertices. This may be expressed as follows:

$$\Lambda_\mu(p, p') = \Lambda_\mu^*[S', D'_{\alpha\beta}, \Gamma_\nu; e_0^2, p, p'], \tag{9.30}$$

where Λ_μ^* is regarded as a functional of the propagator functions and the vertex function, with the bare squared charge e_0^2 and the momenta p, p' appearing as parameters of the functional. In this sense, the graphical series in Fig. 9.2 (b) corresponds to $\Lambda_\mu^*[S, D_{\alpha\beta}, \gamma_\nu; e_0^2, p, p']$, where S and $D_{\alpha\beta}$ are free propagators.

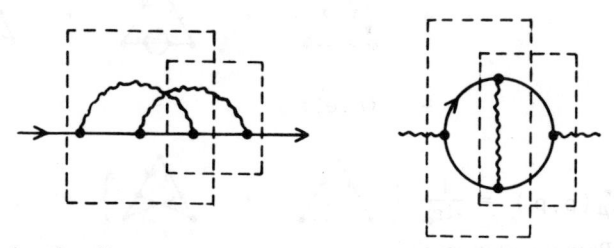

Fig. 9.2 Overlapping divergences

---

[4] N. N. Bogolubov and O. Parasuik, *Acta Math.*, **97**, 227 (1957); K. Hepp, *Comm. Math. Phys.* **2**, 301 (1966).

The functional Λ_μ^* is logarithmically divergent, *i.e.*, it depends on the cut-off Λ as a parameter, and diverges like $\ln\Lambda$ as $\Lambda \to \infty$. The full vertex function Λ_μ is much more divergent, due to divergences coming from the insertions.

3 Electron propagator

The full electron propagator can be obtained directly from Γ_μ through the Ward-Takahashi identity:[5]

$$[S'(p)]^{-1} = [S'(p_0)]^{-1} + (p - p_0)^\mu \Gamma_\mu(p, p_0). \tag{9.31}$$

The right-hand side is actually independent of p_0, but for the sake of definiteness we take p_0 to be the momentum of an electron on the mass shell. "Mass renormalization" consists of asserting that

$$[S'(p_0)]^{-1} = C(\not{p}_0 - m), \tag{9.32}$$

where m is the physical mass of the electron, and C is a cutoff-dependent constant.

4 Photon propagator

The free photon propagator is

$$D_{\mu\nu}(k) = -\left[g_{\mu\nu} - (1 - \lambda)\frac{k_\mu k_\nu}{k^2}\right]\frac{1}{k^2}, \tag{9.33}$$

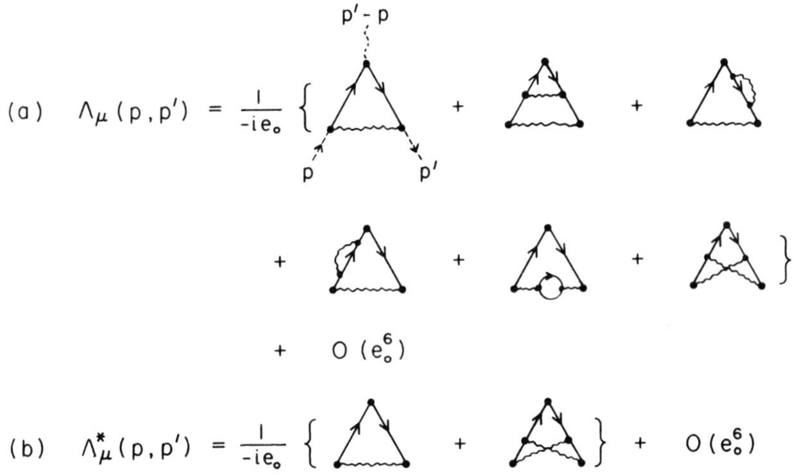

Fig. 9.3 (a) Proper vertex graphs
(b) Skeleton vertex graphs

---

[5] J. C. Ward, *Phys. Rev.*, **78**, 1824 (1950); Y. Takahashi, *N. Cimento*, **6**, 370 (1957).

Renormalization

where λ is the gauge parameter. Regarding $D_{\mu\nu}$ as a matrix with Minkowski metric, we can rewrite (9.33) in matrix form:

$$\mathbb{D}(k) = -\left[\mathbb{P}_T(k) + \lambda \mathbb{P}_L(k)\right]\frac{1}{k^2}, \tag{9.34}$$

where $\mathbb{P}_T$ and $\mathbb{P}_L$ are the transverse and longitudinal projection operators defined in (8.46). The full photon propagator is given in matrix notation by

$$i\mathbb{D}'(k) = i\mathbb{D}(k) + i\mathbb{D}(k)i\mathbb{\Pi}(k)i\mathbb{D}(k) + \cdots \tag{9.35}$$
$$= i\mathbb{D}(k)[1 - i\mathbb{\Pi}(k)\mathbb{D}(k)]^{-1},$$

where $i\Pi_{\mu\nu}$ is the vacuum polarization tensor—the sum of all proper photon SE graphs, with external lines omitted. The graphical expansion for $\Pi_{\mu\nu}$ is shown in Fig. 9.4.

Gauge invariance (or current conservation) requires that

$$k^\mu \Pi_{\mu\nu}(k) = 0. \tag{9.36}$$

Therefore $\Pi_{\mu\nu}$ must have the form[b]

$$\Pi_{\mu\nu}(k) = (g_{\mu\nu}k^2 - k_\mu k_\nu)e_0^2 \Pi(k^2). \tag{9.37}$$

Since, by power counting, $\Pi_{\mu\nu}(k)$ has quadratic skeletal divergences, $\Pi(k^2)$ has only logarithmic skeletal divergences, because two powers of the momentum have been factored out. In matrix notation, (9.37) reads

$$\mathbb{\Pi}(k) = \mathbb{P}_T(k)e_0^2 k^2 \Pi(k^2). \tag{9.38}$$

Substituting this into (9.35), we obtain

$$i\mathbb{D}'(k) = i\mathbb{D}(k)\left[\frac{1}{1 - e_0^2\Pi(k^2)}\mathbb{P}_T(k) + \lambda\mathbb{P}_L(k)\right]. \tag{9.39}$$

Restoring the indices, we write

$$iD'_{\mu\nu}(k) = i\left(g_{\mu\nu} - \frac{k_\mu k_\nu}{k^2}\right)D'(k^2) + \frac{\lambda k_\mu k_\nu}{ik^2}, \tag{9.40}$$

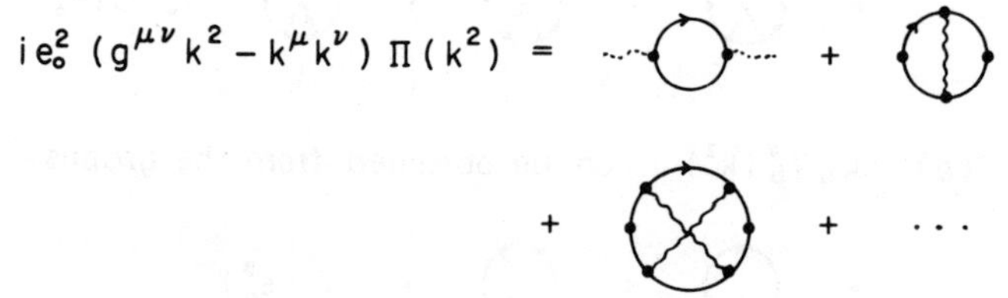

$$ie_0^2(g^{\mu\nu}k^2 - k^\mu k^\nu)\Pi(k^2) = \cdots\bigcirc\cdots + \text{(diagram)} + \text{(diagram)} + \cdots$$

Fig. 9.4 Vacuum polarization tensor

---

[b] Methods to insure gauge invariance in the calculation of $\Pi_{\mu\nu}$ are discussed in Sec. 9.3.

where
$$[iD'(k^2)]^{-1} = ik^2[1 - e_0^2 \Pi(k^2)]. \tag{9.41}$$

This is a gauge-invariant function. The second term in (9.40) depends on the gauge, and is physically irrelevant.

To deal with overlapping divergences, define an auxiliary function $W_\mu(k)$ by

$$W_\mu(k) \equiv \frac{\partial}{\partial k^\mu}[iD'(k^2)]^{-1}. \tag{9.42}$$

Using (9.40), we write

$$W_\mu(k) = 2ik_\mu + ik_\mu T(k^2), \tag{9.43}$$

where

$$T(k^2) = \frac{e_0^2}{k^2} k^\mu \frac{\partial}{\partial k^\mu}[k^2 \Pi(k^2)]. \tag{9.44}$$

To recover D' from W_μ, use the formula

$$[iD'(k^2)]^{-1} = \int_0^1 dx \, k^\mu W_\mu(xk). \tag{9.45}$$

The graphical expansion for W_μ is shown in Fig. 9.5 (a). There are no overlapping divergences. A skeleton expansion for T is obtained by removing

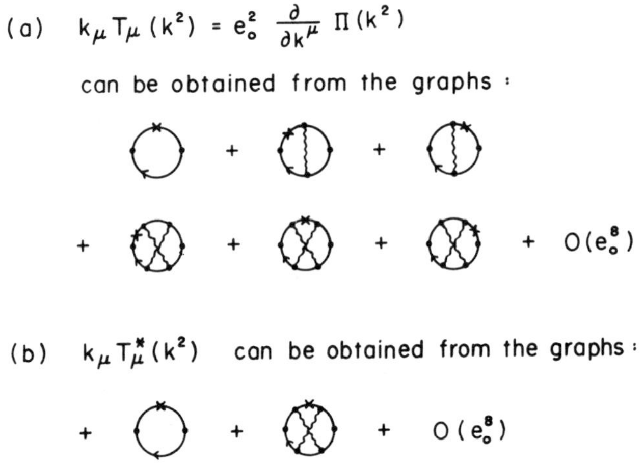

(a) $k_\mu T_\mu(k^2) = e_0^2 \frac{\partial}{\partial k^\mu} \Pi(k^2)$

can be obtained from the graphs:

○ + ⊕ + ⊕

+ ⊗ + ⊗ + ⊗ + $O(e_0^8)$

(b) $k_\mu T_\mu^*(k^2)$ can be obtained from the graphs:

+ ○ + ⊗ + $O(e_0^8)$

Fig. 9.5 (a) Graph for the auxiliary function $T(k^2)$. A cross indicates differentiation with respect to the external momentum k, which is by convention routed exclusively along the upper half of the closed loops.
(b) Skeleton expansion for $T(k^2)$, denoted by $T^*(k^2)$.

Renormalization

from the graphs not only all SE and vertex insertions, but also insertions of W_μ. The expansion, denoted by T^*, is shown in Fig. 9.5 (b). An example illustrating the removal of W_μ is shown in Fig. 9.6.[c] We can now write

$$T(k^2) = T^*[S', D'_{\alpha\beta}, \Gamma_\nu, W_\mu; e_0^2, k^2]. \tag{9.46}$$

The functional T^* is logarithmically divergent.

5 Multiplicative transformations

Graphs defining Λ_μ^* are of even order, and a graph of order $2n$ contains factors of e_0^2, S, $D_{\alpha\beta}$ and γ_ν to various powers, as indicated schematically below:

$$\Lambda_{2n}^* \sim e_0^{2n} S^{2n} (D_{\alpha\beta})^n (\gamma_\nu)^{2n+1}. \tag{9.47}$$

Under the transformation

$$\begin{aligned}\gamma_\nu &\to a\gamma_\nu, \\ D_{\alpha\beta} &\to bD_{\alpha\beta}, \\ S &\to a^{-1}S,\end{aligned} \tag{9.48}$$

where a and b are arbitrary numbers,

$$\Lambda_{2n}^* \to ab^n \Lambda_{2n}^*. \tag{9.49}$$

Therefore,

$$\begin{aligned}a\Lambda_\mu^*[S, D_{\alpha\beta}, \gamma_\nu; e_0^2, p, p'] \\ = \Lambda_\mu^*[a^{-1}S, bD_{\alpha\beta}, a\gamma_\nu; b^{-1}e_0^2, p, p'].\end{aligned} \tag{9.50}$$

For the functional T^*, a graph of order $2n$ has the structure

$$T_{2n}^* \sim e_0^{2n} S^{2n+\sigma} (D_{\alpha\beta})^{n-\sigma} (\gamma_\nu)^{2n+\sigma} (2ik_\mu)^{1-\sigma}, \tag{9.51}$$

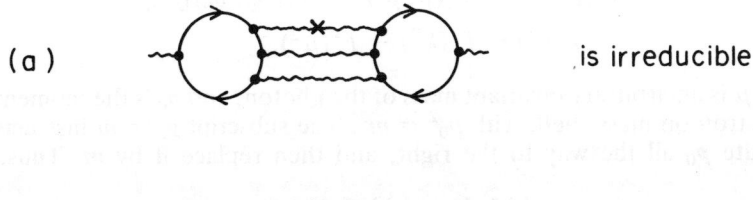

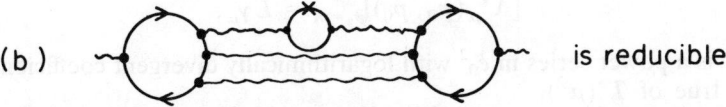

(a)   is irreducible

(b)   is reducible

Fig. 9.6 Illustration of removal of insertions in $T(k^2)$

---

[c] The external momentum k can be routed through a graph in more than one way, and this gives rise to ambiguities in the definition of a skeleton expansion. The difficulty occurs in graphs of W_μ containing at least 3 closed electron loops, and are therefore at least of order e_0^{14}. For a discussion of the difficulty and a convention of momentum-routing that overcomes the difficulty, see T. T. Wu, *Phys. Rev.*, **125**, 1436 (1962).

where σ is an integer that receives an additive contribution 1 from each differentiation of an electron line, and 0 from that of a photon line. Under the transformation (9.48) supplemented by $2ik_\mu \to 2ik_\mu/b$,

$$T^*_{2n} \to b^{n-1} T^*_{2n}. \tag{9.52}$$

Therefore,

$$\begin{aligned}&b^{-1}T^*[S, D_{\alpha\beta}, \gamma_\nu, 2ik_\mu; e_0^2, k^2]\\ &= T^*[a^{-1}S, bD_{\alpha\beta}, a\Gamma_\nu, b^{-1}2ik_\mu; b^{-1}e_0^2, k^2].\end{aligned} \tag{9.53}$$

6 Renormalization

We are interested in renormalizing the functions S', $D'_{\alpha\beta}$, and Γ_ν, which satisfy the following coupled functional equations:

$$\begin{aligned}\Gamma_\mu(p, p') &= \gamma_\mu + \Lambda^*_\mu[S', D'_{\alpha\beta}, \Gamma_\nu; e_0^2, p, p'],\\ W_\mu(k) &= 2ik_\mu + ik_\mu \Gamma^*[S', D'_{\alpha\beta}, \Gamma_\nu, W_\lambda; e_0^2, k^2],\\ [S'(p)]^{-1} &= [S'(p_0)]^{-1} + (p - p_0)^\mu \Gamma_\mu(p, p_0),\\ [D'(k^2)]^{-1} &= \int_0^1 dx\, k^\mu W_\mu(xk).\end{aligned} \tag{9.54}$$

These yield divergent functions as solutions, because the functionals Λ^*_μ and T^* are divergent. However, since they are only logarithmically divergent, they can be rendered convergent through one subtraction.

Using the abbreviations $\Lambda^*_\mu(p, p')$ and $T^*(k^2)$ for the functionals, we define two finite functionals by[d]

$$\begin{aligned}\tilde{\Lambda}_\mu(p, p') &\equiv \Lambda^*_\mu(p, p') - [\Lambda^*_\mu(p_0, p_0)]_{\not{p}_0 = m},\\ \tilde{T}(k^2) &\equiv T^*(k^2) - T^*(\mu^2),\end{aligned} \tag{9.55}$$

where μ is an arbitrary invariant mass of the photon, and p_0 is the momentum of an electron on mass shell, with $p_0^2 = m^2$. The subscript $\not{p}_0 = m$ instructs us to commute $\not{p}_0$ all the way to the right, and then replace it by m. Thus,

$$[\Lambda^*_\mu(p_0, p_0)]_{\not{p}_0 = m} = L\gamma_\mu, \tag{9.56}$$

where L is a power series in e_0^2 with logarithmically divergent coefficients. The same is true of $T^*(\mu^2)$.

We now define finite renormalized functions $\tilde{S}$, $\tilde{D}_{\alpha\beta}$, $\tilde{\Gamma}_\nu$, $\tilde{W}_\mu$ as solutions of the functional equations obtained from (9.54) by replacing Λ^*_μ and T^* by $\tilde{\Lambda}_\mu$ and

---

[d] For simplicity we have chosen to subtract Γ_μ at a mass-shell momentum p_0. Actually the subtraction can be made at any momentum, whose invariant mass would then serve as an extra floating renormalization point in addition to μ.

Renormalization

$\tilde{T}$ respectively, and replacing e_0^2 by an appropriate number e^2:

$$\tilde{\Gamma}_\mu(p, p') = \gamma_\mu + \tilde{\Lambda}_\mu[\tilde{S}, \tilde{D}_{\alpha\beta}, \tilde{\Gamma}_\nu; e^2, p, p'],$$
$$\tilde{W}_\mu(k) = 2ik_\mu + ik_\mu \tilde{T}[\tilde{S}, \tilde{D}_{\alpha\beta}, \tilde{\Gamma}_\nu, \tilde{W}_\lambda; e^2, k^2],$$
$$[\tilde{S}(p)]^{-1} = [\tilde{S}(p_0)]^{-1} + (p - p_0)^\mu \tilde{\Gamma}_\mu(p, p_0), \tag{9.57}$$
$$[\tilde{D}(k^2)]^{-1} = \int_0^1 dx\, k^\mu W_\mu(xk).$$

We fix the normalization of $\tilde{S}$ by the condition

$$[\tilde{S}(p_0)]^{-1} = \not{p}_0 - m. \tag{9.58}$$

Then

$$[\tilde{S}(p)]^{-1} = \not{p} - m + (p - p_0)^\mu \tilde{\Lambda}_\mu(p, p_0), \tag{9.59}$$

with the property

$$[(\not{p} - m)\tilde{S}(p)]_{p=p_0} = 1. \tag{9.60}$$

The normalization of $\tilde{D}$ is such that

$$[ik^2 \tilde{D}(k^2)]_{k^2=\mu^2} = 1. \tag{9.61}$$

We now show that the renormalized quantities are proportional to the unrenormalized ones. Note that $\tilde{\Gamma}_\mu$ can be rewritten as follows:

$$\tilde{\Gamma}_\mu = \gamma_\mu + \Lambda_\mu^* - L Y_\mu = (1 - L)\left(\gamma_\mu + \frac{1}{1-L} \Lambda_\mu^*\right)$$
$$= Z'\left\{\gamma_\mu + \frac{1}{Z'} \Lambda_\mu^*[\tilde{S}, \tilde{D}_{\alpha\beta}, \tilde{\Gamma}_\nu; e^2, p, p']\right\}, \tag{9.62}$$

where

$$Z' = 1 - L. \tag{9.63}$$

This shows that a subtraction is equivalent to re-scaling. Similarly, we can write

$$\tilde{W}_\mu = 2ik_\mu + ik_\mu[T^*(k^2) - T^*(\mu^2)]$$
$$= Z\left\{2ik_\mu + \frac{1}{Z} ik_\mu T^*[\tilde{S}, \tilde{D}_{\alpha\beta}, \tilde{\Gamma}_\nu, \tilde{W}_\lambda; e^2, k^2]\right\}, \tag{9.64}$$

where

$$Z = 1 - \tfrac{1}{2} T^*(\mu^2). \tag{9.65}$$

Using the scaling properties (9.50) and (9.53), we obtain

$$\frac{1}{Z'} \tilde{\Gamma}_\mu = \gamma_\mu + \Lambda_\mu^*\left[Z'\tilde{S}, Z\tilde{D}_{\alpha\beta}, \frac{1}{Z'} \tilde{\Gamma}_\nu; \frac{e^2}{Z}, p, p'\right],$$
$$\frac{1}{Z} \tilde{W}_\mu = 2ik_\mu + ik_\mu T^*\left[Z'\tilde{S}, Z\tilde{D}_{\alpha\beta}, \frac{1}{Z'} \tilde{\Gamma}_\nu, \frac{1}{Z} \tilde{W}_\lambda, \frac{e^2}{Z}, k^2\right]. \tag{9.66}$$

Thus, the system of equations (9.57) can be reduced to (9.54) by putting

$$\Gamma_\mu = \frac{1}{Z'}\widetilde{\Gamma}_\mu,$$
$$W_\mu = \frac{1}{Z}\widetilde{W}_\mu,$$
$$S' = Z'\widetilde{S}, \qquad (9.67)$$
$$D'_{\mu\nu} = ZD_{\mu\nu},$$
$$e_0^2 = \frac{e^2}{Z}.$$

The last statement is known as "charge renormalization".[e]

The renormalized quantities are presumably finite, because they satisfy the finite equations (9.57). To show that they are actually finite, one has to show that (9.57) has finite interactive solutions. That is, the restoration of renormalized insertions into a renormalized skeleton should produce a convergent integral. That this is true has been shown by Weinberg.[6]

9.3 Gauge Invariance and the Photon Mass

As we have stated earlier, gauge invariance (or current conservation) requires that the vacuum polarization tensor $\Pi_{\mu\nu}$ be of the form

$$\Pi_{\mu\nu}(k) = (g_{\mu\nu}k^2 - k_\mu k_\nu)\Pi(k^2). \qquad (9.68)$$

This leads to a photon propagator whose gauge-invariant part reads

$$iD'_{\mu\nu}(k) = \left(g_{\mu\nu} - \frac{k_\mu k_\nu}{k^2}\right)\frac{1}{ik^2[1 + \Pi(k^2)]}. \qquad (9.69)$$

The pole at $k^2 = 0$ corresponds to a massless photon. This pole can be shifted only if $\Pi(k^2)$ develops a pole at $k^2 = 0$, which cannot happen in perturbation theory. Thus, in perturbation theory, the masslessness of the photon is an automatic consequence of gauge invariance, and there is no need for mass renormalization.

However, a naive calculation of $\Pi_{\mu\nu}$ to the lowest order fails to verify (9.68). One finds instead that the leading divergence of $\Pi_{\mu\nu}$ is quadratic and proportional to $g^{\mu\nu}$. This means that naively one would obtain

$$iD'_\mu(k) = \frac{g_{\mu\nu}}{i[k^2 + R(k^2)]}, \qquad (9.70)$$

where $R(k^2)$ appears to be a quadratically divergent mass term.

---

[e] It should be emphasized that the renormalization scheme is based on perturbation theory, and both Z and Z^{-1} must be regarded as power series in α_0 with divergent coefficients. If we terminate the power series to any finite order, as we must do in practice, then both Z and Z^{-1} are divergent, though $ZZ^{-1} = 1$ to the order considered. Formal considerations lead one to believe that $0 \le Z \le 1$, and probably $Z \to 0$ as $\Lambda \to \infty$ [G. Källen, Helv. Phys. Acta, **25**, 417 (1952)].

[6] S. Weinberg, Phys. Rev., **118**, 838 (1960).

Renormalization

The source of the difficulty is the failure of current conservation, due to the singular nature of the current

$$j^\mu(x) = ie\bar\psi(x)\gamma^\mu\psi(x), \qquad (9.71)$$

which involves the product of $\psi(x)$ and its canonical conjugate $\psi^\dagger(x)$ at the same space-time point. There are several ways to rectify the calculation.

A method to overcome the problem, known as the "point-splitting method",[7] is to re-define the current as

$$j^\mu(x) = \lim_{\varepsilon\to 0} ie\bar\psi(x+\tfrac{1}{2}\varepsilon)\gamma^\mu\psi(x-\tfrac{1}{2}\varepsilon)\exp\left[ie\int_{x-\varepsilon/2}^{x+\varepsilon/2} dy^\mu A_\mu(y)\right]. \qquad (9.72)$$

The exponential factor above is necessary to maintain gauge invariance. This factor gives a non-vanishing contribution even in the limit $\varepsilon \to 0$, and cures the problem.

A simpler remedy is to expand the naively calculated $\Pi_{\mu\nu}(k)$ in a Taylor series about $k = 0$, and subtract a sufficient number of leading terms until one gets the desired form. In practice one subtraction suffices. The justification for this procedure is as follows. One argues that the violation of gauge invariance occurs in the real part of $\Pi_{\mu\nu}$ but not in the imaginary part, because the latter describes physical instead of virtual processes. Thus, gauge invariance can be restored by subtracting an appropriate polynomial in k, which has no imaginary part. The practical recipe then is to replace the naive $\Pi_{\mu\nu}(k)$ by $\Pi_{\mu\nu}(k) - \Pi_{\mu\nu}(0)$. One may note that the quadratically divergent constant $\Pi_{\mu\nu}(0)$ can be cancelled by a counter term in the Lagrangian density, of the form $C\,F^{\mu\nu}F_{\mu\nu}$. Thus it can be absorbed by re-scaling $F^{\mu\nu}$. However, such a procedure is purely formal, and does not justify the recipe any better.

The failure of gauge invariance can occur even in finite Feynman graphs, for example, in the scattering of light by light to lowest order. Here, the sum of the 4! graphs, corresponding to all possible permutations of the external photon momenta, gives a convergent integral (although each individual graph is logarithmically divergent). However, the integral is not gauge invariant, and must be modified by the methods mentioned above.

9.4 The Renormalization Group

1 The group invariant

Charge renormalization in quantum electrodynamics is expressed by the relations

$$\widetilde D_{\mu\nu} = \frac{1}{Z(\mu^2)}D'_{\mu\nu},$$

$$e^2(\mu^2) = e_0^2 Z(\mu^2), \qquad (9.73)$$

---

[7] J. Schwinger, *Phys. Rev.*, **82**, 664 (1951); K. Johnson, in *Particles and Field Theory*, eds. S. Deser and K. W. Ford (Prentice-Hall, Englewood Cliffs, N. J., 1965).

which hold for an arbitrary renormalization point (subtraction point) μ^2. A change in μ changes Z but not $D'_{\mu\nu}$ and e_0^2. Hence $\widetilde{D}_{\mu\nu}$ and $e^2(\mu^2)$ change by a multiplicative constant under a change of μ. It is clear that the operations of changing the renormalization point form a multiplicative group, the "renormalization group".

We note that
— unrenormalized functions depend on $\{p, m, \alpha_0, \Lambda\}$,
— renormalized functions depend on $\{p, m, \alpha(\mu^2), \mu\}$,
where p denotes external momenta, Λ denotes the momentum cutoff, μ denotes the renormalization point, and α_0 and $\alpha(\mu^2)$ are respectively the bare and running coupling constants:

$$\alpha_0 = e_0^2/4\pi,$$
$$\alpha(\mu^2) = e^2(\mu^2)/4\pi. \tag{9.74}$$

The experimental fine-structure constant is

$$\alpha \equiv \alpha(0) \cong 1/137. \tag{9.75}$$

We see that in a renormalized function, the original dependences on α_0 and Λ are traded for dependences on $\alpha(\mu^2)$ and μ.

The renormalization constant Z is dimensionless, and does not depend on external momenta:

$$Z = Z\left(\frac{m^2}{\mu^2}, \frac{\mu^2}{\Lambda^2}, \alpha_0\right). \tag{9.76}$$

The gauge-invariant part of the propagators can be written in the forms

$$D'_{\mu\nu} = \left(g_{\mu\nu} - \frac{k_\mu k_\nu}{k^2}\right)\frac{d'}{ik^2},$$
$$\widetilde{D}_{\mu\nu} = \left(g_{\mu\nu} - \frac{k_\mu k_\nu}{k^2}\right)\frac{d}{ik^2}, \tag{9.77}$$

where d' and d are dimensionless scalar functions:

$$d' = d'\left(\frac{k^2}{\Lambda^2}, \frac{m^2}{\Lambda^2}, \alpha_0\right),$$
$$d = d\left(\frac{k^2}{\mu^2}, \frac{m^2}{\mu^2}, \alpha(\mu^2)\right). \tag{9.78}$$

According to (9.61),

$$d\left(1, \frac{m^2}{\mu^2}, \alpha(\mu^2)\right) = 1. \tag{9.79}$$

Renormalization

With all the arguments displayed, (9.73) reads

$$d\left(\frac{k^2}{\mu^2}, \frac{m^2}{\mu^2}, \alpha(\mu^2)\right) = Z^{-1}\left(\frac{m^2}{\mu^2}, \frac{\mu^2}{\Lambda^2}, \alpha_0\right) d'\left(\frac{k^2}{\Lambda^2}, \frac{m^2}{\Lambda^2}, \alpha_0\right),$$

$$\alpha(\mu^2) = \alpha_0 Z\left(\frac{m^2}{\Lambda^2}, \frac{\mu^2}{\Lambda^2}, \alpha_0\right).$$
(9.80)

It is assumed that α_0 depends on m^2/Λ^2 in such a manner as to make the above true in the limit $\Lambda \to \infty$. It follows that

$$\alpha(\mu^2) d\left(\frac{k^2}{\mu^2}, \frac{m^2}{\mu^2}, \alpha(\mu^2)\right) = \alpha_0 d'\left(\frac{k^2}{\Lambda^2}, \frac{m^2}{\Lambda^2}, \alpha_0\right), \quad (9.81)$$

which shows that the left side is independent of μ^2, and is therefore a renormalization group invariant. The invariance property can be trivially reexpressed in the form

$$\alpha(\mu^2) d\left(\frac{k^2}{\mu^2}, \frac{m^2}{\mu^2}, \alpha(\mu^2)\right) = \alpha(\nu^2) d\left(\frac{k^2}{\nu^2}, \frac{m^2}{\nu^2}, \alpha(\nu^2)\right), \quad (9.82)$$

where μ and ν are two arbitrary renormalization points.

2 Running coupling constant

The standard renormalized propagator d_c corresponds to the choice $\mu = 0$:

$$\alpha \, d_c\left(\frac{k^2}{m^2}, \alpha\right) \equiv \lim_{\mu \to 0} \alpha(\mu^2) d\left(\frac{k^2}{\mu^2}, \frac{m^2}{\mu^2}, \alpha(\mu^2)\right). \quad (9.83)$$

It is so normalized that

$$d_c(0, \alpha) = 1. \quad (9.84)$$

Since the right side of (9.83) is actually independent of μ, we can write, for all μ,

$$\alpha \, d_c\left(\frac{k^2}{m^2}, \alpha\right) = \alpha(\mu^2) d\left(\frac{k^2}{\mu^2}, \frac{m^2}{\mu^2}, \alpha(\mu^2)\right). \quad (9.85)$$

By setting $\mu^2 = k^2$ and using (9.79), we obtain

$$\alpha(k^2) = \alpha \, d_c\left(\frac{k^2}{m^2}, \alpha\right). \quad (9.86)$$

Using this, we can rewrite (9.85) as what appears to be a functional equation for the running coupling constant:

$$\alpha(k^2) = \alpha(\mu^2) d\left(\frac{k^2}{\mu^2}, \frac{m^2}{\mu^2}, \alpha(\mu^2)\right). \quad (9.87)$$

However, this equation has no content by itself.

To see this, choose a value $\mu_0(\alpha)$ of the renormalization point such that

$$\alpha(\mu_0^2) = 1. \quad (9.88)$$

Since μ_0^2 has the dimension of squared mass, it must be of the form

$$\mu_0^2(\alpha) = m^2/\phi(\alpha). \tag{9.89}$$

With this, (9.87) reduces to

$$\alpha(k^2) = d\left(\frac{k^2}{m^2}\phi(\alpha), \phi(\alpha), 1\right), \tag{9.90}$$

which does not impose any restriction on the functional form of $\alpha(k^2)$. Nevertheless, (9.87) serves as a convenient vehicle for further input, by which we can find out something about the asymptotic behavior of $\alpha(k^2)$.

We observe that in (9.55), the subtracted term $T^*(\mu^2)$ must vanish in the limit $\mu^2 \to \Lambda^2 \to \infty$. That is, there is no subtraction in that limit. Hence the bare fine-structure constant α_0 is recovered in the limit $\mu^2 \to \infty$:

$$\alpha(\mu^2) \xrightarrow[|\mu^2|\to\infty]{} \alpha_0. \tag{9.91}$$

We place absolute value signs around μ^2, on the assumption that there is no essential singularity in $\alpha(\mu^2)$ at $\mu^2 = \infty$.

3 Gell-Mann-Low function[8]

It seems plausible that the electron mass can be neglected in the renormalized propagator, when the external momentum and the renormalization point both become large. Accordingly we assume that the following limit exists:

$$\lim_{m\to 0} d\left(\frac{k^2}{\mu^2}, \frac{m^2}{\mu^2}, \alpha(\mu^2)\right) = f\left(\frac{k^2}{\mu^2}, \alpha(\mu^2)\right), \tag{9.92}$$

$$f(1, \alpha(\mu^2)) = 1.$$

This statement has been verified to order α^2 in perturbation theory[9]. Note that we cannot set $m = 0$ in the function $d_c(k^2/m^2, \alpha)$, because the renormalization point is $\mu = 0$, and we need m to set the momentum scale.

With (9.92), the identity (9.87) becomes a functional equation with content:

$$\alpha(k^2) = \alpha(\mu^2) f\left(\frac{k^2}{\mu^2}, \alpha(\mu^2)\right) \tag{9.93}$$

$$(k^2/m^2 \gg 1, \mu^2/m^2 \gg 1).$$

Choosing again $\mu = \mu_0(\alpha)$ [see (9.88) and (9.89)], we obtain

$$\alpha(k^2) = f\left(\frac{k^2}{m^2}\phi(\alpha), 1\right) \equiv F\left(\frac{k^2}{m^2}\phi(\alpha)\right) \quad (k^2 \gg m^2). \tag{9.94}$$

This states that, for $k^2 \gg m^2$, the functional form of $\alpha(k^2)$ becomes independent of the value of the fine-structure constant—the latter merely sets the scale.

---

[8] M. Gell-Mann and F. E. Low, *Phys. Rev.*, **95**, 1300 (1954).
[9] This is implicitly shown in R. Jost and J. M. Luttinger, *Hebr. Phys. Acta*, **23**, 20 (1950).

We can use (9.94) to extend the lowest-order calculation of $\alpha(k^2)$ as follows. From (9.18) we have

$$\frac{1}{\alpha(k^2)} = \left(\frac{1}{\alpha} - \frac{1}{3\pi}\ln\frac{k^2}{m^2}\right) + O(\alpha). \tag{9.95}$$

The terms in the braces can be put in the form $F(\phi(\alpha)k^2/m^2)$ by choosing

$$\phi(\alpha) = e^{-3\pi/\alpha}. \tag{9.96}$$

Thus (9.95) is consistent with (9.94) to the order calculated. Since the term $O(\alpha)$ must be a function of $\phi(\alpha) k^2/m^2$ only, its power series expansion in α must have the form

$$c_1\left(\frac{1}{\alpha} - \frac{1}{3\pi}\ln\frac{k^2}{m^2}\right)^{-1} + c_2\left(\frac{1}{\alpha} - \frac{1}{3\pi}\ln\frac{k^2}{m}\right)^{-2} + \cdots$$

$$= \sum_{n=1}^{\infty} c_n \alpha^n \left(1 - \frac{\alpha}{3\pi}\ln\frac{k^2}{m^2}\right)^{-n}.$$

Therefore, for $k^2 \gg m^2$,

$$\left[d_c\left(\frac{k^2}{m^2}, \alpha\right)\right]^{-1} = 1 - \frac{\alpha}{3\pi}\ln\frac{k^2}{m^2} + \sum_{n=1}^{\infty} c_n \alpha^{n+1}\left(1 - \frac{\alpha}{3\pi}\ln\frac{k^2}{m^2}\right)^{-n}. \tag{9.97}$$

Thus, without further calculations, we can assert that the sum of all "leading logarithms" gives

$$d_c\left(\frac{k^2}{m^2}, \alpha\right) = \frac{1}{1 - \frac{\alpha}{3\pi}\ln\frac{k^2}{m^2}}. \tag{9.98}^f$$

We now formulate a more efficient way to get information from (9.93). First, rewrite it in the form

$$\alpha(x) = \alpha(y) f\left(\frac{x}{y}, \alpha(y)\right). \tag{9.99}$$

Differentiating both sides with respect to x at fixed y, and then setting $y = x$, we obtain

$$x \frac{d}{dx} \alpha(x) = \beta(\alpha(x)), \tag{9.100}$$

where

$$\beta(\alpha) \equiv \alpha \left[\frac{\partial f(x, \alpha)}{\partial x}\right]_{x=1} \tag{9.101}$$

---

[f] The pole occurring at $k^2/m^2 = \exp(3\pi/\alpha)$ is called the "Landau ghost", because Landau was alarmed by the pathological behavior of the theory implied by such a pole. [L. D. Landau, in *Niels Bohr and the Development of Physics*, Ed. W. Pauli (McGraw-Hill, New York, 1955)]. In the current view, this formula is of interest only by comparison with the corresponding one in quantum chromodynamics, in which the minus sign is replaced by a plus sign (asymptotic freedom).

is called the "Gell-Mann-Low function" (GL function). The solution to (9.100) is

$$\ln \frac{x}{y} = \int_{\alpha(y)}^{\alpha(x)} \frac{dz}{\beta(z)}. \tag{9.102}$$

The Taylor expansion of $\beta(z)$ can be obtained in perturbation theory, with the result (for quantum electrodynamics)

$$\beta(z) = \frac{z^2}{3\pi} \left(1 + \frac{3}{4\pi} z + \dots \right). \tag{9.103}$$

The first term above can be obtained easily from (9.18). The result (9.98) can be obtained directly from (9.102) by using the first term in the Taylor series for $\beta(z)$, and assuming that we can put $y = m^2$, which can be justified to the order of the calculation.

4 Fixed points

We parametrize the renormalization point by writing

$$\mu^2 = \mu_0^2 + e^t \quad (-\infty < t < \infty), \tag{9.104}$$

where μ_0 is a given reference point. The running coupling constant will be redesignated as $\alpha(t)$. Then (9.102) reads

$$t = \int_{\alpha(0)}^{\alpha(t)} \frac{d\alpha}{\beta(\alpha)}. \tag{9.105}$$

The integral above must diverge when $|t| \to \infty$. This means that, in the limit $|t| \to \infty$, either $\alpha(t)$ diverges or it approaches a finite zero of the GL function $\beta(\alpha)$. We further analyze the latter possibility.

A finite zero of $\beta(\alpha)$ is called a *fixed point*. It is called an ultraviolet (UV) or infrared (IR) fixed point according to whether $\beta(\alpha)$ is a decreasing ($\downarrow$) or an increasing ($\uparrow$) function in the neighborhood of the zero:

$$\begin{aligned}
&\text{Fixed point:} && \beta(\alpha_0) = 0, \\
&\text{UV fixed point:} && \beta(\alpha) \downarrow \text{ near } \alpha = \alpha_0, \\
&\text{IR fixed point:} && \beta(\alpha) \uparrow \text{ near } \alpha = \alpha_0.
\end{aligned} \tag{9.106}$$

Some examples of fixed points are shown in Fig. 9.7. The reason for the terminology is that

$$\begin{aligned}
\alpha(t) &\xrightarrow[t \to \infty]{} \text{UV fixed point}, \\
\alpha(t) &\xrightarrow[t \to -\infty]{} \text{IR fixed point}.
\end{aligned} \tag{9.107}$$

The reader is invited to verify these statements by inspection of Fig. 9.7, in conjunction with (9.105). In Fig. 9.7, we indicate by arrows the directions in which $\alpha(t)$ would move as $t \to \infty$, depending on the starting point $\alpha(0)$.

Fig. 9.7 (a) represents what $\beta(\alpha)$ might look like for quantum electrodynamics. The only property we are sure of is that the origin is an IR fixed point, in the neighborhood of which conventional perturbation theory is valid. Since $\alpha(t)$

Renormalization

does not approach that point as $t \to \infty$, we cannot hope to discover the true asymptotic behavior of $\alpha(t)$ by using perturbation theory. A naive extrapolation of perturbation theory would indicate that $\alpha(t) \to \infty$. But another possibility is that $\beta(\alpha)$ has another zero $\alpha_0 \neq 0$ that is an UV fixed point. In that case, $\alpha(t) \to \alpha_0$, which means that α_0 is the bare fine-structure constant. Thus, if quantum electrodynamics is a finite theory, α_0 must be a root of the GL function. This possible root has so far eluded diligent search. Adler[10] has shown that if it exists, it cannot be a root of any finite order. Others argue that no root can exist.[11]

Fig. 9.7 (b) represents what $\beta(\alpha)$ might look like for an asymptotically free theory, such as quantum chromodynamics. It is characterized by the fact that the origin is an UV fixed point. In this case, perturbation theory is valid in the limit $t \to \infty$ which is the domain relevant to electron-nucleon deep inelastic scattering[12]. However, perturbation theory fails for $t \to -\infty$, which corresponds to the long-distance limit in which the phenomenon of quark confinement is supposed to occur.

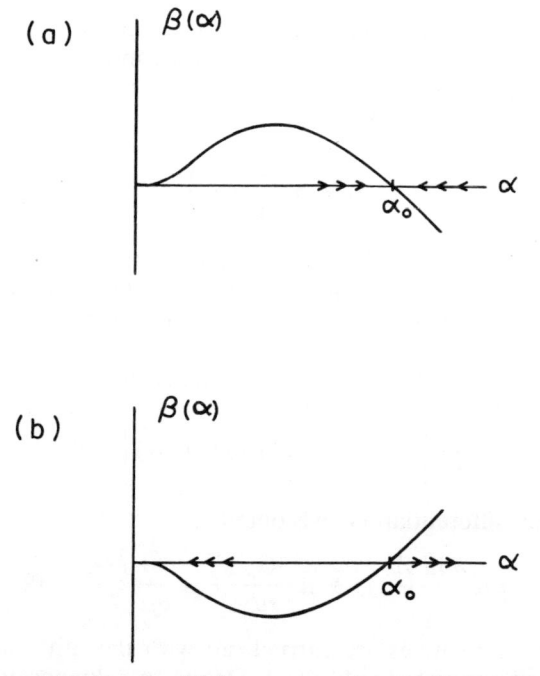

Fig. 9.7 Examples of GL function (a) for a non-asymptotically free theory, and (b) for an asymptotically free theory

---

[10] S. L. Adler, *Phys. Rev.*, D5, 3021 (1972).
[11] S. L. Adler, C. G. Callan, D. J. Gross, and R. Jackiw, *Phys. Rev.*, D6, 2982 (1972); N. Christ, *Phys. Rev.*, D9, 946 (1973); M. Baker and K. Johnson, *Physica*, 96A, 120 (1979).
[12] H. D. Politzer, *Phys. Reports*, 14, 129 (1974); A. H. Mueller, *Phys. Reports*, 73, 239 (1981)

9.5 Callan-Symanzik Equation

It is always hard work to actually renormalize a renormalizable field theory, i.e., to derive relations between unrenormalized and renormalized functions. But if the relations are already given, then one can introduce the running coupling constant and the GL function without ever referring to the details of the renormalization procedure. Callan and Symanzik[13] put this idea into practice.[g]

Consider a renormalizable theory with only one bare coupling constant α_0, which is dimensionless. A connected n-point Green's function G'_n is related to its renormalized version through

$$G'_n(p; \Lambda, \alpha_0) = Z^n\left(\frac{\Lambda}{\mu}, \alpha_0\right) G_n(p; \mu, \alpha), \qquad (9.108)$$

where Z is a positive constant, p stands for all external momenta, Λ denotes the cutoff momentum, and μ is the renormalization point. More generally, the factor Z^n should be supplemented by possible factors of other renormalization constants; but we shall ignore the latter for simplicity. We have set all masses equal to zero, which means that we are considering the asymptotic domain where all external momenta are large, and that the momentum scale is set by μ.

The renormalized coupling constant α depends on Λ, μ, and α_0. Since it is dimensionless, it must have the form

$$\alpha = \alpha\left(\frac{\Lambda}{\mu}, \alpha_0\right). \qquad (9.109)$$

When $\Lambda \to \infty$, this is supposed to approach a function of μ alone, the running coupling constant $\alpha(\mu)$. The entire content of (9.108) rests with the choice of arguments in the various functions, and the assertion that G_n and α approach finite limits when $\Lambda \to \infty$, while G'_n, Z, and α_0 may diverge.

We observe that the right side of (9.108) must be independent of μ, and hence

$$\frac{d}{d\mu}\left[Z^n\left(\frac{\Lambda}{\mu}, \alpha_0\right) G_n(p; \mu, \alpha)\right] = 0. \qquad (9.110)$$

Carrying out the differentiation, we obtain

$$n\mu \frac{\partial \ln Z}{\partial \mu} G_n + \mu \frac{\partial G_n}{\partial \mu} + \mu \frac{\partial \alpha}{\partial \mu} \frac{\partial G_n}{\partial \alpha} = 0, \qquad (9.111)$$

where all partial derivatives are carried out with the other arguments of the function being differentiated held fixed. Define two dimensionless functions β

---

[13] C. G. Callan Jr., *Phys. Rev.*, D2, 1541 (1970); K. Symanzik, *Comm. Math. Phys.*, **18**, 227 (1970).

[g] Historically the Callan-Symanzik equation grew out of considerations of the violation of formal scale invariance in perturbation theory. For references in addition to Ref. 13, see S. Coleman and R. Jackiw, *Annals of Phys.*, **67**, 552 (1971); J. C. Collins, A. Duncan, and S. D. Joglekar, *Phys. Rev.*, D16, 438 (1977).

Renormalization

and γ by

$$\beta(\alpha(x)) \equiv \mu \frac{\partial}{\partial \mu} \alpha\left(\frac{\Lambda}{\mu}, \alpha_0\right),$$

$$\gamma(\mu) \equiv \mu \frac{\partial}{\partial \mu} \ln Z\left(\frac{\Lambda}{\mu}, \alpha_0\right),$$

(9.112)

where $\beta(\alpha(\mu))$ is the GL function, and $\gamma(\mu)$ is called the "anomalous dimension". (It differs by a sign from treatments in which Z is replaced by Z^{-1})[h]. We emphasize that β must be expressed as a function of $\alpha(\mu)$. With these definitions, (9.111) can be written in the form

$$\left[\mu \frac{\partial}{\partial \mu} + \beta(\alpha) \frac{\partial}{\partial \alpha} + n\gamma(\mu)\right] G_n(p; \mu, \alpha) = 0.$$

(9.113)

This is the Callan-Symanzik (CS) equation.

Consider now a dimensionless ratio of Green's functions F, such that the Z factors are divided out. Then F is invariant under the renormalization group. A simple example is

$$F\left(\frac{p}{\mu}, \alpha\right) = \frac{G_n(p; \mu, \alpha)}{G_n(\mu; \mu, \alpha)},$$

(9.114)

for some fixed n. Here $G_n(\mu; \mu, \alpha)$ means $G_n(p; \mu, \alpha)$ with all external momenta set equal to a common momentum with squared invariant mass μ^2. It is easily seen that in this case the CS equation simplifies to

$$\left[\mu \frac{\partial}{\partial \mu} + \beta(\alpha) \frac{\partial}{\partial \alpha}\right] F\left(\frac{p}{\mu}, \alpha\right) = 0.$$

(9.115)

It is convenient to parametrize the renormalization point μ and the running coupling constant α by writing

$$\mu(t) \equiv \mu_0 e^t,$$

$$\alpha(t) \equiv \alpha(\mu(t)),$$

(9.116)

where μ_0 is some fixed reference point. Then the invariance of F can be stated as

$$\frac{d}{dt} F\left(\frac{p}{\mu(t)}, \alpha(t)\right) = 0.$$

(9.117)

Carrying out the differentiation, we obtain

$$\left(\mu \frac{\partial}{\partial \mu} + \frac{d\alpha}{dt} \frac{\partial}{\partial \alpha}\right) F\left(\frac{p}{\mu}, \alpha\right) = 0.$$

(9.118)

---

[h] For simplicity, we use μ instead of μ^2 as independent variable in this section. This does not affect the running coupling constant, but changes $\beta(\alpha)$ by a factor of 2 as compared to the last section.

Comparison with (9.115) yields
$$\frac{d\alpha(t)}{dt} = \beta(\alpha(t)), \qquad (9.119)$$
with solution
$$\int_{\alpha(0)}^{\alpha(t)} \frac{d\alpha}{\beta(\alpha)} = t. \qquad (9.120)$$

Since F is independent of t, we have
$$F\left(\frac{p}{\mu(t)}, \alpha(t)\right) = F\left(\frac{p}{\mu(0)}, \alpha(0)\right). \qquad (9.121)$$

Let us put $p = \mu(t)$, i.e., put all external momenta equal to a common momentum whose squared invariant mass is $\mu^2(t)$. We then have
$$F(1, \alpha(t)) = F\left(\frac{\mu(t)}{\mu(0)}, \alpha(0)\right). \qquad (9.122)$$

The right side is actually a constant and may be normalized to 1. Hence
$$F(1, \alpha(t)) = 1. \qquad (9.123)$$
This may be taken as a definition of $\alpha(t)$.

We rewrite (9.121) trivially as
$$\begin{aligned} F\left(\frac{p}{\mu(t)}, \alpha(t)\right) &= F\left(\frac{e^t p}{e^t \mu(0)}, \alpha(0)\right) \\ &= F\left(\frac{e^t p}{\mu(t)}, \alpha(0)\right). \end{aligned} \qquad (9.124)$$

Putting $p = k\mu(t)/\mu(0)$, and then redesignating k by p, we have
$$F\left(\frac{e^t p}{\mu(0)}, \alpha(0)\right) = F\left(\frac{p}{\mu(0)}, \alpha(t)\right). \qquad (9.125)$$

This shows that a change in all the external momenta of F by a common factor is equivalent to a change in the coupling constant.

For the Green's function G_n, the effects of changing the renormalization point is governed by the CS equation (9.113), which we can rewrite as
$$\left[\frac{d}{dt} + n\gamma(t)\right] G_n(p; \mu(t), \alpha(t)) = 0, \qquad (9.126)$$

where $\gamma(t) \equiv \gamma(\mu(t))$. This is obtained by recognizing that if we regard both μ and α as functions of t, then
$$\mu \frac{\partial}{\partial \mu} + \beta(\alpha) \frac{\partial}{\partial \alpha} = \frac{d}{dt}. \qquad (9.127)$$

Renormalization

The solution to (9.126) is

$$G_n(p; \mu(t), \alpha(t)) = e^{-nD(t)} G_n(p; \mu(0), \alpha(0)),$$

$$D(t) \equiv \int_0^t dt' \, \gamma(t'). \tag{9.128}$$

Now let p be replaced by $e^t p$:

$$G_n(e^t p; \mu(t), \alpha(t)) = e^{-nD(t)} G_n(e^t p; \mu(0), \alpha(0)). \tag{9.129}$$

The left side can be rewritten as

$$G_n(e^t p; e^t \mu(0), \alpha(t)).$$

We note that all the arguments of the dimension of mass are multiplied by a common factor e^t. If G_n itself is of the dimension (mass)^d, then

$$G_n(e^t p; e^t \mu(0), \alpha(0)) = e^{td} G_n(p; \mu(0), \alpha(0)). \tag{9.130}$$

(Recall the assumption that G_n does not contain intrinsic mass parameters). Equating this to the right side of (9.129), we obtain

$$G_n(e^t p; \mu(0), \alpha(0)) = e^{td + nD(t)} G_n(p; \mu(0), \alpha(t)). \tag{9.131}$$

This is the generalization of (9.125). Note that the overall scale factor is not just the naive e^{td} expected from dimensional analysis, but contains an extra factor $\exp\left[n \int dt \, \gamma(t)\right]$ (hence the name "anomalous dimension"). The latter arises from the fact that a scale change also changes the renormalization point, and G_n is not invariant under this operation.

9.6 Example: massless ϕ^4 theory

To illustrate the calculation of the GL function using (9.112), we do it to lowest order in massless ϕ^4 theory. The Lagrangian density is

$$\mathcal{L}(x) = \tfrac{1}{2} \partial_\mu \phi \, \partial^\mu \phi - \alpha_0 \phi^4. \tag{9.132}$$

There is no mass parameter and hence no intrinsic scale. The renormalized charge is defined by the 4-point Green's function, which to second order is given by the Feynman graphs in Fig. 9.8:

$$G_4'(p; \Lambda, \alpha_0) = -i\alpha_0 + \tfrac{1}{2}(-i\alpha_0)^2 [I(p_1 + p_2) \\ + I(p_2 + p_3) + I(p_3 + p_1)], \tag{9.133}$$

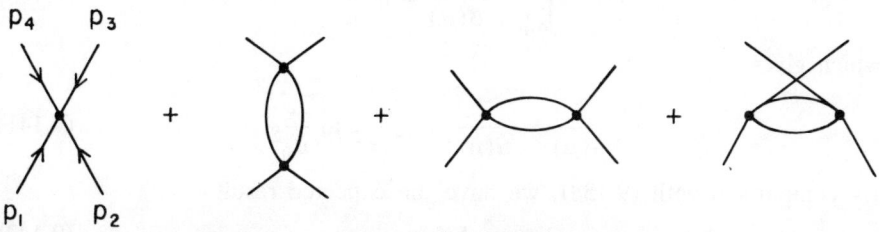

Fig. 9.8 Feynman graphs for 4-point function in ϕ^4 theory

where the factor $\frac{1}{2}$ in the second term comes from the symmetry number, and

$$I(p) = \int \frac{d^4k}{(2\pi)^4} \frac{1}{(k^2 + i\varepsilon)[(k + p)^2 + i\varepsilon]}. \tag{9.134}$$

The integral is logarithmically divergent, and is to be rotated into Euclidean momentum space and cut-off at momentum Λ. By this procedure we obtain

$$I(p) = \frac{i}{16\pi^2} \ln \frac{\Lambda^2}{-p^2 + i\varepsilon}, \tag{9.135}$$

which has been rotated back into Minkowski space. Using this, we obtain

$$G'_4(p; \Lambda, \alpha_0) = -i\alpha_0 + \frac{i\alpha_0^2}{32\pi^2} \left(\ln \frac{\Lambda^2}{s} + \ln \frac{\Lambda^2}{t} + \ln \frac{\Lambda^2}{u} \right),$$

$$s = -(p_1 + p_2)^2 + i\varepsilon, \tag{9.136}$$

$$t = -(p_2 + p_3)^2 + i\varepsilon,$$

$$u = -(p_3 + p_1)^2 + i\varepsilon.$$

The renormalized coupling constant α is defined by

$$\alpha\left(\frac{\Lambda}{\mu}, \alpha_0\right) \equiv iG'_4(\mu; \Lambda, \alpha_0), \tag{9.137}$$

where the argument μ in G' indicates $p_1^2 = p_2^2 = p_3^2 = p_4^2 = -\mu^2$. (Note that we are renormalizing at a Euclidean point). Using (9.136), we obtain

$$\alpha\left(\frac{\Lambda}{\mu}, \alpha_0\right) = \alpha_0 - \frac{3\alpha_0^2}{32\pi^2} \ln \frac{\Lambda^2}{\mu^2}, \tag{9.138}$$

from which we obtain the GL function by using (9.112):

$$\beta(\alpha) = \frac{3\alpha^2}{16\pi^2}. \tag{9.139}$$

Since this is positive, the theory is not asymptotically free. We can obtain the running coupling constant $\alpha(\mu)$ through

$$\int_{\alpha(\mu_0)}^{\alpha(\mu)} \frac{d\alpha}{\beta(\alpha)} = \ln \frac{\mu}{\mu_0}, \tag{9.140}$$

which gives

$$\frac{1}{\alpha(\mu)} = \frac{1}{\alpha(\mu_0)} - \frac{3}{32\pi^2} \ln \frac{\mu^2}{\mu_0^2}. \tag{9.141}$$

By comparison with (9.138), we have the expected result

$$\lim_{\Lambda \to \infty} \alpha(\Lambda) = \alpha_0. \tag{9.142}$$

Renormalization

The renormalization constant is given by

$$Z\left(\frac{\Lambda}{\mu}, \alpha_0\right) = \alpha\left(\frac{\Lambda}{\mu}, \alpha_0\right) \bigg/ \alpha_0 = 1 + \frac{3\alpha_0}{32\pi^2} \ln \frac{\Lambda^2}{\mu^2}. \qquad (9.143)$$

By (9.112) this leads to

$$\gamma(\mu) = -\frac{3}{16\pi} \alpha(\mu). \qquad (9.144)$$

To the order calculated, G'_4 can be written in the form

$$[iG'_4]^{-1} = \frac{1}{\alpha_0} + \frac{1}{32\pi^2} \left(\ln \frac{\Lambda^2}{s} + \ln \frac{\Lambda^2}{t} + \ln \frac{\Lambda^2}{u} \right). \qquad (9.145)$$

Substituting $1/\alpha_0$ from (9.141) with $\mu_0 = \Lambda$, we obtain

$$\begin{aligned}[iG'_4]^{-1} &= \frac{1}{\alpha(\mu)} + \frac{1}{32\pi^2} \left(\ln \frac{\mu^2}{s} + \ln \frac{\mu^2}{t} + \ln \frac{\mu^2}{u} \right) \\ &\equiv [iG_4(p; \mu, \alpha(\mu))]^{-1}.\end{aligned} \qquad (9.146)$$

The factor Z is absent to this order because $Z = 1 + O(\alpha)$.

CHAPTER X
METHOD OF EFFECTIVE POTENTIAL

10.1 Spontaneous Symmetry Breaking

In this chapter we shall discuss renormalization in the presence of spontaneous symmetry breaking. To put things in perspective, it should be pointed out that spontaneous symmetry breaking is a purely theoretical concept that has nothing to do with experiments. The vacuum expectation value of the Higgs field, for example, is not a directly observable quantity. It can be deduced indirectly from observational data only within the framework of a theoretical model. There is no experimental way in which we can tell whether masses arise from "mechanical mass terms" or from spontaneous symmetry breaking.

One might then wonder why we should use this concept at all. The reason, as we have learned in the Weinberg-Salam model, is that it enables us to construct a renormalizable theory of massive vector bosons with which we can do practical calculations. Thus, we should know something about renormalization with spontaneous symmetry breaking, which is most conveniently done with the help of the "effective potential".

To introduce the technique, we shall discuss the mathematical example of a self-coupled scalar field. The relevance to physics comes when the scalar field is coupled to other fields, such as a gauge field. A simple example of the latter will be described, but we forego a full treatment of renormalization in a realistic theory[1].

10.2 The Effective Action[2]

Consider a real scalar field with Lagrangian density

$$\mathcal{L}(x) = \tfrac{1}{2}\partial_\mu \phi(x)\, \partial^\mu \phi(x) - V(\phi(x)),$$

$$V(\phi) = \frac{\alpha_0}{4!}\phi^4 - \frac{m_0^2}{4}\phi^2. \qquad (10.1)$$

The classical value of the vacuum field is at the minimum ρ_0 of $V(\phi)$, but the vacuum expectation value $\langle \phi \rangle$ of the quantum field is not necessarily the same,

---

[1] See B. W. Lee and J. Zinn-Justin, *Phys. Rev.*, **D5**, 3137 (1974); **D7**, 1049 (1973).
[2] J. Goldstone, A. Salam, and J. Weinberg, *Phys. Rev.*, **127**, 965 (1962); G. Jona-Lasinio, *N. Cimento*, **34**, 1790 (1964).

Method of Effective Potential

being defined by

$$\langle \phi \rangle \equiv \lim_{J \to 0} \frac{\langle 0^+ | \phi_{\text{op}} | 0^- \rangle_J}{\langle 0^+ | 0^- \rangle_J}, \tag{10.2}$$

where the notation is that of Sec. 7.3.

Let us recall the definition of the generating functional $W[J]$:

$$\exp \frac{i}{\hbar} W[J] = \mathcal{N} \int (D\phi) \exp \frac{i}{\hbar} \{S[\phi] + (J, \phi)\},$$

$$S[\phi] = \int d^4x \, \mathcal{L}(x), \quad (J, \phi) = \int d^4x \, J(x)\phi(x). \tag{10.3}$$

As a book-keeping device, we keep $\hbar$ as displayed in (10.3), but set $\hbar = c = 1$ everywhere else. By expanding in powers of $\hbar$, we obtain an expansion in terms of the number of closed loops in the Feynman graphs. If we restore all $\hbar$ and c, but keep $\mathcal{L}(x)$ exactly as written in (10.1), we would find that $\hbar c \alpha$ is the only dimensionless parameter in the theory. Thus the loop expansion is actually an expansion in powers of $\hbar c \alpha$, but it performs an infinite re-summation of Feynman graphs.

The vacuum expectation value of the field in the presence of an external source $J(x)$ is given by

$$\phi_c(x) \equiv \frac{\langle 0^+ | \phi_{\text{op}}(x) | 0^- \rangle_J}{\langle 0^+ | 0^- \rangle_J} = \frac{\delta W[J]}{\delta J(x)}. \tag{10.4}$$

The vacuum expectation value $\langle \phi \rangle$ is the limit of $\phi_c(x)$ as $J \to 0$. Here, ϕ_c is determined by the external source function J. We now ask the question: What source function J will produce a given function ϕ_c? To answer the question, we shall reformulate the problem so that it becomes convenient to use ϕ_c as the independent variable, instead of J.

We define an "effective action" $\Gamma[\phi_c]$ by making a Legendre transformation:

$$\Gamma[\phi_c] \equiv W[J] - (J, \phi_c), \tag{10.5}$$

where J is to be eliminated in terms of ϕ_c, through solving (10.4). We call $\Gamma[\phi_c]$ an effective action, because it is a functional of a classical field ϕ_c, and hence akin to $S[\phi]$. We note that

$$\frac{\delta \Gamma[\phi_c]}{\delta \phi_c(x)} = \frac{\delta W[J]}{\delta \phi_c(x)} - J(x) - \int d^4y \, \frac{\delta J(y)}{\delta \phi_c(x)} \phi_c(y),$$

$$\frac{\delta W[J]}{\delta \phi_c(x)} = \int d^4y \, \frac{\delta W[J]}{\delta J(y)} \frac{\delta J(y)}{\delta \phi_c(x)} = \int d^4y \, \frac{\delta J(y)}{\delta \phi_c(x)} \phi_c(y).$$

Substituting the latter equation into the former, we obtain

$$\frac{\delta \Gamma[\phi_c]}{\delta \phi_c(x)} = -J(x). \tag{10.6}$$

If we put $J = 0$, ϕ_c should become a constant, by translational invariance.

Hence $\langle\phi\rangle$ is the root of the equation

$$\left.\frac{d\Gamma[\phi_c]}{d\phi_c}\right|_{\phi_c=\langle\phi\rangle} = 0. \tag{10.7}$$

We recall that $W[J]$ is the generating functional for connected Green's functions. We now show that $\Gamma[\phi_c]$ is the generating functional for proper, or one-particle irreducible (1-PI), Green's functions. We can always expand $\Gamma[\phi_c]$ in powers of ϕ_c, as follows:

$$\Gamma[\phi_c] = \sum_{n=0}^{\infty} \frac{1}{n!} \int d^4x_1 \ldots d^4x_n \, \Gamma_n(x_1, \ldots, x_n) \phi_c(x_1) \ldots \phi_c(x_n). \tag{10.8}$$

We now show that Γ_n is the n-point 1-PI Green's function, i.e., the sum of all connected 1-PI Feynman graphs with n external legs (with external propagators omitted).

Consider the quantity

$$\mathcal{N} \int (D\phi) \exp \frac{i}{a} \{\Gamma[\phi] + (J, \phi)\}. \tag{10.9}$$

As $a \to 0$, the integrand is dominated by its value at a saddle point, which occurs at $\phi = \phi_c$, for (10.6) is precisely the saddle-point condition. Thus

$$\mathcal{N} \int (D\phi) \exp \frac{i}{a} \{\Gamma[\phi] + (J, \phi)\} \xrightarrow[a \to 0]{} \mathcal{N} \exp \frac{i}{a} \{\Gamma[\phi_c] + (J, \phi_c)\}. \tag{10.10}$$

By (10.5), this statement is the same as

$$\exp \frac{i}{a} W[J] \xrightarrow[a \to 0]{} \mathcal{N} \int (D\phi) \exp \frac{i}{a} \{\Gamma[\phi] + (J, \phi)\}. \tag{10.11}$$

In the limit $a \to 0$, the right side is the sum of all tree graphs of a new theory whose action is $\Gamma[\phi]$. A vertex in one of these tree graphs is some Γ_n, by the definition (10.8). This fact is illustrated in Fig. 10.1 (a). On the other hand, $W[J]$ is by definition the sum of all connected Green's functions of the original theory, and a connected Green's function can be dissected into 1-PI components, as shown in Fig. 10.1 (b). Therefore, Γ_n is a 1-PI Green's function of the original theory. ∎

10.3 The Effective Potential

We may alternatively expand $\Gamma[\phi_c]$ in terms of ϕ_c and its derivatives, as follows:

$$\Gamma[\phi_c] = \int d^4x \, [-U(\phi_c(x)) + \tfrac{1}{2}\partial_\mu \phi_c(x) \, \partial^\mu \phi_c(x) Z(\phi_c(x)) + \cdots]. \tag{10.12}$$

The function U is called the "effective potential". It may be extracted from (10.12) by putting $\phi_c(x) = \rho$ (constant). Under this condition, all terms in

Method of Effective Potential

(10.12) vanish except the first, and we have

$$\Gamma[\rho] = -\Omega U(\rho), \qquad (10.13)$$

where Ω is the total volume of space-time.

Let the Fourier transforms of ϕ_c and Γ_n be denoted by

$$\tilde{\phi}_c(k) = \int d^4x\, e^{ik\cdot x}\, \phi_c(x) \qquad (10.14)$$

$$\tilde{\Gamma}_n(k_1, \ldots, k_n)\, \delta^4(k_1 + \cdots + k_n)$$
$$= \int d^4x_1 \ldots d^4x_n\, \Gamma_n(x_1, \ldots, x_n)\, \exp(ik_1\cdot x_1 + \cdots + ik_n \cdot x_n).$$

In terms of these, we can rewrite (10.8) as

$$\Gamma[\phi_c] = \sum_{n=0}^{\infty} \frac{1}{n!} \int \frac{d^4k_1}{(2\pi)^4} \cdots \frac{d^4k_n}{(2\pi)^4}\, \delta^4(k_1 + \cdots + k_n)\, \tilde{\Gamma}_n(k_1, \ldots, k_n)$$
$$\cdot \tilde{\phi}_c(k_1) \ldots \tilde{\phi}_c(k_n). \qquad (10.15)$$

Now put

$$\phi_c(x) = \rho \quad \text{(constant)},$$
$$\tilde{\phi}_c(k) = \Omega\rho, \qquad (10.16)$$

where Ω is the total volume of space-time:

$$\Omega = (2\pi)^4\, \delta^4(0). \qquad (10.17)$$

Then

$$\Gamma[\rho] = \Omega \sum_{n=0}^{\infty} \frac{\rho^n}{n!}\, \tilde{\Gamma}_n(0), \qquad (10.18)$$

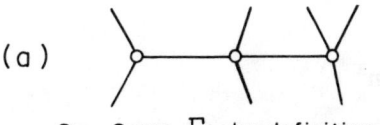

(a)

o = Some Γ_n, by definition of Γ_n

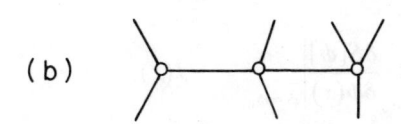

(b)

o = Some 1-PI Green's function, by definition of $W[J]$

Fig. 10.1 Aid for proving that the effective action generates 1P-I Green's functions.

where
$$\tilde{\Gamma}_n(0) \equiv \tilde{\Gamma}_n(0, 0, \ldots, 0). \tag{10.19}$$

Comparing (10.18) with (10.13), we obtain
$$U(\rho) = -\sum_{n=0}^{\infty} \frac{\rho^n}{n!} \tilde{\Gamma}_n(0). \tag{10.20}$$

Thus we can make the identification
$$-\frac{d^n U(\rho)}{d\rho^n}\bigg|_{\rho=0} = \tilde{\Gamma}_n(0). \tag{10.21}$$

We should remember that Γ_n is defined in terms of the field $\phi(x)$. When there is spontaneous symmetry breaking, it is more appropriate to use the shifted field
$$\eta(x) = \phi(x) - \langle\phi\rangle. \tag{10.22}$$

The 1-PI Green's functions $\Gamma_n^{(s)}$ defined with respect to $\eta(x)$ are linear combinations of Γ_n, and may be obtained from the shifted version of (10.20):
$$U(\rho - \langle\phi\rangle) = -\sum_{n=0}^{\infty} \frac{[\rho - \langle\phi\rangle]^n}{n!} \tilde{\Gamma}_n^{(s)}(0). \tag{10.23}$$

Physically useful quantities can be obtained directly from the effective potential:
$$\begin{aligned}
[dU(\rho)/d\rho]_{\rho=\langle\phi\rangle} &= 0 \quad &\text{(condition for } \langle\phi\rangle\text{)},\\
[d^2 U(\rho)/d^2\rho]_{\rho=\langle\phi\rangle} &= m^2 \quad &\text{(renormalized mass)},\\
-[d^4 U(\rho)/d^4\rho]_{\rho=\langle\phi\rangle} &= \alpha \quad &\text{(renormalized coupling constant)}.
\end{aligned} \tag{10.24}$$

The quantities m^2 and α are free parameters of the renormalized theory. Since we must require $m^2 \geq 0$, the first two conditions say that ϕ is at a minimum of $U(\rho)$, just as ρ_0 is at a minimum of $V(\rho)$. We shall proceed to calculate $U(\rho)$ explicitly in the one-loop approximation.

10.4 The Loop Expansion[3]

The loop expansion of $W[J]$ to one-loop order may be obtained from (10.3) by the method of saddle-point integration. The saddle-point is at $\phi = \phi_0$, with ϕ_0 satisfying
$$\frac{\delta S[\phi]}{\delta \phi(x)}\bigg|_{\phi=\phi_0} = -J(x). \tag{10.25}$$

Note that $\phi_0(x)$ is a function of x, and is also a functional of J. When $J \to 0$, $\phi_0(x)$ becomes a solution to the classical equations of motion.

---
[3] R. Jackiw, *Phys. Rev.*, D9, 1686 (1974).

Method of Effective Potential

Expanding $S[\phi]$ about ϕ_0, we have

$$S[\phi + \phi_0] = S[\phi_0] - \int d^4x \, \phi(x)\{\delta S[\phi]/\delta\phi(x)\}_{\phi=\phi_0}$$
$$+ \frac{1}{2}\int d^4x \, d^4y \, \phi(x)\phi(y)\{\delta^2 S[\phi]/\delta\phi(x)\,\delta\phi(y)\}_{\phi=\phi_0}$$
$$+ S_2[\phi, \phi_0], \qquad (10.26)$$

where S_2 includes all higher terms. We introduce a propagator function $\Delta[\phi_0]$, which is a functional of ϕ_0, by the definition

$$\langle x|i\Delta^{-1}[\phi_0]|y\rangle \equiv \left.\frac{\delta^2 S[\phi]}{\delta\phi(x)\,\delta\phi(y)}\right|_{\phi=\phi_0}. \qquad (10.27)$$

Then

$$S[\phi + \phi_0] = S[\phi_0] - (J, \phi) + \tfrac{1}{2}(\phi, i\Delta^{-1}[\phi_0]\phi) + S_2[\phi, \phi_0]. \qquad (10.28)$$

Now we write

$$\exp\frac{i}{\hbar} W[J] = \mathcal{N}\int (D\phi)\exp\frac{i}{\hbar}\{S[\phi + \phi_0] + (\phi + \phi_0, J)\}$$
$$= \mathcal{N}\int (D\phi)\exp\frac{i}{\hbar}\{S[\phi_0] + \tfrac{1}{2}(\phi, i\Delta^{-1}[\phi_0]\phi)$$
$$+ S_2[\phi, \phi_0] + (\phi_0, J)\}$$
$$= \exp\frac{i}{\hbar}\{S[\phi_0] + (\phi_0, J)\}\cdot\left\{\mathcal{N}\int (D\phi)\exp\frac{i}{2\hbar}(\phi, i\Delta^{-1}[\phi_0]\phi)\right\}$$
$$\cdot\left\langle\exp\frac{i}{\hbar}S_2\right\rangle, \qquad (10.29)$$

where $\langle\;\rangle$ denotes functional average with weighting function

$$\exp\frac{i}{2\hbar}(\phi, i\Delta^{-1}[\phi_0]\phi).$$

The middle factor within curly brackets in (10.29) is equal to $\{\det i\Delta^{-1}[\phi_0]\}^{-1/2}$. Thus

$$W[J] = S[\phi_0] + (\phi_0, J) + \frac{i\hbar}{2}\ln\det\{i\Delta^{-1}[\phi_0]\} + W_2[J], \qquad (10.30)$$

where

$$W_2[J] = -i\hbar\ln\left\langle\exp\frac{i}{\hbar}S_2\right\rangle. \qquad (10.31)$$

We can show that W_2 is of higher order in $\hbar$ than first. To do this, rescale ϕ by putting $\phi = \hbar^{1/2}\tilde{\phi}$. Then

$$\hbar^{-1}S_2[\phi, \phi_0] = \hbar^{-1}\phi\phi\phi[\delta^2 S/\delta\phi\,\delta\phi\,\delta\phi]_{\phi=\phi_0} + \cdots$$
$$= \hbar^{1/2}\tilde{\phi}\tilde{\phi}\tilde{\phi}[\delta^2 S/\delta\phi\,\delta\phi\,\delta\phi]_{\phi=\phi_0} + \cdots. \tag{10.32}$$

Thus,

$$W_2 = -i\hbar \ln[1 + O(\hbar^{1/2})] \sim O(\hbar^{3/2}). \tag{10.33}$$

In fact $W_2 \sim O(\hbar^2)$, because we know from Sec. 7.7 that $W[J]$ admits a loop expansion in powers of $\hbar$, and there are no fractional powers. Thus the first two terms in (10.30) represent the tree approximation, and the second term is the complete one-loop correction.

Substitution of (10.30) into (10.5) will give the loop expansion of the effective action $\Gamma[\phi_c]$. However, we must first express $S[\phi_0]$ as a functional of ϕ_c. It is easily seen that $\phi_c = \phi_0$ in the tree approximation. Hence $(\phi_c - \phi_0) \sim O(\hbar)$:

$$\phi_c(x) = \frac{\delta W[J]}{\delta J(x)} \equiv \phi_0(x) + \phi_1(x), \quad \phi_1(x) \sim O(\hbar). \tag{10.34}$$

Now we expand $S[\phi_0]$ in the following manner:

$$S[\phi_0] = S[\phi_c - \phi_1]$$
$$= S[\phi_c] - \int d^4x\,\phi_1(x)\{\delta S[\phi]/\delta\phi(x)\}_{\phi=\phi_0} + O(\hbar^2) \tag{10.35}$$
$$= S[\phi_c] + (\phi_1, J) + O(\hbar^2),$$

where in the last step we have used (10.25). Substituting the last relation into (10.30), and the latter into (10.5), we obtain

$$\Gamma[\phi_c] = S[\phi_c] + (\phi_c, J) + \frac{i\hbar}{2}\ln\det\{i\Delta^{-1}[\phi_c]\} + O(\hbar^2). \tag{10.36}$$

Now we put $J = 0$. Then $\phi_c(x) = \rho$, and

$$\Gamma[\rho] = S[\rho] + \frac{i\hbar}{2}\ln\det\{i\Delta^{-1}[\rho]\} + O(\hbar^2), \tag{10.37}$$

where ρ is a constant. We know, on the other hand, that

$$\Gamma[\rho] = -\Omega U(\rho),$$
$$S[\rho] = \left[\int d^4x\,\mathcal{L}(x)\right]_{\phi=\rho} = -\Omega V(\rho). \tag{10.38}$$

Therefore

$$U(\rho) = V(\rho) - \frac{i\hbar}{2}\ln\det\{i\Delta^{-1}[\rho]\} + O(\hbar^2). \tag{10.39}$$

To calculate the second term, we note that
$$\ln \det\{i\Delta^{-1}[\rho]\} = \text{Tr } \ln\{i\Delta^{-1}[\rho]\}$$
$$= \int \frac{d^4k}{(2\pi)^4} \ln\langle k|i\Delta^{-1}[\rho]|k\rangle. \quad (10.40)$$

Hence
$$U(\rho) = V(\rho) - \frac{i\hbar}{2} \int \frac{d^4k}{(2\pi)^4} \Omega^{-1} \ln\langle k|i\Delta^{-1}[\rho]|k\rangle + O(\hbar^2). \quad (10.41)$$

10.5 One-Loop Effective Potential

In the Lagrangian density, put $\phi(x) = \rho + \phi_1(x)$, where ρ is a constant. Then
$$S[\phi] = \int d^4x [\tfrac{1}{2}\partial_\mu \phi_1 \, \partial^\mu \phi_1 - V(\phi_1 + \rho)]$$
$$= -\frac{1}{2} \int d^4x \, \phi_1 [\Box^2 + V''(\rho)] \phi_1 + \cdots, \quad (10.42)$$

where the omitted terms involve higher powers of ϕ_1. By (10.27), we have
$$\langle x|i\Delta^{-1}[\rho]|y\rangle = \frac{\delta^2 S[\phi]}{\delta\phi_1(x)\,\delta\phi_2(y)}\bigg|_{\phi_1=0}$$
$$= -[\Box^2 + V''(\rho)]\,\delta^4(x-y). \quad (10.43)$$

Hence
$$\Omega^{-1}\langle k|i\Delta^{-1}(\rho)|k\rangle = k^2 - V''(\rho). \quad (10.44)$$

Substituting this into (10.41), we obtain
$$U(\rho) = V(\rho) + \frac{\hbar}{2} \int \frac{d^4k_E}{(2\pi)^4} \ln[k_E^2 + V''(\rho)] + O(\hbar^2), \quad (10.45)$$

where we have rotated into Euclidean momentum space. The integral above is divergent. We cut it off at $k_E^2 = \Lambda^2$, getting
$$\int \frac{d^4k_E}{(2\pi)^4} \ln(k_E^2 + V'') = \pi^2 \left[\Lambda^4 \left(\ln \Lambda - \frac{1}{4}\right) + \Lambda^2 V'' \right.$$
$$\left. + \frac{1}{2} (V'')^2 \left(\ln \frac{V''}{\Lambda^2} - \frac{1}{2}\right)\right] + O\left(\frac{1}{\Lambda^2}\right). \quad (10.46)$$

Using this result, we have
$$U(\rho) = V(\rho) + \hbar V_1(\rho) + O(\hbar^2), \quad (10.47)$$

where

$$V_1(\rho) = \frac{\Lambda^2}{32\pi^2} V''(\rho) + \frac{[V''(\rho)]^2}{64\pi^2}\left[-\frac{1}{2} + \ln\frac{V''(\rho)}{\Lambda^2}\right]. \quad (10.48)$$

We have dropped a constant of the order of $\Lambda^4 \ln \Lambda$, but this has no effect on the location of the minimum of $U(\rho)$.

To separate $V_1(\rho)$ into its divergent and convergent parts, we have to introduce an arbitrary scale parameter μ, so that we can write

$$\ln\frac{V''}{\Lambda^2} = \ln\frac{V''}{\mu^2} + \ln\frac{\mu^2}{\Lambda^2}.$$

With this, we have

$$V_1(\rho) = \frac{1}{32\pi^2}\left[\Lambda^2 V'' - \frac{1}{2}(V'')^2 \ln\frac{\Lambda^2}{\mu^2}\right] + \frac{(V'')^2}{64\pi^2}\left(-\frac{1}{2} + \ln\frac{V''}{\mu^2}\right). \quad (10.49)$$

The first term is divergent and the second term is convergent.

10.6 Renormalization

1 General scheme

The divergences in $V_1(\rho)$ have to be eliminated through renormalization. For this to be possible, the divergent terms must have the same forms as those present in the Lagrangian, so that they can be absorbed. Put another way, the theory is renormalizable if the counter terms needed to cancel the divergences are of the same forms as terms present in the Lagrangian.

For a general $V(\rho)$ that is a polynomial of degree n, $V''(\rho)$ is a polynomial of degree $n - 2$, and the divergent part of $V_1(\rho)$ is a polynomial of degree $2n - 4$. Thus, the theory is renormalizable if $2n - 4 \leq n$, or $n \leq 4$. This condition is satisfied by our choice of $V(\rho)$.

To display the counter terms, we rewrite the parameters in $\mathcal{L}(x)$ in the following forms:

$$\begin{aligned}\alpha_0 &= \alpha_1 + \delta\alpha, \\ m_0^2 &= m_1^2 + \delta m^2,\end{aligned} \quad (10.50)$$

where $\delta\alpha$ and δm^2 may be divergent in perturbation theory, but α_1 and m_1^2 are required to be finite parameters. Since $\delta\alpha$ and δm^2 arise only in quantum theory, we assume for the purpose of the loop expansion that

$$\delta\alpha \sim \delta m^2 \sim O(\hbar). \quad (10.51)$$

We now rewrite

$$\mathcal{L}(x) = \frac{1}{2}\partial_\mu\phi\,\partial^\mu\phi + V(\phi) + \frac{\delta m^2}{4}\phi^2 - \frac{\delta\alpha}{4!}\phi^4,$$

$$V(\phi) = \frac{\alpha_1}{4!}\phi^4 - \frac{m_1^2}{4}\phi^2. \quad (10.52)$$

In the corresponding classical theory, the vacuum value of the field is given by
$$\rho_0 = m_1(3/\alpha_1)^{1/2}. \tag{10.53}$$

The last two terms in $\mathcal{L}(x)$ are the counter terms[a]. Since they are of order $\hbar$, they may be simply added to the one-loop effective potential. Thus we have

$$U(\rho) = V(\rho) + \hbar V_1(\rho) + \frac{\delta\alpha}{4!}\rho^4 - \frac{\delta m^2}{4}\rho^2. \tag{10.54}$$

We choose $\delta\alpha$ and δm^2 so as to cancel the divergent part of $V_1(\rho)$:

$$\delta\alpha = \frac{3\alpha_1^2\hbar}{32\pi^2}\ln\frac{\Lambda^2}{\mu^2},$$
$$\delta m^2 = \frac{\alpha_1\hbar}{32\pi^2}\left(\Lambda^2 + \frac{m_1^2}{2}\ln\frac{\Lambda^2}{\mu^2}\right). \tag{10.55}$$

The right side of these equations are of course ambiguous up to additive finite terms, but they can always be absorbed into the scale parameter μ. Thus we have

$$U(\rho) = V(\rho) + \frac{\hbar}{64\pi^2}[V''(\rho)]^2\left[-\frac{1}{2} + \ln\frac{V''(\rho)}{\mu^2}\right]. \tag{10.56}$$

The minimum occurs at $\rho = \langle\phi\rangle$, where
$$U'(\langle\phi\rangle) = 0. \tag{10.57}$$

We write
$$\langle\phi\rangle = \rho_0 + \rho_1, \tag{10.58}$$

where ρ_0 is defined by (10.53). The renormalized mass m and the renormalized coupling constant α are defined respectively by

$$m^2 \equiv U''(\rho_0 + \rho_1),$$
$$\alpha \equiv U''''(\rho_0 + \rho_1). \tag{10.59}$$

For reference we record formulas for the derivatives of U:

$$\begin{aligned}
U' &= V' + kV''V''' \ln(V''/\mu^2), \\
U'' &= V'' + k\{[(V''')^2 + \alpha_1 V''] \ln(V''/\mu^2) + (V''')^2\}, \\
U''' &= V''' + k\{3\alpha_1 V''' \ln(V''/\mu^2) + [(V''')^3/V''] + 3\alpha_1 V'''\}, \\
U'''' &= \alpha_1 + \alpha_1^2\{3\ln(V''/\mu^2) + (6\alpha_1\rho^2/V'') - [\alpha_1^2\rho^4/(V'')^2] + 3\},
\end{aligned} \tag{10.60}$$

where
$$k \equiv \hbar/32\pi^2, \tag{10.61}$$

---

[a] We have left out wave function renormalization, whereby the kinetic term is replaced by $\tfrac{1}{2}Z\partial_\mu\phi\partial^\mu\phi$. This provides a counter term needed to render finite the overall normalization of the full propagator. The graphs to be cancelled by the counter term come from the momentum-dependent part of the proper self-energy, and contain at least two closed loops. Hence $Z = 1 + O(\hbar^2)$.

and

$$V' = \frac{\alpha_1}{6}\rho^3 - \frac{m_1^2}{2}\rho,$$

$$V'' = \frac{\alpha_1}{2}\rho^2 - \frac{m_1^2}{2}, \tag{10.62}$$

$$V''' = \alpha_1\rho,$$

$$V'''' = \alpha_1.$$

2 Massive case

Suppose $m_1 > 0$. Then we may assume $\rho_1 \ll \rho_0$. Keeping ρ_1/ρ_0 only to first order, we obtain (setting $\hbar = 1$)

$$\frac{\rho_1}{\rho_0} = -\frac{\alpha_1}{32\pi^2}\ln\frac{m_1^2}{\mu^2} + O(\alpha_1^2). \tag{10.63}$$

Using this, we obtain

$$\langle\phi\rangle = m_1\left(\frac{3}{\alpha_1}\right)^{1/2}\left[1 - \frac{\alpha_1}{32\pi^2}\ln\frac{m_1^2}{\mu^2} + O(\alpha_1^2)\right],$$

$$m^2 = m_1^2\left[1 + \frac{3\alpha_1}{32\pi^2}\left(1 + \frac{1}{3}\ln\frac{m_1^2}{\mu^2}\right) + O(\alpha_1^2)\right], \tag{10.64}$$

$$\alpha = \alpha_1\left[1 + \frac{3\alpha_1}{32\pi^2}\left(1 + \ln\frac{m_1^2}{\mu^2}\right) + O(\alpha_1^2)\right].$$

3 Massless case

For the case $m_1 = 0$, there is no intrinsic mass scale, and the coupling constant must be defined at a floating renormalization point μ. From (10.60) we find

$$U''''(\rho) = \alpha + \frac{\alpha^2\hbar}{64\pi^2}\left(11 + 3\ln\frac{\alpha\rho^2}{2\mu^2}\right). \tag{10.65}$$

Let us define a new scale parameter μ_0 by

$$U''''(\mu_0) \equiv \alpha. \tag{10.66}$$

Its value is given by

$$\ln\frac{\alpha\mu_0^2}{2\mu^2} = -\frac{11}{3}. \tag{10.67}$$

We now define the running coupling constant as

$$\alpha(\rho) \equiv U''''(\rho) = \alpha + \frac{3\hbar\alpha^2}{32\pi^2}\ln\frac{\rho^2}{\mu_0^2}. \tag{10.68}$$

Method of Effective Potential

Thus, the parameter α is the value of the running coupling constant at $\rho = \mu_0$. Rewriting (10.68) in the form

$$\frac{1}{\alpha(\rho)} = \frac{1}{\alpha(\mu_0)} - \frac{3}{32\pi^2} \ln \frac{\rho^2}{\mu_0^2}, \tag{10.69}$$

we recognize that this is just (9.141), re-derived by a different method.

The really interesting question in the massless case is whether spontaneous symmetry breaking can occur. In other words, can $\langle\phi\rangle$ be non-zero, purely due to radiative corrections? If the answer is yes, we would have a model of dynamical mass generation, and that would be physically interesting.

From (10.60) we have

$$U'(\rho) = \frac{\rho^4}{4!} \left[\alpha + \frac{3\hbar\alpha^2}{32\pi^2} \left(\ln \frac{\rho^2}{\mu^2} + \text{const.} \right) \right]. \tag{10.70}$$

It appears that there is a root $\langle\phi\rangle$, satisfying

$$\hbar\alpha \ln \frac{\langle\phi\rangle^2}{\mu^2} = -\frac{32\pi^2}{3}. \tag{10.71}$$

However, this requires that a quantity of order $\hbar$ in the loop expansion be equated with a quantity of order 1. Hence this root must be rejected, since it lies beyond the validity of the approximation used.

The technical reason for the failure of (10.71) is that the two terms of (10.70) are of order α and α^2 respectively, and thus cannot cancel each other within the region of validity of the formula. However, if we could modify the model in such a manner that α and α^2 are replaced by two independent coupling constants, then we might have a cancellation, which would then lead to spontaneous symmetry breaking by radiative corrections. Such a model can indeed be found, the simplest example being "massless" scalar electrodynamics.

10.7 Dimensional Transmutation

Coleman and E. Weinberg[4] showed that, in the renormalized version of "massless" scalar electrodynamics, the scalar field has an *arbitrary* vacuum expectation value $\langle\phi\rangle$, even though classically $\langle\phi\rangle = 0$. The consequences are dramatic: $\langle\phi\rangle$ emerges as a spontaneous generated physical mass scale, and both the scalar and the vector particle develop dynamic masses proportional to $\langle\phi\rangle$. Thus, there is actually no such thing as massless scalar electrodynamics. They named this phenomenon "dimensional transmutation".

We shall merely give a sketch of the derivation of the Coleman-Weinberg result. The Lagrangian density is

$$\mathcal{L} = -\frac{1}{4} F^{\mu\nu}F_{\mu\nu} + [(\partial_\mu - ieA_\mu)\phi^*][(\partial^\mu + ieA^\mu)\phi] - \frac{\lambda_0}{6}(\phi^*\phi)^2. \tag{10.72}$$

---

[4] S. Coleman and E. Weinberg, *Phys. Rev.*, D7, 1888 (1973).

The self-coupling λ_0 is necessary to renormalize the scalar-scalar scattering amplitude. We shall immediately transform away the phase of ϕ by going to unitary gauge, and discard the phase as being irrelevant to our purpose. Thus we work with the Lagrangian density

$$\mathscr{L} = -\frac{1}{4} F^{\mu\nu} F_{\mu\nu} + \frac{1}{2} e^2 A_\mu A^\mu \phi^2 + \frac{1}{2} \partial_\mu \phi \, \partial^\mu \phi - \frac{\lambda_0}{4!} \phi^4, \quad (10.73)$$

where we have rescaled ϕ by a factor of $2^{-1/2}$, in order to facilitate comparison with (10.1). The vector field is subject to the subsidiary condition

$$\partial_\mu A^\mu = 0. \quad (10.74)$$

We take e to be a renormalized coupling constant, ignoring counter terms for its renormalization. The coupling constant λ_0 is a bare coupling constant and is to be renormalized.

The action of the theory is

$$S[A, \phi] = \int d^4x \left[\frac{1}{2} A^\mu (\Box^2 + e^2 \phi^2) A_\mu - \frac{1}{2} \phi \Box^2 \phi - \frac{\lambda_0}{4!} \phi^4 \right]. \quad (10.75)$$

We shall integrate out the vector fields in the path integral of the theory, to obtain an action for ϕ alone:

$$\exp iS[\phi] = \mathcal{N} \int (DA) \, \delta[\partial_\mu A^\mu] \exp i S[A, \phi]. \quad (10.76)$$

Each component of the vector field integrated out contributes a factor

$$[\det(\Box^2 + e^2 \phi^2)]^{-1/2} = \exp[-\tfrac{1}{2} \ln \det(\Box^2 + e^2 \phi^2)] \\
= \exp[-\tfrac{1}{2} \operatorname{Tr} \ln(\Box^2 + e^2 \phi^2)]. \quad (10.77)$$

Since there are three independent components, we have

$$S[\phi] = -\int d^4x \left(\frac{1}{2} \phi \Box^2 \phi + \frac{\lambda_0}{4!} \phi^4 \right) - \frac{3}{2} \operatorname{Tr} \ln(\Box^2 + e^2 \phi^2). \quad (10.78)$$

Thus, the effective potential for ϕ, to lowest order in λ_0, is given by

$$U(\rho) = \frac{\lambda_0}{4!} \rho^4 + \frac{3}{2} \int \frac{d^4 k_E}{(2\pi)^4} \ln(k_E^2 + e^2 \rho^2). \quad (10.79)$$

The second term corresponds to a sum of all one-loop graphs with external ϕ lines, with only closed vector boson loops. The scalar loops are regarded as higher order effects.

We can now take over the development following (10.45). The renormalized effective potential (up to an additive constant) can be read off (10.56):

$$U(\rho) = \rho^4 \left[\frac{\lambda}{4!} + \frac{3e^4}{64\pi^2} \left(-\frac{1}{2} + \ln \frac{e^2 \rho^2}{\mu^2} \right) \right], \quad (10.80)$$

Method of Effective Potential

where λ is the running coupling constant at the renormalization point μ. Even though the e^4 term is a one-loop result, while the λ term is a zero-loop result, the two terms can be comparable, because λ and e are independent coupling constants.

Differentiating (10.80), we have

$$U'(\rho) = 4\rho^3 \left(\frac{\lambda}{4!} + \frac{3e^4}{64\pi^2} \ln \frac{e^2\rho^2}{\mu^2} \right), \tag{10.81}$$

which has a non-zero root $\langle\phi\rangle$, satisfying

$$\frac{3e^4}{64\pi^2} \ln \frac{e^2\langle\phi\rangle}{\mu^2} = -\frac{\lambda}{4!}. \tag{10.82}$$

This relation holds within the validity of our approximation. We can use it to eliminate μ from the theory. The mass scale is then set by $\langle\phi\rangle$. Like any renormalized parameter, $\langle\phi\rangle$ is a free parameter of the theory. The effective potential can now be written as

$$U(\rho) = \frac{3e^4\rho^4}{64\pi^2} \left(-\frac{1}{2} + \ln \frac{\rho^2}{\langle\phi\rangle^2} \right). \tag{10.83}$$

The mass of the scalar particle is then given by

$$m_s^2 \equiv U''(\langle\phi\rangle) = \frac{3e^4}{8\pi^2} \langle\phi\rangle^2. \tag{10.84}$$

The vector field develops mass through the Higgs mechanism in the usual way:

$$m_V^2 = e^2 \langle\phi\rangle^2. \tag{10.85}$$

There is nothing unusual about the emergence of a mass scale in renormalization, even when the theory has no intrinsic mass parameter. As we know, in order for the unrenormalized theory to make sense, we have to introduce a cutoff. The cutoff momentum represents a hidden scale of the theory. Upon renormalization, this hidden scale is replaced by an arbitrary finite scale, in the form of a floating renormalization point. In a "massless" theory, one might expect that all renormalization points are equivalent. What is remarkable here is that there is a preferred renormalization point, by virtue of the fact that the particles develop mass dynamically. The term "dimensional transmutation" refers to the emergence of a *preferred* renormalization point in a theory without intrinsic mass scale: the infinite cutoff momentum has been "transmuted" into physical mass.

It is widely believed that dimensional transmutation takes place in quantum chromodynamics with massless quarks. The reasoning is as follows. The "massless" quantum chromodynamics should give a fair account of the observed mass spectrum of the ordinary hadrons, which are bound states of the nearly massless quarks u and d. Since there is no intrinsic mass scale in that theory, dimensional transmutation must take place.

10.8 A Non-Relativistic Example

An instructive example of dimensional transmutation in non-relativistic quantum mechanics is given by Thorn[5]. Consider the two-dimensional Schrödinger equation with an attractive δ-function potential:

$$[-\nabla^2 - \lambda_0\, \delta^2(\mathbf{x})]\psi(\mathbf{x}) = E\psi(\mathbf{x}). \tag{10.86}$$

By dimensional analysis, λ_0 is dimensionless. Thus the Hamiltonian does not contain an intrinsic energy scale. Nevertheless, it is possible for the system to have a bound state. To see this, let $\phi(\mathbf{k})$ be the Fourier transform of $\psi(\mathbf{x})$:

$$\psi(\mathbf{x}) = \int \frac{d^2k}{(2\pi)^2}\, e^{i\mathbf{k}\cdot\mathbf{x}} \phi(\mathbf{k}). \tag{10.87}$$

By Fourier analyzing both sides of (10.86), we have

$$(k^2 + B)\phi(\mathbf{k}) = \lambda_0 \psi(0), \quad B \equiv -E. \tag{10.88}$$

The solution is

$$\phi(\mathbf{k}) = \frac{\lambda_0 \psi(0)}{k^2 + B}. \tag{10.89}$$

Integrating both sides over $\mathbf{k}$, we obtain the eigenvalue condition for the binding energy B:

$$1 = \frac{\lambda_0}{4\pi^2} \int \frac{d^2k}{k^2 + B}. \tag{10.90}$$

The integral on the right-hand side is divergent. Introducing a cutoff at $|\mathbf{k}| = \Lambda$, we obtain

$$1 = \frac{\lambda_0}{4\pi} \ln\left(\frac{\Lambda^2}{B} + 1\right). \tag{10.91}$$

Thus, for large Λ, we have

$$B = \Lambda^2\, e^{-4\pi/\lambda_0}. \tag{10.92}$$

If λ_0 is considered to be an unrenormalized coupling constant, then we can demand that it depends on Λ in such a way that B remains finite as $\Lambda \to \infty$. The infinite cutoff is hereby transmuted into an *arbitrary* binding energy B.

The binding energy now fixes the energy scale of the renormalized system. Consider, for example, the scattering solution

$$\phi(\mathbf{k}) = 4\pi^2\, \delta^2(\mathbf{k} - \mathbf{k}_0) + \frac{\lambda_0}{4\pi^2} \frac{\psi(0)}{k^2 - k_0^2 + i\varepsilon}. \tag{10.93}$$

---

[5] C. Thorn, *Phys. Rev.*, **D6**, 39 (1979), Eq. (4.11).

Method of Effective Potential

Integrating both sides over **k**, we have the condition

$$\psi(0) = 1 + \frac{\lambda_0 \psi(0)}{4\pi^2} \int \frac{d^2k}{k^2 - k_0^2 + i\varepsilon}. \tag{10.94}$$

Again introducing the cutoff Λ, we find

$$\psi(0) = -\frac{\ln \Lambda^2}{\ln(-k_0^2 + i\varepsilon)}. \tag{10.95}$$

Eliminating Λ with the help of (10.92), we obtain

$$\lambda_0 \psi(0) = -4\pi \left[\ln \frac{-k_0^2 + i\varepsilon}{B}\right]^{-1}. \tag{10.96}$$

Thus, the renormalized wave function is

$$\phi(\mathbf{k}) = 4\pi^2 \delta^2(\mathbf{k} - \mathbf{k}_0) - \frac{1}{\pi}\left[(k^2 - k_0^2 + i\varepsilon)\ln\left(\frac{-k_0^2 + i\varepsilon}{B}\right)\right]^{-1}, \tag{10.97}$$

from which we can obtain the total scattering cross section at energy $E = k_0^2$:

$$\sigma_{tot}(E) = \frac{64}{\pi\sqrt{E}}\left[\left(\ln \frac{E}{B}\right)^2 + \pi^2\right]^{-1}. \tag{10.98}$$

Naive dimensional analysis would lead us to expect $\sigma_{tot} \propto E^{-1/2}$; but actually $\sigma_{tot} \propto E^{-1/2} (\ln E)^{-2}$ at high energies.

The nice things about this model is that we can understand the meaning of the cutoff Λ in a physical way. Let us consider the δ-function potential to be the limit of a suitably chosen square well:

$$[-\nabla^2 + V(r)]\psi(\mathbf{x}) = E\psi(\mathbf{x}) \tag{10.99}$$

where $r = |\mathbf{x}|$, and

$$V(r) = \begin{cases} -\lambda_0/\pi a^2 & (r < a), \\ 0 & (r > a). \end{cases} \tag{10.100}$$

There is a bound state in the square well, with binding energy

$$B = \frac{4}{\gamma^2 a^2} e^{-4\pi/\lambda_0} \quad (\ln \gamma = 0.5772 \cdots). \tag{10.101}$$

To maintain a bound state at fixed binding energy B as $a \to 0$, the well depth must increase according to

$$\frac{\lambda_0}{\pi a^2} = \frac{4}{a^2}\left(\ln \frac{4}{\gamma^2 a^2 B}\right)^{-1}. \tag{10.102}$$

Thus, the parameter λ_0 must approach zero, so that the well depth increases more slowly than a^{-2}. Otherwise, the binding energy would diverge, and the Hamiltonian would not be bounded from below. The cutoff Λ in the previous

10.9 Application to Weinberg-Salam Model

If the Higgs field in the Weinberg-Salam model is taken to be a dynamical field, then the Higgs potential will have radiative corrections. Write the Higgs potential in the Weinberg-Salam Lagrangian density in the form

$$V(\rho) = \frac{\lambda_0}{4!} \rho^4 + \frac{m_0^2}{2} \rho^2 + \text{const.}, \tag{10.104}$$

where the parameter m_0^2 is allowed to have either sign. [We are in unitary gauge, and the field ρ is $\sqrt{2}$ times that appearing in (6.32)]. We shall include radiative corrections coming from all one-loop graphs containing only gauge vector boson loops. The effective potential can be obtained immediately from (10.80):

$$U(\rho) = \frac{\lambda}{4!} \rho^4 + \frac{m^2}{2} \rho^2 + \frac{3\rho^4}{64\pi^2} \sum_V e_V^4 \left(-\frac{1}{2} + \ln \frac{e_V^2 \rho^2}{\mu^2} \right), \tag{10.105}$$

where the sum over V extends over all vector bosons coupled to the Higgs field, namely, W^+, W^-, and Z. Each vector boson develops mass through the Higgs mechanism according to

$$m_V^2 = e_V^2 \langle \phi \rangle^2. \tag{10.106}$$

This relation can be used to eliminate the coupling constant e_V. We can also re-define μ to absorb all ρ^4 terms in the effective potential. Thus we write

$$U(\rho) = \frac{m^2}{2} \rho^2 + K\rho^4 \ln \frac{\rho^2}{\mu^2},$$

$$K = \frac{3}{64\pi^2 \langle \phi \rangle^4} (2m_W^4 + m_Z^4). \tag{10.107}$$

The derivatives of $U(\rho)$ are given by

$$U'(\rho) = m^2 \rho + 4B\rho^3 \left(\frac{1}{2} + \ln \frac{\rho^2}{\mu^2} \right),$$

$$U''(\rho) = m^2 + 12B\rho^2 \left(\frac{7}{6} + \ln \frac{\rho^2}{\mu^2} \right). \tag{10.108}$$

If radiative corrections are ignored, we must take $m^2 < 0$ in order to have spontaneous symmetry breaking. When radiative corrections are taken into account, spontaneous symmetry breaking occurs even if $m^2 = 0$, and the Higgs field mass is given by the generalization of (10.84). Now, imagine that we

Method of Effective Potential

increase m^2 from zero. For sufficiently small positive m^2, we would still have spontaneous symmetry breaking, although the Higgs mass should become smaller. If we keep increasing m^2, there will come a point when spontaneous symmetry breaking disappears, and at this point the Higgs mass will have its smallest possible value. The qualitative features of $U(\rho)$ are shown in Fig. 10.2, for different values of m^2.

From (10.108) we find that $U'(\rho)$ has a non-zero root at $\langle\phi\rangle$, satisfying

$$\langle\phi\rangle^2\left(\frac{1}{2} + \ln\frac{\langle\phi\rangle^2}{\mu^2}\right) = -\frac{m^2}{4K}. \tag{10.109}$$

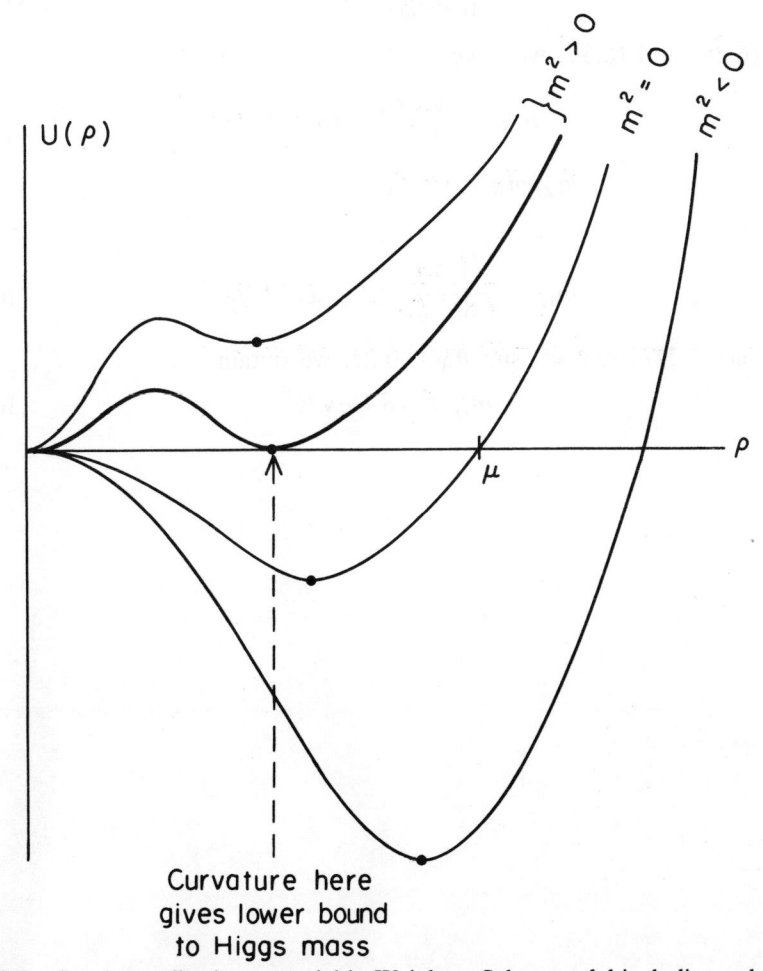

Fig. 10.2 One-loop effective potential in Weinberg-Salam model including only effects of closed gauge boson loops. The potentials are shown for various choices of the mass parameter m^2 in the potential term in the Lagrangian.

At this root we have

$$U(\langle\phi\rangle) = -K\langle\phi\rangle^4\left(1 + \ln\frac{\langle\phi\rangle^2}{\mu^2}\right),$$

$$m_H^2 \equiv U''(\langle\phi\rangle) = 8K\langle\phi\rangle^2\left(\frac{3}{2} + \ln\frac{\langle\phi\rangle^2}{\mu^2}\right), \tag{10.110}$$

where m_H is the mass of the Higgs boson. It is clear that $\langle\phi\rangle$ will cease to be the lowest minimum if $U(\langle\phi\rangle)$ becomes positive. Hence we must have $\ln\langle\phi\rangle^2/\mu^2 \leq -1$. This gives a theoretical lower bound to the Higgs mass:[6]

$$m_H^2 \geq \frac{3}{16\pi^2\langle\phi\rangle^2}(2m_W^4 + m_Z^4). \tag{10.111}$$

From (6.28) and (6.36) we have

$$m_W^2 = \frac{\pi\alpha\langle\phi\rangle^2}{\sin^2\theta_W} \quad (\alpha \cong 1/137),$$

$$m_Z/m_W = \sin\theta_W. \tag{10.112}$$

Hence

$$m_H \geq \frac{\sqrt{3}\,\alpha\langle\phi\rangle}{4\sin^2\theta_W}(2 + \sec^4\theta_W)^{1/2}. \tag{10.113}$$

Using $\langle\phi\rangle = 247$ GeV/c², $\sin^2\theta_W = 0.22$, we obtain

$$m_H \geq 6\cdot 8 \text{ GeV}/c^2. \tag{10.114}$$

---

[6] S. Weinberg, *Phys. Rev. Lett.*, **36**, 294 (1976).

CHAPTER XI

THE AXIAL ANOMALY

11.1 Origin of the Axial Anomaly

We are familiar with the fact that, for a free Dirac particle, the left and right chiral states L and R decouple from each other in the massless limit [see (6.5)]. Since the electromagnetic interaction does not couple L to R [see (6.3)], one might expect that such a decoupling also takes place for a charged Dirac particle. This is of course true if one starts with a massless theory; but if one starts with a massive theory, then L and R remain coupled to each other in the massless limit[1]. The reason is that, even in the massless limit, a charged Dirac particle of given *helicity* can make a transition into a virtual state of the opposite helicity, by emitting a real photon. This effect, which we shall discuss in detail later, is the physical origin of the axial anomaly.

The masslessness of a Dirac field theory is formally expressed by the invariance of the Lagrangian density under a chiral transformation

$$\psi(x) \to e^{-i\omega\gamma_5}\psi(x). \tag{11.1}$$

According to Noether's theorem, chiral invariance implies the existence of a conserved axial-vector current, usually taken to be the gauge-invariant chiral current[a]

$$j_5^\mu = \bar\psi\gamma^\mu\gamma_5\psi. \tag{11.2}$$

Using the equations of motion for Heisenberg fields, one can derive the formal operator identity

$$\partial_\mu j_5^\mu = 2mj_5, \tag{11.3}$$

where m is the mass, and j_5 is the chiral density:

$$j_5 = i\bar\psi\gamma_5\psi. \tag{11.4}$$

---

[a] Definitions of currents do not include the factors representing the unrenormalized coupling constants. We remind the reader that $\gamma_5 = -i\gamma^0\gamma^1\gamma^2\gamma^3$, and $\varepsilon^{0123} = -\varepsilon_{0123} = 1$. These differ by a sign from the convention used by Adler (ref. 2).

[1] T. D. Lee and M. Nauenberg, *Phys. Rev.*, **133B**, 1549 (1964). These authors also point out that massless spinor electrodynamics is a pathological theory: the S-matrix elements of the theory are infinite, due to "mass singularities". To cancel these singularities, one has to define the S-matrix elements with respect to appropriately chosen statistical ensembles of degenerate states (rather than pure states). For this reason, a massless Dirac theory should always be considered to be the limit of a massive theory.

Thus, one expects $\partial_\mu j_5^\mu \to 0$ when $m \to 0$. However, this is false. As we shall show later, a more careful analysis gives instead the "anomalous" result

$$\partial_\mu j_5^\mu = 2mj_5 + \frac{\alpha_0}{2\pi} \widetilde{F}^{\mu\nu} F_{\mu\nu}, \qquad (11.5)$$

where $\widetilde{F}^{\mu\nu} = \frac{1}{2}\varepsilon^{\mu\nu\alpha\beta} F_{\alpha\beta}$, and α_0 is the unrenormalized fine-structure constant. [The operators in (11.5) are unrenormalized field operators]. Thus, j_5^μ is not conserved in the massless limit. This fact does not contradict Noether's theorem, for $\widetilde{F}^{\mu\nu} F_{\mu\nu}$ is the 4-divergence of a vector, and hence it is possible to define a new axial vector current that is conserved. Unfortunately, the new current is not gauge invariant, and hence cannot be coupled to physical fields. (We expand on this point later). The last term in (11.5) is the axial anomaly. Its presence reflects a conflict between gauge invariance and chiral invariance.

Why should there be a difference between the results of formal and "careful" reasoning? In our discussion of vacuum polarization in Chap. IX, we learned that currents like $j^\mu = \bar{\psi}\gamma^\mu\psi$ and $j_5^\mu = \bar{\psi}\gamma^\mu\gamma_5\psi$ are singular operators, due to the fact that ψ is coupled to its canonical conjugate $\psi^\dagger$ at the same space-time point. As a consequence, formal statements about currents may be ambiguous. For example, through formal use of the equations of motion, one could seemingly show that all matrix elements of $\partial_\mu j^\mu$ vanish; but what one really gets is an indeterminate quantity like $\infty - \infty$. To obtain unambiguous results, the theory must be made well-defined, through extra requirements (such as the condition $k_\mu \Pi^{\mu\nu} = 0$ for the vacuum polarization tensor). The conflict between gauge invariance and chiral invariance arises from the fact that no physically acceptable requirement can be found that would maintain $\partial_\mu j^\mu = 0$ and $\partial_\mu j_5^\mu = 0$ simultaneously. The axial anomaly arises when one insists upon gauge invariance, through the definition $\partial_\mu j^\mu \equiv 0$. The justification for the latter ultimately rests with experiments.

The manipulation of singular operators is a delicate matter. Special tools have been devised for that purpose, including such items as "point splitting method", "Schwinger term", "seagull term", and "T* product". We avoid their use to keep things simple, and refer the interested reader to the reviews by Adler[2] and Jackiw.[3]

11.2 The Triangle Graph

The most elementary manifestation of the axial anomaly is the triangle graph: a closed fermion loop with one axial-vector vertex and two vector vertices, as shown in Fig. 11.1. The two graphs there differ only in the labelling of the external photons, and are collectively referred to as "the triangle graph". Adler[4] and Bell and Jackiw[5] were the first to discuss the anomaly arising therefrom, hence also the name "Adler-Bell-Jackiw anomaly".

---

[2] S. Adler, in *Lectures on Elementary Particles and Quantum Field Theory*, eds. S. Deser, M. Grisaru, and H. Pendleton (MIT Press, Cambridge, 1970).
[3] R. Jackiw, in S. B. Treiman, R. Jackiw, and D. J. Gross, *Lectures on Current Algebra and Its Applications* (Princeton University Press, Princeton, 1972).
[4] S. Alder, *Phys. Rev.*, **177**, 2426 (1969).
[5] J. S. Bell and R. Jackiw, *N. Cimento*, **60A**, 47 (1969).

The Axial Anomaly

The triangle graph can occur in different physical theories. In quantum electrodynamics, it is the lowest-order description of the creation of two photons by an external axial-vector source. In the Weinberg-Salam model, it gives the lowest-order amplitude for the virtual process $Z \to 2\gamma$, when one sums over all possible internal fermion loops. (The amplitude of the corresponding real process vanishes, because a spin 1 state cannot decay into two real photons[6]). In both cases, the graph is given by the same mathematical expression. The difference lies in the values of the coupling constants, which give an overall multiplicative factor, and the external propagators or wave functions that one chooses to attach to the graph.

Omitting all coupling constants and external propagators, we denote the triangle graph by

$$t_{\alpha\beta\mu}(k_1, k_2) = s_{\alpha\beta\mu}(k_1, k_2) + s_{\beta\alpha\mu}(k_2, k_1), \tag{11.6}$$

where

$$s_{\alpha\beta\mu}(k_1, k_2)$$
$$= -i \int \frac{d^4p}{(2\pi)^4} \, \text{Tr}\left(\frac{1}{\slashed{p} + \slashed{k}_1 - m} \gamma_\alpha \frac{1}{\slashed{p} - m} \gamma_\beta \frac{1}{\slashed{p} - \slashed{k}_2 - m} \gamma_\mu \gamma_5 \right). \tag{11.7}$$

This is a linearly divergent integral; but, as we shall see later, the sum of the two terms in (11.6) is convergent.

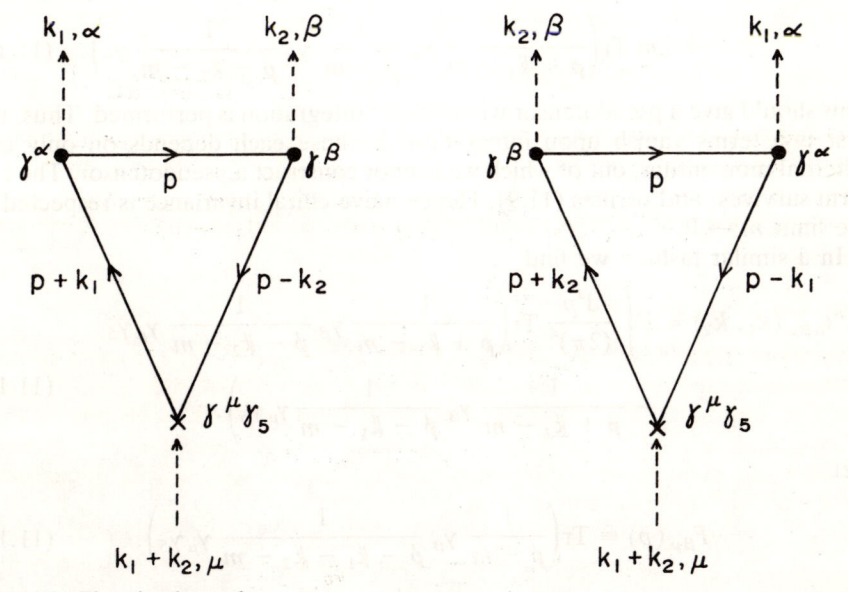

Fig. 11.1 The triangle graph

---

[6] C. N. Yang. *Phys. Rev.*, **77**, 242 (1950); L. D. Landau, *Dokl. Akad. Nauk.* (USSR) **60**, 207 (1948).

Gauge invariance requires that
$$k_1{}^\alpha t_{\alpha\beta\mu}(k_1, k_2) = k_2{}^\beta t_{\alpha\beta\mu}(k_1, k_2) = 0. \tag{11.8}$$
The naive statement (11.3) leads us to expect
$$(k_1 + k_2)^\mu t_{\alpha\beta\mu}(k_1, k_2) = 2m v_{\alpha\beta}(k_1, k_2), \tag{11.9}$$
where
$$v_{\alpha\beta}(k_1, k_2) = \int \frac{d^4p}{(2\pi)^4} \operatorname{Tr}\left(\frac{1}{\not{p} + \not{k}_1 - m} \gamma_\alpha \frac{1}{\not{p} - m} \gamma_\beta \frac{1}{\not{p} - \not{k}_2 - m} \gamma_5\right)$$
$$+ (k_1 \rightleftarrows k_2, \alpha \rightleftarrows \beta). \tag{11.10}$$

We shall now check these statements.

Contracting (11.7) with $(k_1 + k_2)^\mu$, we have
$$(k_1 + k_2)^\mu s_{\alpha\beta\mu}(k_1, k_2)$$
$$= \int \frac{d^4p}{(2\pi)^4} \operatorname{Tr}\left[\frac{1}{\not{p} + \not{k}_1 - m} \gamma_\alpha \frac{1}{\not{p} - m} \gamma_\beta \frac{1}{\not{p} - \not{k}_2 - m} (\not{k}_1 + \not{k}_2)\gamma_5\right]. \tag{11.11}$$

The trace above can be rewritten as
$$\operatorname{Tr}\left(\frac{1}{\not{p} + \not{k}_1 - m} \gamma_\alpha \frac{1}{\not{p} - m} \gamma_\beta \gamma_5\right) + \operatorname{Tr}\left(\gamma_\alpha \frac{1}{\not{p} - m} \gamma_\beta \frac{1}{\not{p} - \not{k}_2 - m} \gamma_5\right)$$
$$+ 2m \operatorname{Tr}\left(\frac{1}{\not{p} + \not{k}_1 - m} \gamma_\alpha \frac{1}{\not{p} - m} \gamma_\beta \frac{1}{\not{p} - \not{k}_2 - m} \gamma_5\right). \tag{11.12}$$

This should give a pseudotensor when the p^μ integration is performed. Thus, the first two terms vanish upon integration, because each depends on only one external momentum, out of which we cannot construct a pseudotensor. The last term survives, and verifies (11.9). Hence naive chiral invariance is respected in the limit $m \to 0$.

In a similar fashion we find
$$k_1{}^\alpha t_{\alpha\beta\mu}(k_1, k_2) = i \int \frac{d^4p}{(2\pi)^4} \operatorname{Tr}\left(\frac{1}{\not{p} + \not{k}_1 - m} \gamma_\beta \frac{1}{\not{p} - \not{k}_2 - m} \gamma_\mu \gamma_5\right.$$
$$\left. - \frac{1}{\not{p} + \not{k}_2 - m} \gamma_\beta \frac{1}{\not{p} - \not{k}_1 - m} \gamma_\mu \gamma_5\right). \tag{11.13}$$

Let
$$F_{\beta\mu}(p) \equiv \operatorname{Tr}\left(\frac{1}{\not{p} - m} \gamma_\beta \frac{1}{\not{p} - \not{k}_1 - \not{k}_2 - m} \gamma_\mu \gamma_5\right). \tag{11.14}$$

Then
$$k_1{}^\alpha t_{\alpha\beta\mu}(k_1, k_2) = i \int \frac{d^4p}{(2\pi)^4} \left[F_{\beta\mu}(p + k_1) - F_{\beta\mu}(p + k_2)\right]. \tag{11.15}$$

The Axial Anomaly

This would vanish, if the two terms on the right-hand side were at worst logarithmically divergent, for then we could make independent shifts of the integration variables. But the two terms are in fact linearly divergent. We note that

$$\int d^4p \, F_{\beta\mu}(p + a) = \int d^4p \left[F_{\beta\mu}(p) + a^\lambda \frac{\partial F_{\beta\mu}(p)}{\partial p^\lambda} + \cdots \right], \quad (11.16)$$

where the omitted terms give vanishing surface integrals; but the second term does not vanish, because asymptotically $F_{\beta\mu} \sim p^{-3}$. Thus,

$$k_1{}^\alpha t_{\alpha\beta\mu}(k_1, k_2) = \frac{i}{(2\pi)^4} (k_2 - k_1)^\lambda \int d^4p \, \frac{\partial F_{\beta\mu}(p)}{\partial p^\lambda}. \quad (11.17)$$

The integral above can be calculated by transforming to Euclidean momentum space:

$$I \equiv a^\lambda b^\beta c^\mu \int d^4p \, \frac{\partial F_{\beta\mu}(p)}{\partial p^\lambda} = i a_E{}^\lambda b_E{}^\beta c_E{}^\mu \int d^4p_E \, \frac{\partial F_{\beta\mu}(p_E)}{\partial p_E{}^\lambda}$$

$$= i a_E{}^\lambda b_E{}^\beta c_E{}^\mu \int dS_E{}^\lambda F_{\beta\mu}(p_E), \quad (11.18)$$

where

$$dS_E{}^\lambda = \frac{p_E{}^\lambda}{p_E} dS_E, \quad (11.19)$$

$$dS_E = p_E{}^3 \, d\Omega,$$

$$\int d\Omega = 2\pi^2.$$

We shall need only the asymptotic form of $F_{\beta\mu}(p_E)$:

$$b_E{}^\beta c_E{}^\mu F_{\beta\mu}(p_E) \xrightarrow[p_E \to \infty]{} \frac{4i}{p_E{}^4} b_E{}^\beta c_E{}^\mu p_E{}^\alpha (k_1 + k_2)_E{}^\nu \varepsilon_{\alpha\beta\mu\nu}. \quad (11.20)$$

Thus,

$$I = -4 a_E{}^\lambda b_E{}^\beta c_E{}^\mu (k_1 + k_2)_E{}^\nu \varepsilon_{\alpha\beta\mu\nu} \int d\Omega \, \frac{p_E{}^\lambda p_E{}^\alpha}{p_E{}^2}. \quad (11.21)$$

Noting that

$$\int d\Omega \, \frac{p_E{}^\lambda p_E{}^\alpha}{p_E{}^2} = \frac{1}{4} \delta_{\lambda\alpha} \int d\Omega = \frac{\pi^2}{2} \delta_{\lambda\alpha},$$

we have

$$I = -2\pi^2 a_E{}^\alpha b_E{}^\beta c_E{}^\mu (k_1 + k_2)_E{}^\nu \varepsilon_{\alpha\beta\mu\nu}. \quad (11.22)$$

Using this in (11.17), and transforming back to Minkowski space, we finally obtain

$$k_1{}^\alpha t_{\alpha\beta\mu}(k_1, k_2) = -\frac{i}{4\pi^2} k_1{}^\alpha k_2{}^\nu \varepsilon_{\alpha\beta\mu\nu}. \quad (11.23)$$

Similarly,

$$k_2^\beta t_{\alpha\beta\mu}(k_1, k_2) = \frac{i}{4\pi^2} k_1^\alpha k_2^\beta \varepsilon_{\alpha\beta\mu\nu}. \tag{11.24}$$

We see that gauge invariance is violated.

As in the case of vacuum polarization, we can enforce gauge invariance by subtracting a suitable polynomial in the external momenta (see Sec. 9.3). It is clear that the following redefined amplitude satisfies the requirements of gauge invariance:

$$T_{\alpha\beta\mu}(k_1, k_2) \equiv t_{\alpha\beta\mu}(k_1, k_2) - \frac{i}{4\pi^2}(k_1 - k_2)^\nu \varepsilon_{\alpha\beta\mu\nu}. \tag{11.25}$$

We take this to be the correct expression of the triangle graph.

With (11.25), and noting (11.9), we find

$$(k_1 + k_2)^\mu T_{\alpha\beta\mu}(k_1, k_2) = 2m v_{\alpha\beta}(k_1, k_2) + \frac{i}{2\pi^2} k_1^\mu k_2^\nu \varepsilon_{\alpha\beta\mu\nu}, \tag{11.26}$$

which does not vanish in the limit $m \to 0$. This shows that chiral invariance is violated by the triangle graph, when gauge invariance is enforced. The last term in (11.26) is the axial anomaly.

Note that $T_{\alpha\beta\mu}$ is symmetric under the interchange of the two external photons (Bose symmetry), and that the anomaly is independent of m.

The triangle graph depends only on two independent momenta k_1, k_2. Let

$$q^\mu \equiv (k_1 + k_2)^\mu. \tag{11.27}$$

By Lorentz invariance and Bose symmetry, $T_{\alpha\beta\mu}$ must be an invariant function times a pseudotensor, which can only be one of the following:

$$\begin{aligned}
&\varepsilon_{\alpha\beta\lambda\nu} k_1^\lambda k_2^\nu q_\mu, \\
&\varepsilon_{\alpha\beta\mu\nu}(k_1 - k_2)^\nu, \\
&(\varepsilon_{\alpha\mu\lambda\sigma} k_{1\beta} - \varepsilon_{\beta\mu\lambda\sigma} k_{2\alpha}) k_1^\lambda k_2^\sigma, \\
&(\varepsilon_{\alpha\mu\lambda\sigma} k_{2\beta} - \varepsilon_{\beta\mu\lambda\sigma} k_{1\alpha}) k_1^\lambda k_2^\sigma.
\end{aligned} \tag{11.28}$$

All but the first are ruled out by gauge invariance, i.e.,

$$k_1^\alpha T_{\alpha\beta\mu} = k_2^\beta T_{\alpha\beta\mu} = 0.$$

Hence

$$T_{\alpha\beta\mu}(k_1, k_2) = i\varepsilon_{\alpha\beta\lambda\nu} k_1^\lambda k_2^\nu q_\mu R(q^2, k_1^2, k_2^2). \tag{11.29}$$

The fact that (11.23) is finite shows, through (11.25), that R is finite.

By similar arguments we can write

$$v_{\alpha\beta}(k_1, k_2) = i\varepsilon_{\alpha\beta\lambda\nu} k_1^\lambda k_2^\nu S(q^2, k_1^2, k_2^2). \tag{11.30}$$

The Axial Anomaly

Thus, (11.26) gives the relation

$$q^2 R(q^2, k_1^2, k_2^2) = 2mS(q^2, k_1^2, k_2^2) + \frac{1}{2\pi^2}. \tag{11.31}$$

In the limit $m \to 0$, R becomes independent of k_1^2, k_2^2:

$$q^2 R(q^2, k_1^2, k_2^2) \xrightarrow[m \to 0]{} \frac{1}{2\pi^2}. \tag{11.32}$$

Therefore

$$T_{\alpha\beta\mu}(k_1, k_2) \xrightarrow[m \to 0]{} \frac{i}{2\pi^2} \varepsilon_{\alpha\beta\lambda\nu} k_1^\lambda k_2^\nu \frac{q_\mu}{q^2 + i\varepsilon}. \tag{11.33}$$

In the massless limit, the entire contribution to the triangle graph comes from the anomaly, which is manifested as a pole at $q^2 = 0$ (the "anomaly pole"). We shall give a physical explanation of this result later. Note that, according to (11.31), the anomaly pole at $q^2 = 0$ is always present, regardless of m. But it is below the physical threshold $q^2 = 4m^2$, and touches the physical region only when $m \to 0$.

11.3 Radiative Corrections

The second-order radiative corrections to the triangle graph are represented by the Feynman graphs shown in Fig. 11.2. Adler and Bardeen[7] have shown that their net contribution to the axial anomaly vanishes. They argue that higher radiative corrections are also absent. This leads to the remarkable result that the

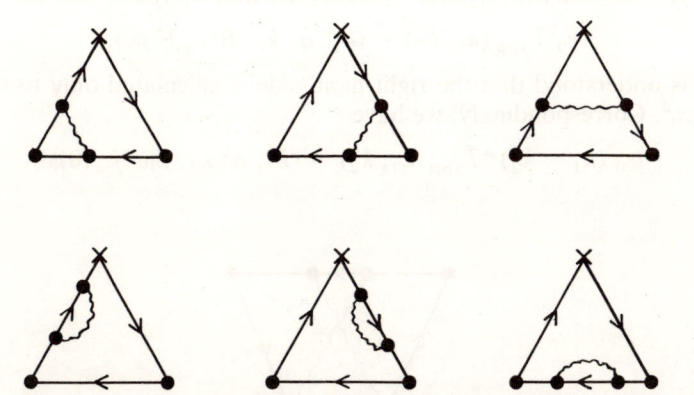

Fig. 11.2 Second-order radiative corrections to the triangle graph. Their anomalies have been shown to cancel one another.

---

[7] S. L. Adler and W. A. Bardeen, *Phys. Rev.*, **182**, 1517 (1969). See also A. Zee, *Phys. Rev. Lett.*, **29**, 1198 (1972).

last term in (11.26) represents the exact anomaly, to all orders of perturbation theory.

The argument for the absence of radiative corrections is as follows. Any radiative correction to the basic triangle graph must contain at least one internal photon line, and two extra vertices on the basic triangular loop. If one regulates the *photon* propagator (thus preserving chiral invariance at the expense of gauge invariance), then one would have a finite integral, there being at least two extra electron propagators along the basic triangular loop to make the integral converge. Therefore, no anomaly occurs, because one could freely shift integration variables. The only graph without internal photon lines is the basic triangle graph, which is solely responsible for the axial anomaly.

In view of the importance of the conclusion, it should be pointed out that the argument is not air-tight, even within the framework of perturbation theory. There is room for escape, because the argument relies on a cut-off procedure that does not respect gauge invariance. It is not clear whether the result is independent of cut-off procedure. The second-order calculation does not really vindicate the argument, because the corrections are special, consisting solely of SE or vertex insertions. Thus, the absence of radiative corrections to second order may be no more than a reflection of the renormalizability of quantum electrodynamics. For a real test, one has to go to fourth order, where one encounters for the first time a new skeleton graph, an example of which is shown in Fig. 11.3. Explicit fourth-order calculations have not been done.

11.4 Anomalous Divergence of the Chiral Current

We can deduce (11.5) from (11.26) as follows. The gauge-invariant amplitude $T_{\alpha\beta\mu}$ may be expressed as the matrix element of the unrenormalized Heisenberg operator j_5^μ between the physical vacuum state and a physical two-photon state:

$$e_0^2 T_{\alpha\beta\mu}(k_1, k_2) = \langle k_1, \alpha; k_2; \beta | j_{5\mu}(0) | 0 \rangle, \qquad (11.34)$$

where it is understood that the right-hand side is calculated only to the lowest order in e_0^2. Correspondingly, we have

$$-ie_0^2(k_1 + k_2)^\mu T_{\alpha\beta\mu}(k_1, k_2) = \langle k_1, \alpha; k_2, \beta | \partial^\mu j_{5\mu}(0) | 0 \rangle. \qquad (11.35)$$

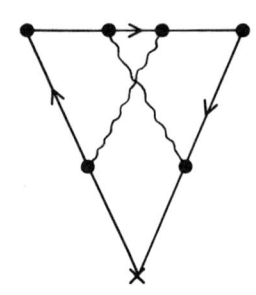

Fig. 11.3 A fourth-order correction to the triangle graph, representing a first skeletal correction.

The Axial Anomaly

The result (11.26) can be reproduced by taking

$$\partial^\mu j_{5\mu}(x) = 2m j_5(x) + \frac{\alpha_0}{2\pi} \tilde{F}^{\mu\nu}(x) F_{\mu\nu}(x), \tag{11.36}$$

where

$$F^{\mu\nu} = \partial^\mu A^\nu - \partial^\nu A^\mu,$$
$$\tilde{F}^{\mu\nu} = \tfrac{1}{2}\varepsilon^{\mu\nu\alpha\beta} F_{\alpha\beta}, \tag{11.37}$$
$$\alpha_0 = e_0^2/4\pi.$$

The unrenormalized fine-structure constant α_0 in (11.36) is replaced by the renormalized one, if all operators there are replaced by renormalized operators. Absence of radiative corrections to (11.26) would mean that (11.36) is an operator identity, valid to all orders of perturbation theory.

Noting that

$$\tilde{F}^{\mu\nu} F_{\mu\nu} = \partial^\mu X_\mu,$$
$$X_\mu = 2\varepsilon_{\mu\alpha\beta\gamma} A^\alpha \partial^\beta A^\gamma, \tag{11.38}$$

we can define a new axial vector current

$$J_5^\mu \equiv \bar{\psi}\gamma^\mu \gamma_5 \psi - \frac{\alpha_0}{2\pi} X^\mu, \tag{11.39}$$

which satisfies

$$\partial_\mu J_5^\mu = 2m j_5, \tag{11.40}$$

and is therefore conserved in the limit $m \to 0$. The new current is not gauge invariant, and hence cannot be coupled to physical fields. However, the corresponding charge Q_5, which by virtue of (11.38) is a constant of motion in the limit $m \to 0$, is gauge invariant:

$$Q_5 \equiv \int d^3x \, J_5^0 = \int d^3x \left(\psi^\dagger \gamma_5 \psi - \frac{\alpha_0}{\pi} \mathbf{A} \cdot \mathbf{\nabla} \times \mathbf{A} \right). \tag{11.41}$$

A gauge transformation $\mathbf{A} \to \mathbf{A} + \mathbf{\nabla}\chi$ gives rise to a surface integral that vanishes. Thus Q_5 can be a physical quantity. In fact,

$$[Q_5, \psi(x)] = -\gamma_5 \psi(x), \tag{11.42}$$

which shows that Q_5 is the generator of infinitesimal chiral transformations:

$$\psi \to (1 - i\omega\gamma_5)\psi = \psi + i\omega[Q_5, \psi]. \tag{11.43}$$

We mention in passing that the above situation changes if A^μ is replaced by a non-Abelian gauge field A_a^μ. In that case (11.36) is generalized, with $\tilde{F}^{\mu\nu} F_{\mu\nu}$ replaced by $\tilde{F}_a^{\mu\nu} F_{a\mu\nu}$.[8] The quantity X_μ in (11.38) is replaced by (5.6). If we continue to use (11.39), then Q_5 is not gauge invariant under "large" gauge transformations (see Sec. 8.6), owing to the existence of the topological charge. Physical consequences of this fact will be discussed in Sec. 12.5.

---

[8] W. A. Bardeen, *Phys. Rev.*, **184**, 1848 (1969).

11.5 Physical Explanation of the Axial Anomaly

To understand the origin of the axial anomaly physically, we consider the process:

Axial-vector source → Two real photons,

of which the triangle graph gives the lowest order description. We shall examine its "absorptive part"[9], which is obtained by "cutting" the graph as shown in Fig. 11.4. This means that we replace the cut electron propagators with their imaginary parts, which are δ-functions that force the electrons to go on mass shell. An electron propagating in the opposite sense to the arrow on the propagator is defined to be a positron. Thus, the absorptive part is a product of two amplitudes describing a succession of two physical processes:

1. Source $\to e^+ + e^-$,
2. $e^+ + e^- \to \gamma + \gamma$.

The triangle graph can be obtained from its absorptive part through a dispersion relation[10]. The absorptive part determines the imaginary part of the invariant function $R(q^2)$ in (11.29). For complex values of the argument, $R(s)$ is

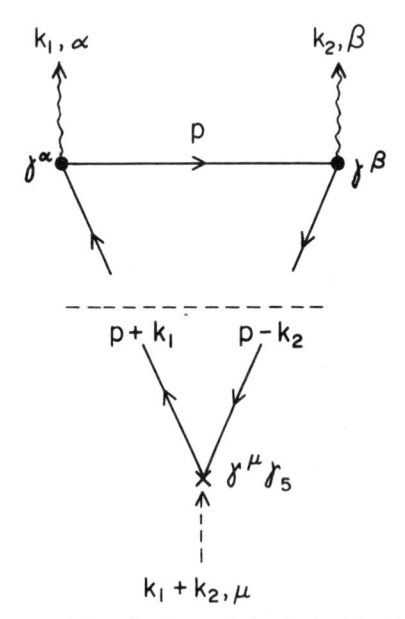

Fig. 11.4 The absorptive part of the triangle graph is obtained by "cutting" the graph.

---

[9] A. D. Dolgov and V. I. Zakharov, *Nucl. Phys.* **B27**, 525 (1971).
[10] For reference on dispersion relations and the analytic properties of Feynman graphs see R. J. Eden, P. V. Landshoff, D. I. Olive, and J. C. Polkinghorne, *The Analytic S-Matrix* (Cambridge University Press, Cambridge, England, 1966).

The Axial Anomaly

given by

$$R(s) = \frac{1}{\pi} \int_{4m^2}^{\infty} ds' \frac{\text{Im } R(s')}{s' - s}, \qquad (11.44)$$

plus possible "subtraction terms" which are polynomials in s. From (11.25) we see that $T_{\alpha\beta\mu}$ and $t_{\alpha\beta\mu}$ share the same absorptive part, which is gauge invariant and unambiguous. The axial anomaly comes from the fact that the absorptive part does not vanish in the limit $m \to 0$. In fact, we see from (11.32) that

$$\text{Im } R(q^2) \xrightarrow[m \to 0]{} -\frac{1}{2\pi} \delta(q^2). \qquad (11.45)$$

Our purpose is to explain this result physically.

It seems at first sight that the absorptive part vanishes in the massless limit, by chiral invariance. The reasoning runs as follows.

Go to the center-of-mass frame of the two final photons. In the initial process, the axial-vector source must make an e^+e^- pair of total spin 0, because a spin 1 state cannot subsequently go into two real photons[6]. The source can manage this via the interaction $\gamma_0\gamma_5$. Thus e^+ and e^- must have the same helicity, and hence opposite chirality in the massless limit. Since $\bar{R}\gamma_0\gamma_5 L = 0$, the initial process is forbidden in the massless limit.

In the final process, the e^+e^- pair annihilates into two photons by going through a virtual intermediate state, one of which is depicted in Fig. 11.5. (Another possible intermediate state is obtained by reversing the helicity of the

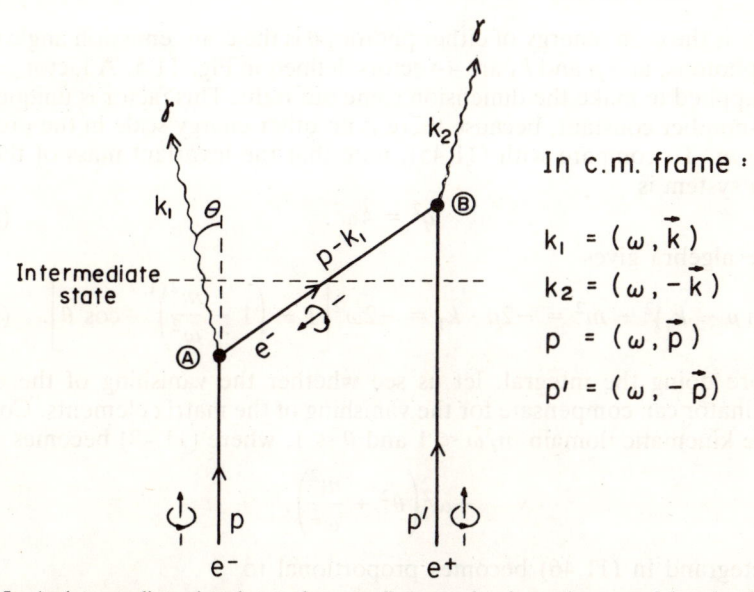

Fig. 11.5 An intermediate virtual state that contributes to the absorptive part of the triangle graph. The anomaly arises from the fact that in the massless limit the virtual state can become real, and gives a non-vanishing contribution at threshold.

virtual electron; still others are obtained by replacing the outgoing virtual electron by an incoming virtual positron). The transition at vertex B is allowed; but that at A is forbidden in the limit $m \to 0$. The reason is that vertex A calls for helicity flip, which is the same as chirality flip in the massless limit, and the electromagnetic interaction conserves chirality. For any other intermediate state, there is always one allowed and one forbidden vertex. Thus the final process is also forbidden.

However, this is not the whole story. As $m \to 0$, the virtual state in the final process approaches a real state, because in that limit the transition $e \to e + \gamma$ is no longer forbidden by energy-momentum conservation. If the electron has non-zero momentum $\mathbf{p}$, the photon can be emitted only in the forward direction; but if $\mathbf{p} \to 0$, the photon can be emitted in any direction. Thus, although the matrix elements vanish, so does the energy denominator, and they may compensate each other. The situation is not as simple as it appears at first glance: we must not take the limit $m \to 0$ too soon.

We shall estimate the absorptive part through the following crude argument. The matrix element of the initial process (source $\to e^+e^-$) carries a factor m. The matrix element of the final process ($e^+e^- \to \gamma\gamma$), when summed over all possible intermediate states, is m times the Feynman propagator of the virtual electron. We multiply together the matrix elements and integrate over all directions of emission of one of the final photons. Thus we have

$$\text{Im } R(\omega, m) \propto \frac{m^2}{\omega^2} \int d(\cos\theta) \, \frac{1}{(p-k_1)^2 - m^2} \quad (\omega > m), \qquad (11.46)$$

$$\text{Im } R(\omega, m) = 0 \quad (\omega < m),$$

where ω is the c. m. energy of either photon, θ is the c. m. emission angle of one of the photons, and p and k_1 are 4-vectors defined in Fig. 11.5. A factor ω^{-2} has been supplied to make the dimension come out right. This factor is unique up to a pure-number constant, because there is no other energy scale in the problem, barring m. To compare with (11.45), note that the invariant mass of the two-photon system is

$$q^2 = 4\omega^2. \qquad (11.47)$$

A little algebra gives

$$(p - k_1)^2 - m^2 = -2p \cdot k_1 = -2\omega^2 \left[1 - \left(1 - \frac{m^2}{\omega^2} \right)^{1/2} \cos\theta \right]. \qquad (11.48)$$

Before doing the integral, let us see whether the vanishing of the energy denominator can compensate for the vanishing of the matrix elements. Consider first the kinematic domain $m/\omega \ll 1$ and $\theta \ll 1$, where (11.48) becomes

$$-\omega^2 \left(\theta^2 + \frac{m^2}{\omega^2} \right).$$

The integrand in (11.46) becomes proportional to

$$\frac{(m/\omega)^2}{\theta^2 + (m/\omega)^2} \xrightarrow[m \to 0]{} \begin{cases} 1 & (\theta = 0) \\ 0 & (\theta \neq 0). \end{cases} \qquad (11.49)$$

The Axial Anomaly

This gives zero when integrated over angles. Hence no compensation occurs here; the absorptive part vanishes as $m \to 0$, for any value of ω above threshold.

Next, consider the threshold value $\omega = m$. Here (11.48) becomes $-2m^2$, independent of θ. Thus, as $m \to 0$, the energy denominator diverges like m^{-2}, and exactly compensates for the vanishing matrix elements. We expect the absorptive part to be peaked at threshold energy, and diverge there in the massless limit (because of the extra factor ω^{-2} from dimensional analysis).

Substituting (11.48) into (11.46), we obtain

$$\text{Im } R(\omega, m) = C \frac{m^2}{\omega^4} \left(1 - \frac{m^2}{\omega^2}\right)^{-1/2} \ln \frac{1-(1-m^2/\omega^2)^{1/2}}{1+(1-m^2/\omega^2)^{1/2}}, \quad (11.50)$$

where C is a constant pure number. A detailed calculation shows that this is in fact the right answer[9]. In Fig. 11.6 we plot Im R as a function of ω^2, for different values of m. We can see that

$$\text{Im } R(\omega, m) \xrightarrow[m \to 0]{} \text{Const. } \delta(\omega^2). \quad (11.51)$$

This explains (11.43).

We can also understand the behavior of Im R from the standpoint of analytic properties. We see from (11.31) that the anomaly is manifested through the fact that R has a fixed pole at $q^2 = 0$, with fixed residue (the anomaly pole). It touches the physical region $q^2 \geq 4m^2$ only in the limit $m \to 0$. If we increase m

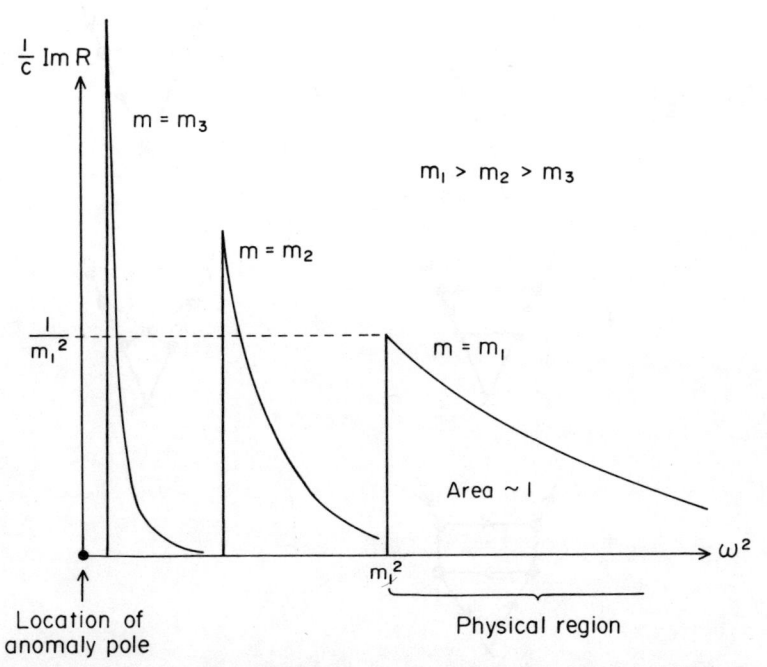

Fig. 11.6 Absorptive part of the triangle graph. The areas under the curves remain finite as $m \to 0$. Hence they approach $\delta(\omega^2)$ as $m \to 0$.

232   *Quarks, Leptons and Gauge Fields*

from 0, the pole remains intact, but the physical region moves away from it. The presence of the pole is then manifested only indirectly, by the fact that Im R increases towards threshold.

11.6 Cancellation of Anomalies

In analogy with the vertex $\Gamma^\mu(p_1, p_2)$ discussed in section 9.2, we define an axial vertex function $\Gamma_5{}^\mu(p_1, p_2)$ to be the sum of all connected proper Feynman graphs (with external propagators omitted) that describe the creation of a fermion-anti-fermion pair by an external axial-vector source. Some of these graphs are shown in Fig. 11.7. We are especially interested in the last graph, which is "anomalous" in the sense that it harbors a triangle subgraph. The graph is logarithmically divergent (even though the triangle subgraph is finite) because the triangle graph grows linearly with its external momenta. The divergence of the anomalous graph makes $\Gamma_5{}^\mu$ non-renormalizable.

We recall that renormalizability depends on a scaling property, such as (9.50) in the case of Γ^μ, which enables us to subtract off the divergent part of a Feynman graph and re-express the operation as infinite rescaling. In quantum electrodynamics, the scaling property of a "normal" Feynman graph is determined by the fact that each vertex touches two electron lines and a photon line, and carries a factor e_0. The triangle graph, however, has a *different* scaling property, as we can see indirectly by referring to (11.36). The first term there is

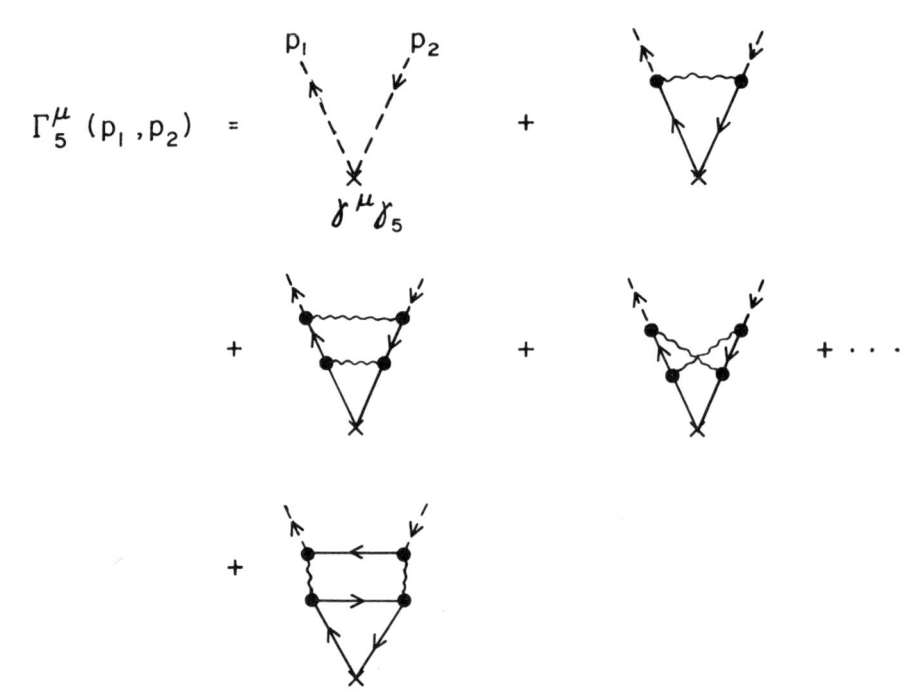

Fig. 11.7 Feynman graphs for the axial vertex function. The last graph is "anomalous", in that it has different scaling properties from the others, leading to non-renormalizability.

The Axial Anomaly

normal, but the anomalous second term corresponds to a Feynman graph in which two photon lines are attached to a single vertex carrying a factor e_0^2. This means that anomalous graphs in Γ_5^μ scale differently from the normal ones, and there is no renormalization procedure that could simultaneously get rid of both the divergence from anomalous graphs and that from normal graphs. Therefore Γ_5^μ is truly divergent.

It should be noted that the mass-dependent part of the triangle graph is normal, and would have been renormalizable in the usual manner all by itself. The anomalous part, which destroys renormalizability, is independent of the mass.

The non-renormalizability of Γ_5^μ is of no consequence in quantum electrodynamics, because Γ_5^μ is not a physical quantity in that theory. However, it does occur in the Weinberg-Salam model, for example in the graph shown in Fig. 11.8, which contributes to $e - \nu$ scattering. Though of high order, this graph by itself would give the disastrous prediction that the scattering cross section is infinite. But we have to sum over all possible fermion triangular loops. As we shall see below, when the contributions of all members in the first family e, ν, u, d are added together, the anomalous parts miraculously cancel one another, leaving a renormalizable mass-dependent residue. The latter is of such a high order in the coupling constants that we can ignore it in practice.

Let the possible fermions in the triangular loop be numbered by n, and let $T^{(n)}_{\alpha\beta\mu}$ denote the contribution of the nth fermion. Then the amplitude for the virtual process $Z \to \gamma\gamma$ is given by

$$M_{\alpha\beta\mu} = \sum_n Q_n^2 \, (Q'_R - Q'_L)_n T^{(n)}_{\alpha\beta\mu}, \tag{11.52}$$

where Q_n is the electric charge, and Q'_{Rn} and Q'_{Ln} the right and left-handed neutral charges of the nth fermion. These quantum numbers have been given in

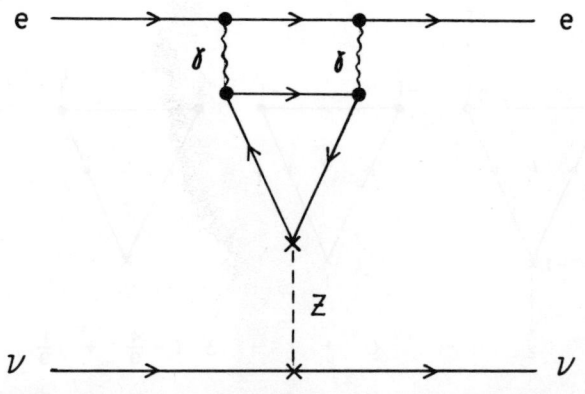

Fig. 11.8 An infinite non-renormalizable contribution to electron-neutrino scattering in the Weinberg-Salam model. Renormalizability requires cancellation of axial anomalies among all possible triangular fermion loops.

Table 6.2. Using (11.26), we obtain

$$q^\mu M_{\alpha\beta\mu} = 2 \sum_n Q_n^2 (Q'_R - Q'_L)_n m_n v_{\alpha\beta}^{(n)} \quad (11.53)$$
$$+ \frac{i}{2\pi^2} k_1^\mu k_2^\nu \varepsilon_{\alpha\beta\mu\nu} \sum_n Q_n^2 (Q'_R - Q'_L)_n.$$

The last term above is the anomaly, which is independent of masses. In Fig. 11.9, we show the relevant vertices for the fermions in the first family, from which follows

$$\sum_n Q_n^2 (Q'_L - Q'_R)_n = \frac{1}{\sin 2\theta_w} [0 + 1 + 3(-\tfrac{4}{9} + \tfrac{1}{9})] = 0. \quad (11.54)$$
$$\phantom{\sum_n Q_n^2 (Q'_L - Q'_R)_n = \frac{1}{\sin 2\theta_w}\ } \nu \quad e \quad\ u \quad d$$

Note that each quark flavor must be counted thrice to take into account the color quantum number. Thus, the anomaly cancels for any value of the Weinberg angle, and $M_{\alpha\beta\mu}$ is renormalizable in the usual manner.

There are possible triangle graphs in the Weinberg-Salam model, corresponding to $Z \to W^+W^-$, $\gamma \to W^+W^-$, etc. The anomalies in all such graphs must cancel, in order that the theory be renormalizable. In these more general graphs, the fermion can change identity as it goes around the triangular loop, so that the fermion propagators along the loop in general have different masses. Again, because the anomalies are independent of the masses, their cancellation depends only on the quantum numbers of the fermions. For a particular process, the condition that the anomalies cancel is

$$\text{Tr}[(V_1 V_2 + V_2 V_1) A] = 0, \quad (11.55)$$

where V_1, V_2 are the charge matrices that occur at the vector vertices, and A is the axial-charge matrix occurring at the axial-vector vertex. Instead of trying to verify the above case by case, we shall formulate a general rule for anomaly cancellation.

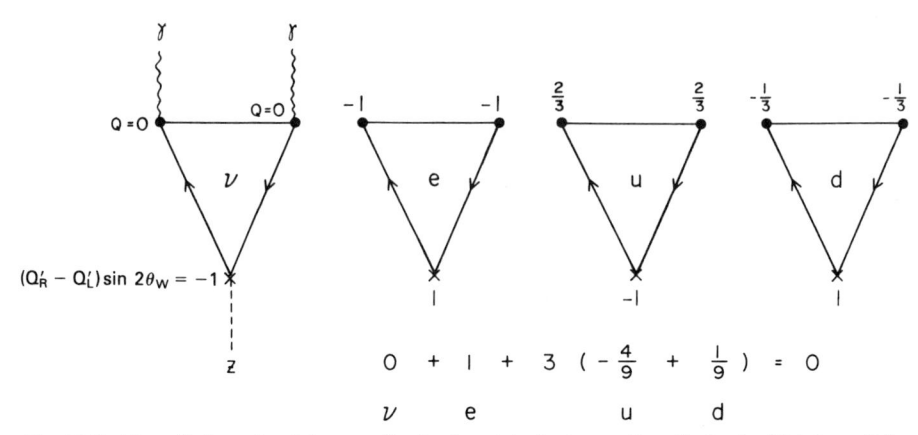

Fig. 11.9 Cancellation of axial anomalies in the virtual process $Z \to 2\gamma$ in the Weinberg-Salam model.

The Axial Anomaly

Consider any gauge theory. Let the fermion multiplet be denoted by the column vector ψ, in which the entries are Dirac spinors. The conserved currents coupled to the gauge fields are

$$j_a^\mu = \bar\psi \gamma^\mu t_a \psi, \tag{11.56}$$

where t_a ($a = 1, \ldots, N$) are the generators of the gauge group, represented by matrices that act on ψ. If the right and left-handed components of ψ transform differently under the gauge group, then t_a will be represented by different matrices t_a^R, t_a^L, when it acts on the right- or left-handed component. We write accordingly

$$j_a^\mu = \tfrac{1}{2}\bar\psi \gamma^\mu t_a^R(1 + \gamma_5)\psi + \tfrac{1}{2}\bar\psi \gamma^\mu t_a^L(1 - \gamma_5)\psi. \tag{11.57}$$

In any triangle graph in which all three vertices involve these currents, the fermion loop must be either completely left-handed or completely right-handed, because there is no coupling between right and left. If a particular left-handed loop is allowed, so is the corresponding right-handed loop, whose contribution differs only by a sign at the $\gamma^\mu \gamma_5$ vertex. Thus, the condition that no anomalies arise from the gauge interactions is that, for all a, b, c,

$$\text{Tr}[\{t_a^L, t_b^L\}t_c^L] - \text{Tr}[\{t_a^R, t_b^R\}t_c^R] = 0. \tag{11.58}$$

Of course, the theory may have other conserved currents (e.g., those related to ungauged global symmetries) that give rise to anomalies. But these do not make the theory unrenormalizable, just as the anomalies in Γ_5^μ have no relevance to the renormalizability of quantum electrodynamics.

We now check (11.58) for the Weinberg-Salam model. There are 4 generators t_0 and t_k ($k = 1, 2, 3$), with

$$[t_0, t_k] = 0, \tag{11.59}$$

$$\text{Tr } t_0 = \text{Tr } t_k = 0.$$

Recall that

$$t_k^R \equiv 0, \tag{11.60}$$

$$t_k^L = \tfrac{1}{2}\tau_k,$$

and that both members of a left-handed doublet have the same t_0. It is easy to verify that for both right and left-handed representations,

$$\text{Tr}[\{t_i, t_j\}t_k] = 0,$$

$$\text{Tr}[\{t_i, t_j\}t_0] = 0, \tag{11.61}$$

$$\text{Tr}(t_i t_0^2) = 0.$$

The only thing left to be checked is the trace of t_0^3. By direct calculation using Table 6.2, we find (remembering that quarks have color),

$$\text{Tr}(t_0^L)^3 = \text{Tr}(t_0^R)^3 = -\tfrac{2}{9}. \tag{11.62}$$

Thus, all anomalies cancel in the Weinberg-Salam model.

The sufficient conditions that enable the cancellation to occur are

— that the anomaly is independent of masses,
— that the anomaly has no radiative corrections.

The cancellations then follows purely from the quantum number assignment within the fermion family.

Since the three fermion families in the Weinberg-Salam model are identical in multiplet structure, all anomalies cancel within each family. The necessity for these cancellations to occur is the best theoretical argument we have for the standard family structure. It strongly suggests that the yet unobserved members of the third family, *i.e.* v'' and t, should exist. More generally, the requirement that all anomalies cancel imposes a severe constraint on possible physical models.[11]

11.7 't Hooft's Principle

With the proliferation of quarks and leptons, it is natural to ask whether they can be reduced to something simpler. One ordinarily thinks in a language one knows. So it is not surprising that the first thought is to make them out of other fermions—"preons", perhaps[12]. The first puzzle we face is how the electron could be so light when it is so small. Suppose the electron is a bound state of massless preons confined within the electron's intrinsic radius, which has an experimental upper bound of $a = 10^{-6}$ cm. Naive intuition would lead us to expect a mass greater than $a^{-1} \sim 10^6 \, m_e$. If the bound-state picture is qualitatively correct, then some principle must be at work to suppress the electron mass. 't Hooft[13] suggests that the principle is chiral invariance, as expressed in the following form:

A composite particle must reproduce the axial anomaly due to its fermionic constituents.

The principle seems obvious. We know that the electron must exhibit the axial anomaly, which in a composite picture has to arise from its more point-like constituents. However, the consequences are non-trivial, as we shall see later in detail. Briefly it works as follows. In the idealized limiting case of massless preons, the anomaly pole (which, regardless of mass, is always located at $q^2 = 0$) occurs in the physical region. If preons are confined and unobservable, then the anomaly pole must appear to be due to a physical bound state, which must therefore have zero mass. If chiral symmetry is not spontaneously broken, we identify the bound state with the electron. The actual observed mass of the electron is then attributed to the fact that preons have mass (for whatever reason). Although the principle does not enable us to calculate the electron mass, it gives a qualitative explanation for its smallness.

---

[11] D. J. Gross and R. Jackiw, *Phys. Rev.* **D6**, 477 (1972).
[12] The name was suggested by J. C. Pati, A. Salam, and J. Strathdee, *Phys. Letters*, **59**B, 265 (1975).
[13] G. 't Hooft, in *Recent Development in Gauge Theories*, eds. G. 't Hooft *et al.* (Plenum Press, New York, 1980).

The Axial Anomaly

To discuss 't Hooft's principle in more detail, we adopt a model for the preons, hoping that the results will be more general than the model. The preons are associated with a Dirac field ψ coupled to gauge fields $G_a{}^\mu$, which correspond to the generators L_a of a simple gauge group, say, $SU(N)$. The Lagrangian density is

$$\mathcal{L} = i\bar{\psi}(\gamma_\mu \partial^\mu + igL_a G_a{}^\mu)\psi - \kappa\bar{\psi}\psi, \qquad (11.63)$$

where the limit $\kappa \to 0$ is to be taken. This model has the same structure as quantum chromodynamics (QCD), with possible differences coming only from the choice of the gauge group, and the flavor multiplet structure of the fermions. We exploit this similarity by drawing upon the folklores of QCD.

Like the quarks in QCD, the preons are assumed to be permanently confined. Thus, dimensional transmutation should take place, giving rise to a confinement scale, namely, the size of an electron. The theory should be asymptotically free, which means that preon flavors are limited. For simplicity take one flavor of preons in the fundamental representation of $SU(N)$.

In this model there are two ungauged symmetries, the "electromagnetic" $U(1)$ and the chiral $U(1)$, which we assume to be unbroken[b]. They give rise to two conserved currents:

$$\begin{aligned} j^\mu &= \bar{\psi}\gamma^\mu\psi, \\ j_5{}^\mu &= \bar{\psi}\gamma^\mu\gamma_5\psi. \end{aligned} \qquad (11.64)$$

Although these currents are not coupled to any dynamical fields in the model, we can imagine turning on external sources that couple to them, and measuring the three-current correlation function

$$C_{\alpha\beta\mu}(x, y) = i\langle 0|Tj_\alpha(x)j_\beta(y)j_{5\mu}(0)|0\rangle. \qquad (11.65)$$

This describes vacuum fluctuations of the system in the absence of external sources. Existence of the axial anomaly implies that there are definite correlations in the vacuum fluctuations of j_α, j_β, and $j_{5\mu}$, at different space-time points. The absorptive part of $C_{\alpha\beta\mu}$ contains information about the physical excitations of the system that can transmit signals from one correlated point to another. Thus, $C_{\alpha\beta\mu}$ is an intrinsic property of the preon system.

We assume that the short-distance behavior of $C_{\alpha\beta\mu}$ can be studied by treating the gauge interactions of the preons in perturbation theory, taking the preons to be free particles in the zeroth-order approximation. This assumption is the same as that underlying the parton picture in QCD. Lacking an understanding of confinement from first principles, we cannot really prove its validity. However, we can offer the following plausibility arguments:

1. Owing to asymptotic freedom, preons should be seen as free particles by an agent that delivers a momentum transfer much higher than that set by the confinement scale.

---

[b] In this respect the model differs from QCD, where chiral symmetry is believed to be spontaneously broken, giving rise to a pseudoscalar Goldstone boson identifiable with the pion (see Sec. 12.4). We have no understanding from first principles why this should happen. Thus we cannot say what constraints must be placed on the preon model in order that the spontaneous breakdown does not happen.

2. Asymptotic perturbation theory in QCD has yielded results consistent with experiments.

According to this assumption, the Fourier transform of $C_{\alpha\beta\mu}$ for asymptotically high momenta is given by the usual triangle graph, with preons in the triangular loop:

$$T_{\alpha\beta\mu}(k_1, k_2) = \frac{iN}{2\pi^2} \varepsilon_{\alpha\beta\lambda\nu} k_1{}^\lambda k_2{}^\nu \frac{q_\mu}{q^2 + i\varepsilon}, \tag{11.66}$$

$$q \equiv k_1 + k_2,$$

where $k_1 \to \infty$, $k_2 \to \infty$. Note that the asymptotic domain includes the point $(k_1 + k_2)^2 = 0$, and hence covers the anomaly pole. A factor N is supplied, corresponding to the number of preon "colors". Assuming that there are no radiative corrections, we take (11.66) to be valid to all orders in the gauge coupling g.

We recall that (11.66) is obtained through enforcing the "electromagnetic" $U(1)$ symmetry, at the expense of chiral $U(1)$. There is no necessity for doing it here. We can add to (11.66) a polynomial in k_1, k_2 without changing its physical content, which resides in the absorptive part:

$$\text{Ab } T_{\alpha\beta\mu} \propto \delta(q^2). \tag{11.67}$$

This is non-zero, because the physical region in the asymptotic domain is $q^2 \geq 0$, the preons being massless. The point $q^2 = 0$ lies both in the asymptotic and the finite-momentum region. Thus, having obtained (11.67) in the asymptotic domain, we can continue to use it in the finite-momentum region. In fact, the amplitude $T_{\alpha\beta\mu}$ is given by (11.66) for all k_1, k_2, because it must have the general form (11.29), by symmetry arguments.

The reasoning from here on[14] relies on the assumptions that
(a) preons are confined,
(b) chiral symmetry is not spontaneously broken.

The confinement assumption means that, in the finite-momentum region, preons cannot exist as physical states, and hence cannot contribute to the absorptive part of $T_{\alpha\beta\mu}$. Therefore, there must be physical bound states that do. Since the absorptive is non-vanishing only at $q^2 = 0$, there must be massless physical bound states.

These massless bound states cannot be spin 0 particles, for if there were such particles in this model, they would have to be Goldstone bosons associated with the spontaneous breakdown of chiral symmetry. We have decreed that the latter does not occur.[c]

The anomaly pole cannot be due to particles of spin 1 or higher, because such particles can be coupled to external sources only via effective derivative

---

[c] If chiral symmetry breaks down spontaneously, then 't Hooft's principle says something about the coupling of the associated Goldstone boson to the currents j^μ and $j_5{}^\mu$. Applied to QCD, it reproduces the well-known calculation of the lifetime of the decay $\pi^0 \to 2\gamma$, which we discuss in Sec. 12.4.

[14] For more detailed discussions see Y. Frishman, A. Schwimmer, T. Banks, and S. Yankielowicz, *Nucl. Phys.* **B177**, 157 (1981); S. Coleman and B. Grossman (to be published).

The Axial Anomaly

couplings. Consequently, their contributions to the absorptive part of the triangle graph vanish at threshold. But the absorptive part is non-vanishing only at threshold.

The only remaining possibility is a massless spin 1/2 bound state, which, as we know, can produce the anomaly pole. This is identified as the electron.

We can represent the matrix elements of j^μ and j_5^μ between electron states in the forms

$$\langle e_2|j^\mu|e_1\rangle = g_V \bar{u}(\mathbf{p}_2, s_2)\gamma^\mu u(\mathbf{p}_1, s_1),$$
$$\langle e_2|j_5^\mu|e_1\rangle = g_A \bar{u}(\mathbf{p}_2, s_2)\gamma^\mu \gamma_5 u(\mathbf{p}_1, s_1), \quad (11.68)$$

where $u(\mathbf{p}, s)$ are Dirac spinors, and g_V and g_A are constants. In order that the electron reproduce the anomaly (11.66), we must have

$$g_V^2 g_A = N, \quad (11.69)$$

which is a concrete result of 't Hooft's principle. In a more realistic model with multi-flavored preons, the above condition generalized to a relation between the multiplet structure constants of the preons and those of the bound states. One may then use it to see what type of preon flavor structures can give rise to bound states having the observed multiplet structure of quarks and leptons.[15]

For a space-time view of 't Hooft's principle, calculate the Fourier transform of (11.66) to obtain the three-current correlation function:

$$C_{\alpha\beta\mu}(x, y) = \frac{1}{2\pi^2} \varepsilon_{\alpha\beta\mu\lambda} \left[\frac{\partial}{\partial r_\lambda} \delta^4(r) \right] \frac{1}{\rho} \frac{d}{d\rho} D_F(\rho),$$
$$r \equiv \frac{1}{2}(x + y), \quad (11.70)$$
$$\rho \equiv x - y.$$

Here $D_F(\rho)$ is the Feynman propagator for a massless particle:

$$D_F(\rho) \equiv -\int \frac{d^4q}{(2\pi)^4} \frac{e^{-iq\cdot\rho}}{q^2 + i\varepsilon}. \quad (11.71)$$

The imaginary part of $iD_F(\rho)$ vanishes outside the light cone $\rho^2 = 0$. Hence the imaginary part of the correlation function is non-zero only if $x = y$ and $x^2 = 0$. This means that there must be physical excitations traveling on the light cone to produce the correlation. They must therefore be massless excitations. Consider a ray on the light cone, as indicated by the line OP in Fig. 11.10. The short-distance domain is a neighborhood about O in which asymptotic freedom tells us that the massless excitations are free preons. If we go sufficiently far away from O along the ray OP, we eventually go beyond the confinement scale. Massless excitations in that region must be bound states—electrons.

If we extend the model by giving mass to the preons, the situation becomes complicated, and reliance on QCD folklore is not enough. The complicating

---

[15] S. Dimopoulos, S. Raby, and L. Susskind, *Nucl. Phys.* **B173**, 208 (1980).

circumstance is that the anomaly pole, which still lies at $q^2 = 0$, is beyond the reach of the physical region, and does not directly contribute to the absorptive part of $T_{\alpha\beta\mu}$. The latter will generally have non-vanishing values for all q^2 above threshold, as depicted qualitatively in Fig. 11.6. To determine it requires a dynamical calculation. In particular, the location of the threshold becomes a dynamical question: It could be higher than the kinematic threshold for free preon creation, because the absorptive part can vanish in that neighborhood, for dynamical reasons. Thus, we cannot derive the electron mass in any simple way.

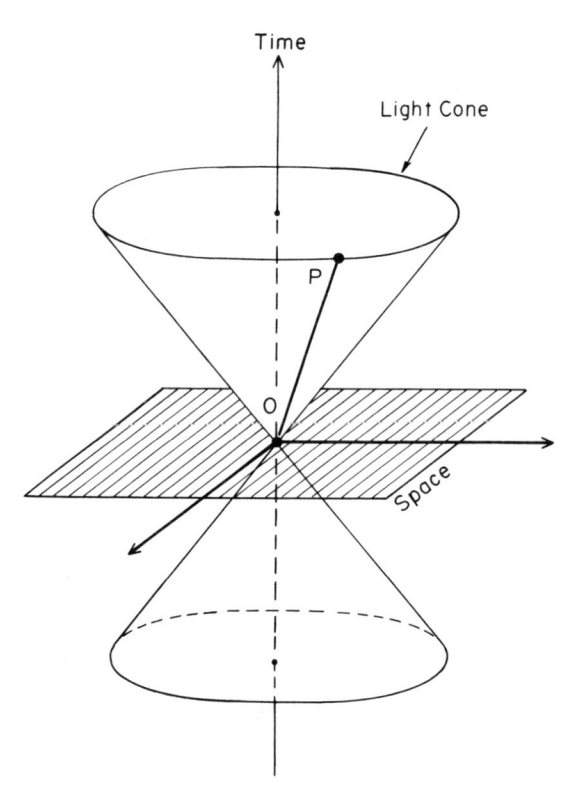

Fig. 11.10 Space-time view of 't Hooft's principle. There must exist massless excitations that move on the light cone, in order to account for the three-current correlations implied by the existence of the axial anomaly. Near O these excitations are preons. At P, beyond the confinement scale, they are massless bound states—electrons.

CHAPTER XII
QUANTUM CHROMODYNAMICS

12.1 General Properties

1 Lagrangian Density

We have discussed in Chapter II the physical motivation for introducing color as a quantum number for quarks. By gauging the symmetry group $[SU(3)]_{\text{color}}$, we obtain quantum chromodynamics (QCD), the currently accepted model of the strong interactions.

The 8 generators of $[SU(3)]_{\text{color}}$ are represented in the fundamental representation by the Gell-Mann matrices $\lambda_a/2$ ($a = 1, \ldots, 8$) listed in Table 2.2, with the normalization.

$$\text{Tr}(\lambda_a \lambda_b) = 2 \delta_{ab}. \tag{12.1}$$

The gauge fields, called "gluon fields," are denoted by G_a^μ ($a = 1, \ldots, 8$). We use the notation (see Chapter IV)

$$G^\mu(x) \equiv \tfrac{1}{2}\lambda_a G_a^\mu(x),$$
$$\mathcal{G}^{\mu\nu}(x) = \partial^\mu G^\nu(x) - \partial^\nu G^\mu(x) + ig_0[G^\mu(x), G^\nu(x)], \tag{12.2}$$

where g_0 is a dimensionless number—the unrenormalized gauge coupling constant.

The matter fields consists of spinor quark fields denoted collectively by $q(x)$, with components $q_\alpha^{fi}(x)$, where

$i = 1, 2, 3$ (color index: red, yellow, green),
$f = 1, \ldots, 6$ (flavor index: u, d, c, s, t, b),
$\alpha = 1, \ldots, 4$ (spinor index).

We usually suppress the spinor index, and sometimes all indices. Instead of using the flavor index, we sometimes denote the quark fields of various flavors by their conventional names:

u	d	c	s	t	b
(up)	(down)	(charmed)	(strange)	(top)	(bottom)

The existence of the top quark has not been confirmed experimentally, but is suggested by the internal consistency of the Weinberg-Salam model of electroweak interactions. There may be other flavors yet undiscovered.

The complete Lagrangian density of QCD is

$$\mathcal{L} = -\tfrac{1}{4}\mathcal{G}_a{}^{\mu\nu}\mathcal{G}_{a\mu\nu} + \bar{q}(i\slashed{D} - M)q,$$
$$D^\mu = \partial^\mu + ig_0 G^\mu, \qquad (12.3)$$

where M is a color-independent mass matrix in the flavor indices, which will be discussed in detail in Sec. 12.6. The theory is renormalizable, but we shall not present the proof.[1]

2 Feynman Rules

It is generally believed that quarks and gluons are confined, i.e., colored states do not exist in the physical sector of the Hilbert space. Thus, the basis states for the S-matrix of QCD consists not of single quark or gluon states, but color-singlet states containing physical hadrons. However, we can always define Green's functions that are vacuum expectation values of time-ordered products of quark and gluon field operators. Quark confinement merely means that mass shells for quarks and gluons do not exist. It is thus possible, at least formally, to define Feynman graph expansions for these Green's functions in terms of quark and gluon propagators and vertices.

The generating functional for connected Green's functions $W[J, K]$ is given by (see Chapter VIII)

$$\exp iW[J, K] = \mathcal{N} \int (DG)\,(D\bar{q})\,(Dq)\,(D\eta^*)\,(D\eta).$$
$$\cdot \exp i \int d^4x [\mathcal{L}_{\text{eff}} + J_a{}^\mu G_{a\mu} + (\bar{K}q) + (\bar{q}K)], \qquad (12.4)$$

where $J_a{}^\mu(x)$ is a c-number vector source, and $K(x)$ is an anticommuting c-number spinorial source. The effective Lagrangian density $\mathcal{L}_{\text{eff}}$ is given in a general Lorentz gauge by

$$\mathcal{L}_{\text{eff}} = \mathcal{L} + \frac{1}{2\alpha}(\partial^\mu G_{a\mu})(\partial^\nu G_{a\nu})$$
$$+ \tfrac{1}{2}\eta_a^*(\delta_{ab}\Box^2 + g_0 f_{abc} G_c{}^\mu \partial_\mu)\eta_b, \qquad (12.5)$$

where λ is the gauge parameter [see (8.62)], η_a^* and η_a are anticommuting c-number ghost fields, and f_{abc} are the $SU(3)$ structure constants given in Table 2.3.

We divide the effective Lagrangian density into a "free" part and an "interaction" part:

$$\mathcal{L}_{\text{eff}} = \mathcal{L}_0 + \mathcal{L}'. \qquad (12.6)$$

The free Lagrangian density is taken to be

$$\mathcal{L}_0 = \tfrac{1}{2}G_{a\mu}\left[g^{\mu\nu}\Box^2 - \left(1 - \frac{1}{\alpha}\right)\partial^\mu\partial^\nu\right]G_{a\nu}$$
$$+ i\bar{q}\gamma^\mu\partial_\mu q + \tfrac{1}{2}\eta_a^*\Box^2 \eta_a. \qquad (12.7)$$

[1] G. 't Hooft, *Nucl. Phys.*, B33, 173 (1971); B35, 167 (1971).

Quantum Chromodynamics

The interaction Lagrangian density is given by

$$\mathcal{L}' = \mathcal{L}'_{\text{gluon}} + \mathcal{L}'_{\text{quark}} + \mathcal{L}'_{\text{ghost}}, \tag{12.8}$$

$$\mathcal{L}'_{\text{gluon}} = \tfrac{1}{2}g_0 f_{abc}(\partial^\mu G_a{}^\nu - \partial^\nu G_a{}^\mu)G_{b\mu}G_{c\nu}$$
$$- \tfrac{1}{4}g_0^2 f_{abc}f_{ab'c'}G_b{}^\mu G_c{}^\nu G_{b'\mu}G_{c'\nu}, \tag{12.9}$$

$$\mathcal{L}'_{\text{quark}} = -\tfrac{1}{2}g_0(\bar{q}\gamma_\mu \lambda_a q)G_a{}^\mu, \tag{12.10}$$

$$\mathcal{L}'_{\text{ghost}} = \tfrac{1}{2}g_0 f_{abc}(\eta_a^* \partial_\mu \eta_b)G_c{}^\mu. \tag{12.11}$$

The Feynman rules for Green's functions can be read off the above formulas, and are given in Fig. 12.1.

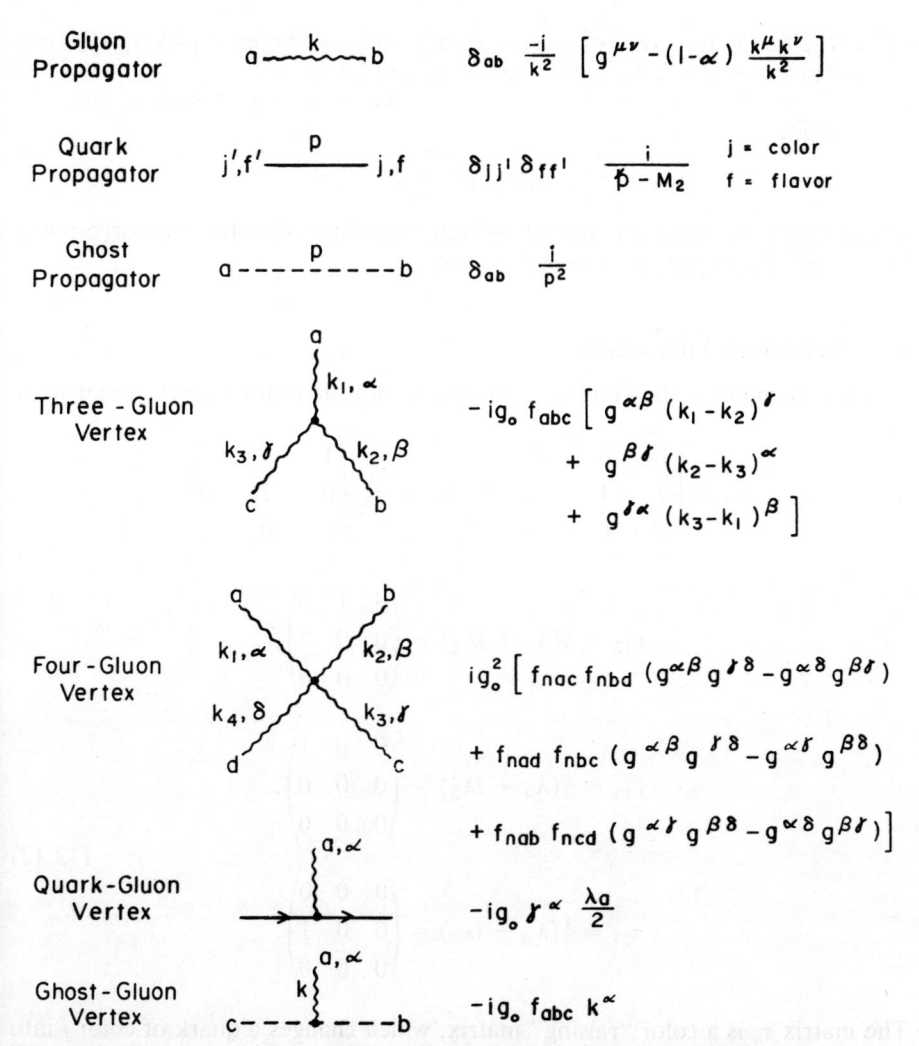

Gluon Propagator	$a \sim\!\!\sim\!\!\sim^{k}\!\!\sim\!\!\sim b$	$\delta_{ab}\,\dfrac{-i}{k^2}\left[g^{\mu\nu} - (1-\alpha)\dfrac{k^\mu k^\nu}{k^2}\right]$
Quark Propagator	$j',f' \xrightarrow{p} j,f$	$\delta_{jj'}\delta_{ff'}\,\dfrac{i}{\slashed{p} - M_2}$ j = color, f = flavor
Ghost Propagator	$a \dashrightarrow^{p} b$	$\delta_{ab}\,\dfrac{i}{p^2}$
Three-Gluon Vertex	$a(k_1,\alpha)$, $b(k_2,\beta)$, $c(k_3,\gamma)$	$-ig_0 f_{abc}\left[g^{\alpha\beta}(k_1-k_2)^\gamma + g^{\beta\gamma}(k_2-k_3)^\alpha + g^{\gamma\alpha}(k_3-k_1)^\beta\right]$
Four-Gluon Vertex	$a(k_1,\alpha)$, $b(k_2,\beta)$, $c(k_3,\gamma)$, $d(k_4,\delta)$	$ig_0^2\big[f_{nac}f_{nbd}(g^{\alpha\beta}g^{\gamma\delta} - g^{\alpha\delta}g^{\beta\gamma})$ $+ f_{nad}f_{nbc}(g^{\alpha\beta}g^{\gamma\delta} - g^{\alpha\gamma}g^{\beta\delta})$ $+ f_{nab}f_{ncd}(g^{\alpha\gamma}g^{\beta\delta} - g^{\alpha\delta}g^{\beta\gamma})\big]$
Quark-Gluon Vertex	a,α	$-ig_0 \gamma^\alpha \dfrac{\lambda_a}{2}$
Ghost-Gluon Vertex	a,α; $c \dashrightarrow b$, momentum k	$-ig_0 f_{abc} k^\alpha$

Fig. 12.1 Feynman rules for QCD. All gluon momenta flow into the vertex.

To illustrate combinatorial considerations in the derivation of the Feynman rules, we discuss the three-gluon vertex, which arises from the first term in (12.9). Suppose the three external gluon lines have the following labels:

Color index: $a, b, c,$
Momentum: $k_1, k_2, k_3,$
Polarization index: $\alpha, \beta, \gamma,$

with all momenta flowing into the graph. The first term in (12.9) can be rewritten as

$$g_0 f_{a'b'c'}(\partial^\mu G_{a'}{}^\nu) G_{b'\mu} G_{c'\nu},$$

due to the fact that f_{abc} is antisymmetric in its indices. We must pick up all terms in the above sum that contribute to the vertex graph. First consider $a' = a$. Then either $b' = b$, $c' = c$, or $b' = c$, $c' = b$. These two possibilities give the contribution

$$-ig_0 f_{abc}(k_1{}^\beta g^{\alpha\gamma} - k_1{}^\gamma g^{\alpha\beta}).$$

The complete vertex is obtained by adding two other contributions corresponding to the alternatives $a' = b$ and $a' = c$.

3 Quark-Gluon Interactions

Let us recombine the Gell-Mann matrices into the following set of matrices:

$$\lambda_3 = \begin{pmatrix} 1 & 0 & 0 \\ 0 & -1 & 0 \\ 0 & 0 & 0 \end{pmatrix}, \quad \lambda_8 = \frac{1}{\sqrt{3}}\begin{pmatrix} 1 & 0 & 0 \\ 0 & 1 & 0 \\ 0 & 0 & -2 \end{pmatrix},$$

$$\tau_{12} \equiv \tfrac{1}{2}(\lambda_1 + i\lambda_2) = \begin{pmatrix} 0 & 1 & 0 \\ 0 & 0 & 0 \\ 0 & 0 & 0 \end{pmatrix},$$

$$\tau_{13} \equiv \tfrac{1}{2}(\lambda_4 + i\lambda_5) = \begin{pmatrix} 0 & 0 & 1 \\ 0 & 0 & 0 \\ 0 & 0 & 0 \end{pmatrix},$$

$$\tau_{23} \equiv \tfrac{1}{2}(\lambda_6 + i\lambda_7) = \begin{pmatrix} 0 & 0 & 0 \\ 0 & 0 & 1 \\ 0 & 0 & 0 \end{pmatrix}.$$

(12.12)

The matrix τ_{ij} is a color "raising" matrix, which changes a quark of color j into one of color i. The hermitian conjugate $\tau_{ij}^\dagger \equiv \tau_{ji}$ does the reverse. Correspond-

Quantum Chromodynamics

ingly, define

$$A^\mu \equiv G_3{}^\mu,$$
$$B^\mu \equiv G_8{}^\mu,$$
$$X^\mu \equiv 2^{-1/2}(G_1{}^\mu + iG_2{}^\mu), \qquad (12.13)$$
$$Y^\mu \equiv 2^{-1/2}(G_4{}^\mu + iG_5{}^\mu),$$
$$Z^\mu \equiv 2^{-1/2}(G_6{}^\mu + iG_7{}^\mu).$$

Then

$$\mathcal{L}'_{\text{quark}} = -\frac{g_0}{2}\bar{q}(\lambda_3 A\!\!\!/ + \lambda_8 B\!\!\!/)q$$

$$-\frac{g}{\sqrt{2}}[\bar{q}(\tau_{21} X\!\!\!\!/ + \tau_{31} Y\!\!\!\!/ + \tau_{32} Z\!\!\!/)q + \text{c.c.}], \qquad (12.14)$$

which shows that the quarks have two kinds of "charges": a color 'isotopic charge" corresponding to an eigenvalue of $\lambda_3/2$, and a color "hypercharge" corresponding to an eigenvalue of $\lambda_8/2$. The gluons X, Y, Z also carry these charges, for a quark can change color by absorbing or emitting one of these gluons, as illustrated in Fig. 12.2. The charge assignments for quarks and gluons are listed in Table 12.1.

4 Gluon Self-Interactions

Before we examine the gluon self-interactions in QCD, it is instructive to study those in a simpler case: $SU(2)$ pure-gauge theory, with gauge fields $G_1{}^\mu$,

Table 12.1 COLOR CHARGES
All charges are in units of g_0.
Q_A = Color isotopic charge (source of A field)
Q_B = Color hypercharge (source of B field)

		Q_A	Q_B
Quarks	1 red	$\frac{1}{2}$	$\frac{1}{2\sqrt{3}}$
	2 yellow	$-\frac{1}{2}$	$\frac{1}{2\sqrt{3}}$
	3 green	0	$-\frac{1}{\sqrt{3}}$
Charged Gluons	X	-1	0
	Y	$-\frac{1}{2}$	$-\frac{\sqrt{3}}{2}$
	Z	$\frac{1}{2}$	$-\frac{\sqrt{3}}{2}$

$G_2{}^\mu$, $G_3{}^\mu$, and field tensor

$$\mathcal{F}_a{}^{\mu\nu} = \partial^\mu G_a{}^\nu - \partial^\nu G_a{}^\mu - e\varepsilon_{abc} G_b{}^\mu G_c{}^\nu \qquad (12.15)$$
$$(a = 1, 2, 3).$$

The analog of (12.13) is

$$\begin{aligned} A^\mu &\equiv G_3{}^\mu, \\ X^\mu &\equiv 2^{-1/2}(G_1{}^\mu + iG_2{}^\mu), \end{aligned} \qquad (12.16)$$

which defines a real "photon" field A^μ, and a charged vector boson field X^μ. Writing the field tensor in terms of A^μ and X^μ, using the shorthand notation (6.51) for 4-tensors, we have

$$\begin{aligned} \mathcal{F}_1 &= \frac{1}{\sqrt{2}}(D \times X + \text{c.c.}), \\ \mathcal{F}_2 &= \frac{1}{i\sqrt{2}}(D \times X - \text{c.c.}), \\ \mathcal{F}_3 &= F + ieX^* \times X, \end{aligned} \qquad (12.17)$$

where

$$\begin{aligned} F^{\mu\nu} &\equiv \partial^\mu A^\nu - \partial^\nu A^\mu, \\ D^\mu &\equiv \partial^\mu - ieA^\mu. \end{aligned} \qquad (12.18)$$

Thus, the field X^μ has charge $-e$. The Lagrangian density is

$$\begin{aligned} \mathcal{L} &= -\tfrac{1}{4}\mathcal{F}_a{}^{\mu\nu}\mathcal{F}_{a\mu\nu} \\ &= -\tfrac{1}{4}F^{\mu\nu}F_{\mu\nu} - ieF^{\mu\nu}X_\mu^* X_\nu \\ &\quad -\tfrac{1}{2}(D^\mu X^\nu - D^\nu X^\mu)^*(D_\mu X_\nu - D_\nu X_\mu) \\ &\quad +\tfrac{1}{2}e^2[(X^* \cdot X)^2 - (X \cdot X)^*(X \cdot X)]. \end{aligned} \qquad (12.19)$$

Apart from the last term, this is the Lagrangian density of massless vector electrodynamics, with an 'anomalous" magnetic moment.

The last term, a quartic self-interaction of the charged field, marks the difference between massless vector electrodynamics and full $SU(2)$ gauge

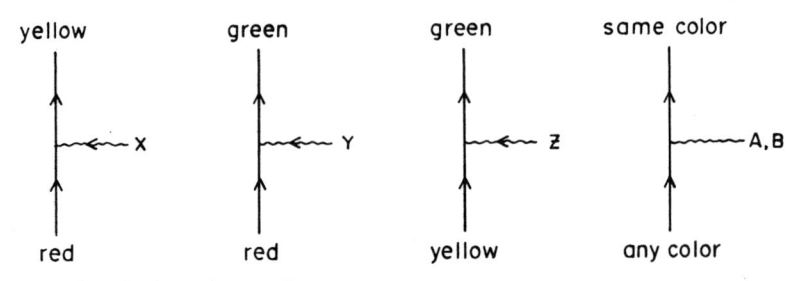

Fig. 12.2 Quark-gluon interactions

Quantum Chromodynamics

theory. Without it, the theory is invariant only under a smaller gauge group $U(1)$. It is also not renormalizable, due to the logarithmic divergences of the skeleton graphs for charge-charge scattering [Fig. 12.3(a)]. The quartic self-interaction [Fig. 12.3(b)] provides the necessary counter term to cancel these divergences, making the theory renormalizable. This interaction is the analog of $\lambda(\phi^*\phi)^2$ in scalar electrodynamics; but its coefficient is not an independent coupling constant, due to spin constraints.

In the Lagrangian density (12.9), the term

$$-ieF^{\mu\nu}X_\mu^* X_\nu \qquad (12.20)$$

is not necessary for $U(1)$ gauge invariance, and appears to be a "non-minimal" interaction from the point of view of electrodynamics, giving rise to an "anomalous" magnetic moment over and above the orbital moment. If the charged vector field were massive, the resulting total magnetic moment would correspond to a gyromagnetic ratio $g = 2$ [see (6.55)]. As we shall show in the next section, this $g - 2 = 0$ is a necessary condition for a massive charged vector theory to have a massless limit. From the point of view of $SU(2)$ gauge invariance, the term (12.20) is part of the minimal interaction, for it arises from the cross product between the two terms of $\mathcal{F}_3{}^{\mu\nu}$ in (12.17). Thus, the value $g - 2 = 0$ is natural, and of kinematical origin (just as in the Dirac equation). Indeed, it is needed for internal consistency: if $g - 2 \neq 0$, then the vacuum polarization tensor of the A field would diverge quadratically rather than logarithmically, spoiling renormalizability.

We now return to QCD. According to (12.13), there are two "photon" fields A and B, coupled respectively to color isotopic charge and color hypercharge. There are three charged vector fields X, Y, Z. To express the gluon field tensor in terms of these new fields, it is helpful to have all the structure constants f_{abc} displayed for easy reference. This is provided by Fig. 12.4. We obtain in a straightforward manner (again using a shorthand tensor notation) the following

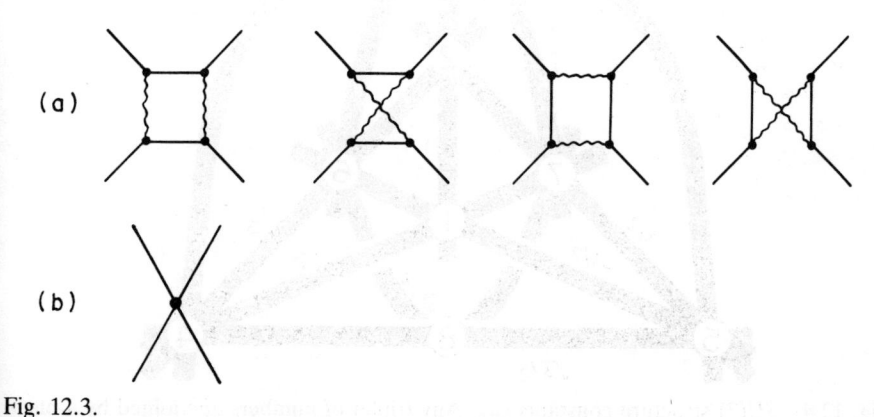

Fig. 12.3.
(a) Logarithmically divergent skeleton graphs for the scattering of charged vector bosons through exchange of photons.
(b) Quartic self-interaction that cancels the divergences in (a).

expressions:

$$\mathcal{F}_3 = \partial \times A + ig_0(X^* \times X + \tfrac{1}{2}Y^* \times Y + \tfrac{1}{2}Z^* \times Z),$$

$$\mathcal{F}_8 = \partial \times B + ig_0 \frac{\sqrt{3}}{2} \cdot (Y^* \times Y + Z^* \times Z),$$

$$\frac{1}{\sqrt{2}}(\mathcal{F}_1 + i\mathcal{F}_2) = D \times X + \frac{g_0}{\sqrt{2}} Y \times Z^*, \qquad (12.21)$$

$$\frac{1}{\sqrt{2}}(\mathcal{F}_4 + i\mathcal{F}_5) = D \times Y + \frac{ig_0}{2} Z \times X,$$

$$\frac{1}{\sqrt{2}}(\mathcal{F}_6 + i\mathcal{F}_7) = D \times Z + \frac{g_0}{\sqrt{2}} X^* \times Y,$$

where the covariant derivatives are defined by

$$DX \equiv (\partial - ig_0 A)X,$$

$$DY \equiv (\partial - \frac{ig_0}{2} A - ig \frac{\sqrt{3}}{2} B)Y, \qquad (12.22)$$

$$DZ \equiv (\partial + \frac{ig_0}{2} A - ig \frac{\sqrt{3}}{2} B)Z,$$

from which we can read off the various charges, confirming the assignments given in Table 12.1.

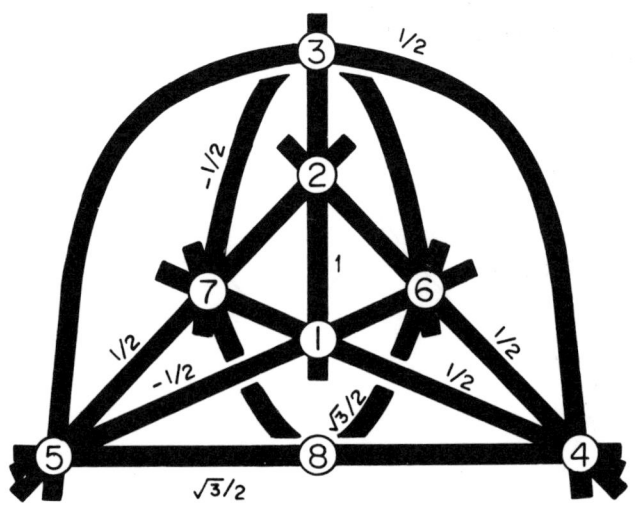

Fig. 12.4 $SU(3)$ structure constants f_{abc}. Any triplet of numbers abc joined by a black belt in the figure corresponds to a non-vanishing element f_{abc}. The number labelling the black belt is the value of f_{abc} for the ordering $a > b > c$. The value of f_{acb} etc. can be obtained by noting that f_{abc} is completely antisymmetric in its indices.

Quantum Chromodynamics

We indicate the structure of the Lagrangian density for pure QCD as follows:

$$\mathcal{L}_{\text{pure QCD}} = -\tfrac{1}{4}\mathcal{F}_a{}^{\mu\nu}\mathcal{F}_{a\mu\nu}$$
$$= \mathcal{L}_{\text{kin}} + \mathcal{L}_{\text{mag. mom.}} + \mathcal{L}_{\text{int}},$$
$$\mathcal{L}_{\text{kin}} = -\tfrac{1}{4}(\partial \times A)^2 - \tfrac{1}{4}(\partial \times B)^2 - \tfrac{1}{2}[|D \times X|^2 + |D \times Y|^2 + |D \times Z|^2],$$
$$\mathcal{L}_{\text{mag. mom.}} = -ig_0(\partial^\mu A^\nu - \partial^\nu A^\mu)(X_\mu^* X_\nu + \tfrac{1}{2}Y_\mu^* Y_\nu + \tfrac{1}{2}Z_\mu^* Z_\nu),$$
$$-ig_0\frac{\sqrt{3}}{2}(\partial^\mu B^\nu - \partial^\nu B^\mu)(Y_\mu^* Y_\nu + Z_\mu^* Z_\nu). \tag{12.23}$$

The term $\mathcal{L}_{\text{kin}}$ taken alone would describe a theory of three charged vector bosons X, Y, Z interacting in minimal electromagnetic fashion with two "photon" fields A and B. The term $\mathcal{L}_{\text{mag. mom.}}$ endows each of the charged vector bosons with color gyromagnetic ratio $g = 2$, with respect to both the A field and the B field. The rest of the Lagrangian density is lumped into $\mathcal{L}_{\text{int}}$, which we shall not write out. It contains interactions involving at least three charged vector bosons at the same point, and makes the theory renormalizable.

12.2 The Color Gyromagnetic Ratio

We discuss the color gyromagnetic ratio, which was mentioned earlier in connection with the nature of gluon interactions, and will be relevant to our discussion of asymptotic freedom in the next section.

As a thought experiment, consider the behavior of the charged gluons X, Y, Z in an external color magnetic field associated with the A or B gluon field. Ignoring all gluon interactions except those with the imposed external field, we reduce the problem to that of massless charged vector bosons interacting independently with an external Abelian magnetic field. The non-Abelian nature of QCD enters only in so far as it endows the vector bosons with charges.

We first consider a *massive* charged spinning particle, of charge e and mass m, interacting with a weak homogeneous external magnetic field **B**. Let ζ be the average spin, defined as the expectation value of the spin operator in the rest frame of a one-particle state. The gyromagnetic ratio g is conventionally defined through the equation

$$\frac{d\zeta}{dt} = g\frac{e}{2m}\zeta \times \mathbf{B}. \tag{12.24}$$

We shall try to transform this equation to an arbitrary Lorentz frame, and then take the limit $m \to 0$. If the limit exists, then the resulting equation defines g in the massless case.

To do this, we define the polarization 4-vector S^μ by

$$S^\mu \equiv (0, \zeta) \quad \text{(in rest frame)}. \tag{12.25}$$

Let u^μ be the 4-velocity (of the center of a wave packet, whose motion can be treated classically):

$$u^\mu \equiv \frac{1}{\sqrt{1-v^2}}(1, \mathbf{v}), \tag{12.26}$$

where $\mathbf{v}$ is the 3-velocity. In every Lorentz frame we have $S \cdot u = 0$, or

$$S^0 = \mathbf{S} \cdot \mathbf{v}. \tag{12.27}$$

The time derivative of this, evaluated in the rest frame, together with (12.24), give the equation of motion for S^μ in the rest frame:

$$\frac{dS^0}{dt} = \mathbf{S} \cdot \frac{d\mathbf{v}}{dt} \quad \text{(in rest frame)},$$

$$\frac{d\mathbf{S}}{dt} = g \frac{e}{2m} \mathbf{S} \times \mathbf{B} \quad \text{(in rest frame)}. \tag{12.28}$$

The covariant generalization of the above is[2]

$$\frac{dS^\mu}{d\tau} = g \frac{e}{2m} [F^{\mu\nu}S_\nu + u^\mu(S \cdot F \cdot u)] - u^\mu \left(\frac{du^\nu}{d\tau} S_\nu\right), \tag{12.29}$$

where τ is the proper time:

$$d\tau = \frac{dt}{\sqrt{1-v^2}}. \tag{12.30}$$

The electromagnetic tensor $F^{\mu\nu}$ is related to the electric and magnetic fields by (3.28). We use the notation $(S \cdot F \cdot u) = S_\alpha F^{\alpha\beta} u_\beta$. We can verify (12.29) by noting that the right-hand side contains all possible 4-vectors that can contribute, and reduces to (12.28) in the rest frame with no electric field. In a homogeneous electromagnetic field, u^μ obey the equation of motion

$$\frac{du^\mu}{d\tau} = \frac{e}{m} F^{\mu\nu} u_\nu. \tag{12.31}$$

Substituting this into (12.29), we obtain

$$\frac{dS^\mu}{d\tau} = \frac{e}{2m} [gF^{\mu\nu}S_\nu + (g-2)u^\mu(S \cdot F \cdot u)]. \tag{12.32}$$

In terms of the time t of a fixed observer, this reads

$$\frac{dS^\mu}{dt} = \frac{g}{2} \frac{e}{E} F^{\mu\nu}S_\nu + \frac{e}{2m}(g-2)v^\mu(S \cdot F \cdot v), \tag{12.33}$$

where $v^\mu \equiv (1, \mathbf{v})$, and E is the energy of the particle:

$$E \equiv \frac{m}{\sqrt{1-v^2}}. \tag{12.34}$$

---

[2] V. Bargman, L. Michel, and V. L. Telegdi, *Phys. Rev. Lett.*, **2**, 435 (1959).

Quantum Chromodynamics

As $m \to 0$, (12.33) approaches a well-defined limit only if

$$g - 2 = 0, \tag{12.35}$$

which is therefore a necessary constraint for a spinning charged massless particle.

In the absence of an electric field, the spatial component of (12.32) reads

$$\frac{d\mathbf{S}}{dt} = \frac{g}{2}\frac{e}{E} \mathbf{S} \times \mathbf{B} - \frac{e}{2m}(g-2)\mathbf{v}(\mathbf{S}\cdot\mathbf{v}\times\mathbf{B}). \tag{12.36}$$

The instantaneous spin $\boldsymbol{\zeta}$ is related to $\mathbf{S}$ through a Lorentz transformation, and can be shown to obey the equation of motion[3]

$$\frac{d\boldsymbol{\zeta}}{dt} = \left[\frac{e}{E} + \frac{e}{2m}(g-2)\right]\boldsymbol{\zeta}\times\mathbf{B} + \frac{e}{2m}(g-2)\frac{E}{E+m}(\mathbf{v}\cdot\mathbf{B})\mathbf{v}\times\boldsymbol{\zeta}. \tag{12.37}$$

For $g - 2 = 0$, $\boldsymbol{\zeta}$ and $\mathbf{S}$ become identical, with

$$\frac{d\mathbf{S}}{dt} = \frac{e}{E}\mathbf{S}\times\mathbf{B}, \tag{12.38}$$

which is the equation describing the spin precession of a massless charged particle.

From the above discussion, we see that the gluons X, Y, Z all have color gyromagnetic ratio $g - 2 = 0$. The quarks also have $g - 2 = 0$, by virtue of the Dirac equation. (The value of $g - 2$ here refers only to the "kinematic" value, and does not include radiative corrections).

12.3 Asymptotic Freedom

1 The Running Coupling Constant

The running coupling constant of QCD is defined through the renormalized gluon propagator[a]. The unrenormalized full propagator may be represented in the form

$$D'^{\mu\nu}_{ab}(k) = \delta_{ab}\left(g^{\mu\nu} - \frac{k^\mu k^\nu}{k^2}\right)\frac{d'(k^2/\Lambda^2, \alpha_0)}{ik^2} \tag{12.39}$$
$$+ \text{(gauge-dependent terms)},$$

where Λ is a cutoff momentum, and

$$\alpha_0 \equiv g_0^2/4\pi. \tag{12.40}$$

---

[a] We refer to Chap. IX for concepts not fully explained here.

[3] V. B. Berestetskii, E. M. Lifshitz, and L. P. Pitaevskii, *Relativistic Quantum Theory*, Part 1 (Pergamon, Oxford, England, 1971), p. 127.

The renormalized propagator is obtained by putting

$$d'\left(\frac{k^2}{\Lambda^2}, \alpha_0\right) = Z\left(\frac{\lambda^2}{\Lambda^2}, \alpha_0\right) d\left(\frac{k^2}{\lambda^2}, \alpha_\lambda\right), \tag{12.41}$$

where d is a finite function, and

$$\alpha_\lambda \equiv \alpha_0 Z\left(\frac{\lambda^2}{\Lambda^2}, \alpha_0\right). \tag{12.42}$$

We call this the running coupling constant corresponding to the renormalization point λ.[b] The definition of d is made unique by the normalization condition

$$d(1, \alpha_\lambda) = 1. \tag{12.43}$$

From (12.41) and (12.42) we see that

$$\alpha_\lambda d\left(\frac{k^2}{\lambda^2}, \alpha_\lambda\right) = \alpha_0 d'\left(\frac{k^2}{\Lambda^2}, \alpha_0\right). \tag{12.44}$$

Thus, the left-hand side is a renormalization group invariant:

$$\alpha_\lambda d\left(\frac{k^2}{\lambda^2}, \alpha_\lambda\right) = \alpha_\nu d\left(\frac{k^2}{\nu^2}, \alpha_\nu\right), \tag{12.45}$$

where λ and ν are two arbitrary renormalization points. Putting $\nu^2 = k^2$, we obtain

$$\alpha_k = \alpha_\lambda d\left(\frac{k^2}{\lambda^2}, \alpha_\lambda\right). \tag{12.46}$$

The Gell-Mann-Low (GL) function is defined by

$$\beta(\alpha_\lambda) \equiv \lambda^2 \frac{d}{d\lambda^2}\left[\alpha_0 Z\left(\frac{\lambda^2}{\Lambda^2}, \alpha_0\right)\right], \tag{12.47}$$

where the right-hand side is to be re-expressed as a function of α_λ with the help of (12.42). Under a change in the renormalization point λ, α_λ changes according to

$$\frac{d\alpha_\lambda}{d \ln(\lambda^2)} = \beta(\alpha_\lambda). \tag{12.48}$$

The right-hand side of (12.44) is a power series in α_0 beginning with the power α_0^2. Hence the GL function has a power series expansion of the form

$$\beta(\alpha) = \beta_0 \alpha^2 + \beta_1 \alpha^3 + \cdots. \tag{12.49}$$

Gross and Wilczek[4], Politzer,[5] were the first to point out that $\beta_0 < 0$, and that

---

[b] The numbers Λ^2 and λ^2 are negative, being the squared invariant masses of Euclidean momenta. The limit $\lambda \to \infty$ is understood to mean $-\lambda^2 \to \infty$.

[4] D. J. Gross and F. Wilczek, *Phys. Rev. Lett.*, **30**, 1343 (1973); *Phys. Rev.*, D**8**, 3633 (1973).
[5] H. D. Politzer, *Phys. Rev. Lett.*, **30**, 1346 (1973).

this implies

$$\alpha_\lambda \xrightarrow[\lambda \to \infty]{} 0. \tag{12.50}$$

This means, for example, that the interaction between two quarks due to one-gluon exchange vanishes in the limit of infinite momentum. This phenomenon is called "asymptotic freedom", and furnishes a basis for the parton model of electron-proton deep inelastic scattering.

Using the lowest-order term in (12.49), we obtain from (12.48)

$$\frac{1}{\alpha_k} = \frac{1}{\alpha_\lambda} - \beta_0 \ln \frac{k^2}{\lambda^2} + O(\alpha_\lambda), \tag{12.51}$$

which is the analog of (9.18) in quantum electrodynamics, except that this is exact in the limit $k \to \infty$. Using (12.46) we can also write

$$d\left(\frac{k^2}{\lambda^2}, \alpha_\lambda\right) = 1 - \beta_0 \alpha_\lambda \ln \frac{k^2}{\lambda^2} + O(\alpha_\lambda^2). \tag{12.52}$$

The GL function can be obtained directly from (12.47), by calculating the Feynman graphs for the gluon propagator[4]. The lowest-order coefficient β_0 can be obtained from the graphs shown in Fig. 12.5, with the result

$$\beta_0 = -\frac{1}{6\pi}\left(\frac{33}{2} - N_f\right), \tag{12.53}$$

where N_f is the number of quark flavors. The quark contribution is $N_f/2$ times the value $(3\pi)^{-1}$ in quantum electrodynamics [see (9.103)][c]. For β_0 to be

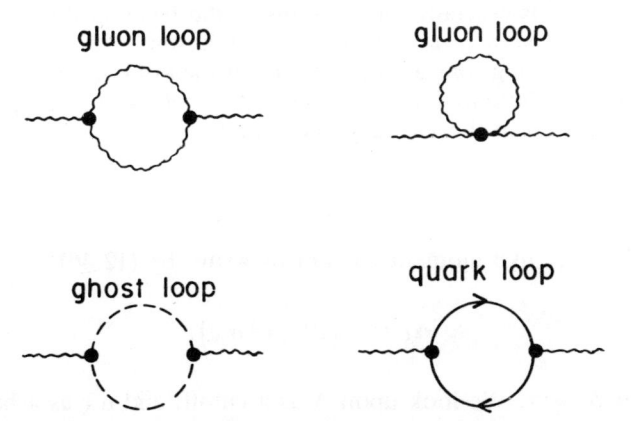

Fig. 12.5 Feynman graphs contributing to the GL function to lowest order.

---

[c] The factor 1/2 arises as follows. To find the quark contribution to β_0, it suffices to consider the self-energy graph of the A field with one quark loop. For any given flavor, we see from Table 12.1 that two colors contribute, and the charges are both 1/2. Thus, $2(1/2)^2 = 1/2$. The same result is obtained by considering the B field, because $\Sigma Q_A^2 = \Sigma Q_B^2$, as we can see from Table 12.1.

negative, it is necessary for the gluon self-interactions to dominate, which requires

$$N_f < 17. \tag{12.54}$$

The next coefficient is given by[6]

$$\beta_1 = -\frac{1}{8\pi^2}\left(51 - \frac{19}{3} N_f\right). \tag{12.55}$$

We note that (12.51) admits the solution

$$\alpha_k = -\frac{1}{\beta_0 \ln(k^2/\bar{\Lambda}^2)}, \tag{12.56}$$

where $\bar{\Lambda}^2$ is an arbitrary scale parameter, which can be determined by fitting QCD predictions to experimental data from electron-proton deep inelastic scattering. Existing data corresponding to $-k^2$ between 10 and 100 $(\text{GeV})^2$ can be fitted with $\bar{\Lambda} \sim 0.5$ Gev.[7] This gives $\alpha_k \cong 0.3$ at $(-k^2) = (10 \text{ GeV})^2$, for $N_f = 6$.

2 The Vacuum as Magnetic Medium

To establish asymptotic freedom, we only need to know β_0, which we shall calculate by an elementary method due to Nielsen[8]. In this method, one views the vacuum state as a magnetic medium, and finds β_c through the response of the medium to an external magnetic field. The method is applicable to any field theory, and brings out the interesting view that asymptotic freedom is the outcome of a competition between Landau diamagnetism and Pauli paramagnetism, and depends crucially on the spins of the fields in the theory. We develop this point of view in a version due to Johnson.[9]

The running coupling constant α_k determines the kth Fourier transform of the color electrostatic potential between two static charges (isotopic charge or hypercharge) placed in the vacuum [see (9.20)]:

$$V(k) = \frac{\alpha_k}{k^2}, \tag{12.57}$$

where k is a space-like momentum. Let us write, by (12.46),

$$\alpha_k = \alpha_\Lambda d\left(\frac{k^2}{\Lambda^2}, \alpha_\Lambda\right), \tag{12.58}$$

and consider $\Lambda \to \infty$. We look upon Λ as a cutoff, and α_Λ as a bare coupling

---

[6] W. E. Caswell, *Phys. Rev. Lett.*, **33**, 244 (1974); D. R. T. Jones, *Nucl. Phys.*, B75, 531 (1974).
[7] A. J. Buras, *Rev. Mod. Phys.*, **52**, 199 (1980).
[8] N. K. Nielsen, *Am. J. Phys.*, **49**, 1171 (1981).
[9] K. Johnson, in *Asymptotic Realms of Physics*, Eds. A. Guth, K. Huang, and R. L. Jaffe (MIT Press, Cambridge, 1983).

Quantum Chromodynamics

constant. The dielectric constant of the vacuum may be defined as

$$\varepsilon_\Lambda(k) \equiv 1/d\left(\frac{k^2}{\Lambda^2}, \alpha_\Lambda\right). \tag{12.59}$$

Introducing the electric polarizability $P_\Lambda(k)$ by

$$\varepsilon_\Lambda(k) = 1 + P_\Lambda(k), \tag{12.60}$$

we find from (12.52) that

$$P_\Lambda(k) = \beta_0 \alpha_\Lambda \ln\frac{\Lambda^2}{k^2} + O(\alpha_\Lambda^2). \tag{12.61}$$

Thus, asymptotic freedom means that the vacuum is a medium with negative electric polarizability.

In a Lorentz-invariant theory, the dielectric constant and the magnetic permeability are inverses of each other[d]. Hence the latter is given by

$$\mu_\Lambda(k) = d\left(\frac{k^2}{\Lambda^2}, \alpha_\Lambda\right). \tag{12.62}$$

Introducing the magnetic susceptibility of the vacuum $\chi_\Lambda(k)$ by

$$\mu_\Lambda(k) = 1 + \chi_\Lambda(k), \tag{12.63}$$

we find

$$\chi_\Lambda(k) = -\beta_0 \alpha_\Lambda \ln\frac{\Lambda^2}{k^2} + O(\alpha_\Lambda^2). \tag{12.64}$$

Paramagnetism corresponds to $\beta_0 < 0$ (asymptotic freedom), and diamagnetism corresponds to $\beta_0 > 0$.

To calculate χ_Λ, we first calculate the vacuum energy in the presence of an external color magnetic field associated with the neutral gluon field A or B. To lowest order in α_Λ, all quarks and charged gluons can be treated as free particles, except for their interactions with the external field. The non-Abelian nature of the theory comes in only through the fact that gluons carry charge. The magnetic susceptibility will receive independent additive contributions from the quarks of different colors and flavors, and from the charged gluons.

We only need to consider the following prototype system: a charged massless free field (boson or fermion), of charge e and spin S, interacting with a weak homogeneous external magnetic field. The single-particle energies are the eigenvalues of the single-particle Hamiltonian H, whose square is given by

$$H^2 = |\mathbf{p} - e\mathbf{A}|^2 - 2e\mathbf{S} \cdot \mathbf{B}, \tag{12.65}$$

where $\mathbf{p} = -i\nabla$, $\mathbf{B} = \nabla \times \mathbf{A}$ is the external magnetic field, and the components of $\mathbf{S}$ are spin matrices, each having only two possible eigenvalues $\pm S$. (For example, $\mathbf{S} = \boldsymbol{\sigma}/2$ for the spin 1/2 case). The last term in (12.65) is designed to

---

[d] This is because the velocity of light in the QCD vacuum is $[\varepsilon_\Lambda(k)\mu_\Lambda(k)]^{-1/2}$.

give the equation of motion
$$[\mathbf{S}, H^2] = 2ie\mathbf{S} \times \mathbf{B}, \tag{12.66}$$
whose one-particle expectation leads to (12.38).

Let us choose $\mathbf{B}$ to be a homogeneous field pointing along the z-axis:
$$\mathbf{B} = \hat{z}B. \tag{12.67}$$

The two terms in (12.65) commute with each other, and the first term is of the form of a non-relativistic Hamiltonian (not squared Hamiltonian) of a particle in a magnetic field. Its eigenvalues have the well-known Landau spectrum:[10]

$$\text{Landau eigenvalue} = p_z^2 + 2eB(n + \tfrac{1}{2}) \quad (n = 0, 1, 2, \ldots),$$
$$\text{Degeneracy} = \Omega^{2/3}eB/2\pi, \tag{12.68}$$

where p_z is the momentum along the direction of the magnetic field, and Ω is the total spatial volume. Each value of the quantum number n corresponds to a circular classical orbit of the particle. The degeneracy arises from the fact that the center of the orbit could be anywhere in a plane normal to $\mathbf{B}$. The single-particle energies are labelled by the three quantum numbers p_z, n, S_z:

$$E(p_z, n, S_z) = [p_z^2 + 2eB(n + \tfrac{1}{2} - S_z)]^{1/2},$$
$$n = 0, 1, 2, \ldots, \tag{12.69}$$
$$S_z = \pm S.$$

The vacuum energy per unit volume is

$$\mathscr{E}_{\text{vac}} = (-1)^{2S} \frac{eB}{2\pi} \int \frac{dp_z}{2\pi} \sum_n \sum_{S_z} \left[p_z^2 + 2eB(n + \tfrac{1}{2} - S_z) \right]^{1/2}, \tag{12.70}$$

where the factor $(-1)^{2S}$ takes into account the connection between spin and statistics[e]. The magnetic susceptibility χ is defined by

$$\mathscr{E}_{\text{vac}} = -\tfrac{1}{2}\chi B^2. \tag{12.71}$$

For $S > 1/2$, (12.69) has complex eigenvalues, corresponding to the unstable modes studied by Nielsen and Olesen[11]. However, they do not contribute to χ.

We cut off the divergent expression (12.70) by restricting the values of p_z and n such that
$$E(p_z, n, S_z) < \Lambda. \tag{12.72}$$

---

[e] We take the vacuum energy to be the zero-point energy of the system. For a single boson field, this is 1/2 the sum of all single-particle energies. For a charged boson field, the factor 1/2 is cancelled by the fact that the charged field has two components. For a fermion field, the vacuum energy is the sum of all negative single-particle energies.

---

[10] K. Huang, *Statistical Mechanics* (Wiley, New York, 1963), Chap. 11. The right-hand side of (11.88) should be multiplied by a factor of two.
[11] N. K. Nielsen and P. Olesen, *Nucl. Phys.*, **B144**, 376 (1978).

Quantum Chromodynamics

In the limit $\Lambda \to \infty$ and $B \to 0$, (12.70) can be evaluated by approximating the higher terms in the n-sum by the Euler formula[12]

$$\sum_{n=N}^{K} f(n + \tfrac{1}{2}) \cong \int_{N}^{K} dx\, f(x) - \frac{1}{24}[f'(K) - f'(N)], \qquad (12.73)$$

where N is chosen to be large enough for the approximation to be valid. Its precise value is unimportant; the terms for $n < N$ will affect only the scale of Λ. The calculation is done in great detail in Ref. 7, and will not be repeated here. We only note that the continuum approximation to the n-sum does not contribute to χ; only the second term in (12.73) is relevant. This is related to the fact that a classical system does not exhibit diamagnetism[10]. The result for the vacuum energy density is

$$\mathcal{E}_{\text{vac}} = -(-1)^{2S} \frac{e^2 B^2}{16\pi^2} [(2S)^2 - \tfrac{1}{3}] \ln \frac{\Lambda^2}{ceB} + \text{const.}, \qquad (12.74)$$

where c is some number unimportant for our purpose. From this we obtain

$$\chi = (-1)^{2S} \frac{e^2}{8\pi^2} [(2S)^2 - \tfrac{1}{3}] \ln \frac{\Lambda^2}{ceB}, \qquad (12.75)$$

which is the vacuum magnetic susceptibility in a weak static external magnetic field ($k^2 = ceB \to 0$).

3 The Nielsen-Hughes Formula

By comparing (12.75) with (12.64), we find

$$\beta_0 = -\frac{(-1)^{2S}}{2\pi}[(2S)^2 - \tfrac{1}{3}]. \qquad (12.76)$$

This is valid for a massless charged field of any spin S (for which $g - 2 = 0$). It has also been derived by Hughes[13] from a rather different point of view. We call it the "Nielsen-Hughes formula".

When (12.76) is used in (12.49) and (12.48), we have the rate of change of the running coupling constant $\alpha = e^2/4\pi$, where e is the charge with respect to the "electromagnetic" field used in deriving the magnetic susceptibility. If the running coupling constant is defined as $\alpha = g^2/4\pi$, and the charge referred to above is $e = Qg$, then (12.76) should be multiplied by Q^2.

The term $(2S)^2$ in (12.76) represents the effect due to an enhancement of the external magnetic field coming from spin alignment, while the term $-1/3$ represents the effect of orbital motion, which produces a magnetic field that tends to cancel the imposed field. These terms are respectively the analogs of Pauli paramagnetism and Landau diamagnetism in solid-state physics, where the respective susceptibilities also bear the ratio (for $S = 1/2$)[10]

$$\chi_{\text{Landau}}/\chi_{\text{Pauli}} = -\tfrac{1}{3}. \qquad (12.77)$$

---

[12] M. Abramowitz and I. A. Stegun, *Handbook of Mathematical Functions* (National Bureau of Standards, Washington, 1964), p. 806.
[13] R. J. Hughes, *Phys. Lett.* **97b**, 246 (1980); *Nucl. Phys.*, **B186**, 376 (1981).

In the present case, an additional effect arises due to fact that fermions have negative vacuum energy, as expressed by the factor $(-1)^{2S}$. Thus, quantum electrodynamics would have been asymptotically free, were it not for the necessity for "hole theory".

To apply the Nielsen-Hughes formula to QCD, we add the contributions from the quarks of various colors and flavors, and from the gluons X, Y, Z. Taking the external magnetic field to be that associated with a neutral gluon field (A or B), we find with the help of Table 12.1 that

$$(\beta_0)_{\text{quarks}} = N_f/6\pi, \qquad (12.78)$$
$$(\beta_0)_{\text{gluons}} = -11/4\pi.$$

Therefore

$$\beta_0 = -\frac{1}{6\pi}\left(\frac{33}{2} - N_f\right), \qquad (12.79)$$

as stated in (12.53).

12.4 The Pion as Goldstone Boson

1 The Low-Energy Domain

The scale $\bar{\Lambda} \sim 0.5$ GeV associated with asymptotic freedom defines a high-momentum (or short-distance) regime

$$\ln(k^2/\bar{\Lambda}^2) \gg 1,$$

in which quarks and gluons can be treated as weakly interacting particles in perturbative QCD. At the other end of the scale ($k^2/\bar{\Lambda}^2 \lesssim 1$) is low-energy hadronic physics, in which the interacting units are not individual quarks and gluons, but hadrons. So far we cannot solve QCD in this domain. However, from the vast amount of experimental information available, some highly fruitful ideas had been developed long before QCD was invented. Translated into expected properties of QCD, some of these ideas become concrete statements that are much easier to comprehend. They also serve as clues to the mathematical structure of QCD. We are referring to what is known as "current algebra".[14] In this section we translate a piece of that wisdom called "PCAC" (partially conserved axial-vector current hypothesis).

2 Chiral Symmetry: an Idealized Limit

The mass matrix M in the QCD Lagrangian density is a phenomenological quantity whose origin is unknown. It is the same as the quark mass matrix in the Weinberg-Salam model discussed in Chap. VI. If we adopt the latter to describe the electroweak interactions, then renormalizability requires that M arise from

---

[14] See S. B. Treiman, in S. B. Treiman, R. Jackiw, and D. J. Gross, *Lectures in Current Algebra and Its Applications* (Princeton University Press, Princeton, 1972); S. L. Adler and R. F. Dashen, *Current Algebras* (Benjamin, New York, 1968).

the vacuum expectation value of the Higgs field; but it also contains arbitrary constants, and remains a phenomenological quantity. Whatever the origin, M can always be brought to diagonal form in the manner indicated in (6.71), through flavor-mixing transformations.[f] Thus, the mass term in the Lagrangian density can be written as

$$\mathcal{L}_{\text{mass}} = -\sum_{i=1}^{3}\sum_{f=1}^{6} m_f \bar{q}^{fi} q^{fi}. \qquad (12.80)$$

If QCD leads to quark confinement, as we shall assume, then the mass parameters m_f are not observable quantities. As we shall see later, however, they can be determined in terms of observable hadronic masses through "current algebra" methods. They are called "current" quark masses, to distinguish them from "constituent" quark masses, which are parameters used in phenomenological quark models of hadronic structure.

The approximate unitary symmetry of the strong interactions implies that $m_u \cong m_d \cong m_s$. Since isospin conservation is a much better symmetry than the whole of flavor $SU(3)$, the relation $m_u \cong m_d$ should hold to a higher degree of accuracy than $m_d \cong m_s$. The mass parameters for c, b, t should all be much larger than those for u, d, s, because we see no trace of flavor $SU(4)$ or higher symmetries in the hadronic spectrum.

If we put $m_u = m_d$, then isospin will become exactly conserved. In this limit, n and p will have the same mass, and so will π^+, π^-, and π^0. But this does not explain the smallness of the pion mass $(m_\pi/m_n = 0.14)$, which makes the pion very special among hadrons. To understand this, Nambu[15] and Chou[16] suggested that there is a limit in which the pion is a massless Goldstone boson associated with spontaneous symmetry breaking. This limit is of course an idealized theoretical construct. As an example, Nambu and Jona-Lasinio[17] gave a non-renormalizable model in which the fundamental fields are massless nucleon fields, and thus possesses chiral symmetry, whose spontaneous breakdown gives rise to nucleon mass, and a massless Goldstone boson identified with the pion.

To translate these ideas into QCD, we consider the idealized limit corresponding to

$$m_u = m_d = 0. \qquad (12.81)$$

The Lagrangian density exhibiting the chiral symmetry in this limit is

$$\mathcal{L}_{\text{chiral}} = -\tfrac{1}{4} G_a^{\mu\nu} G_{a\mu\nu} + \bar{\psi} i \slashed{D} \psi, \qquad (12.82)$$

where

$$\psi = \begin{pmatrix} u \\ d \end{pmatrix}. \qquad (12.83)$$

The quark color index has been suppressed.

---

[f] We temporarily ignore the complications relating to CP violation discussed in Sec. 12.6.

[15] Y. Nambu, *Phys. Rev. Lett.*, **4**, 380 (1960).
[16] Chou Kuang-chao, *Soviet Phys.* JETP, **12**, 492 (1961).
[17] Y. Nambu and G. Jona-Lasinio, *Phys. Rev.*, **122**, 345 (1961); **124**, 246 (1961).

The full QCD Lagrangian density can be written in the form

$$\mathcal{L} = \mathcal{L}_{\text{chiral}} - (m_u \bar{u} u + m_d \bar{d} d) + \mathcal{L}_{scbt}, \quad (12.84)$$

where $\mathcal{L}_{scbt}$ contains terms pertaining only to the quarks s, c, b, t. We shall regard $\mathcal{L}_{\text{chiral}}$ as an "unperturbed" Lagrangian density. The idealized limit of chiral symmetry corresponds to the unperturbed problem.

As we shall show later, m_u and m_d are in fact small compared to the nucleon mass. Hence $\mathcal{L}_{\text{chiral}}$ should, by itself, give a reasonably good description of "ordinary" hadronic physics, in which strangeness and higher flavors do not play a direct role. This assumption implies that hadronic masses arise from dimensional transmutation, since there are no intrinsic mass parameters in $\mathcal{L}_{\text{chiral}}$.

Our "unperturbed" system, as described by $\mathcal{L}_{\text{chiral}}$, is invariant under the following global symmetry groups:

$$\begin{aligned} [SU(2)]_V: & \ \psi \to e^{-i\boldsymbol{\tau}\cdot\boldsymbol{\omega}/2} \psi, \\ [U(1)]_V: & \ \psi \to e^{-i\alpha} \psi, \\ [SU(2)]_A: & \ \psi \to e^{-i\boldsymbol{\tau}\cdot\boldsymbol{\theta}/2} \gamma_5 \psi, \\ [U(1)]_A: & \ \psi \to e^{-i\beta} \gamma_5 \psi, \end{aligned} \quad (12.85)$$

where $\boldsymbol{\omega}$, $\boldsymbol{\theta}$, α, β are arbitrary real constants. The subscripts V and A stand respectively for "vector" and "axial-vector". The associated Noether currents are respectively

$$\begin{aligned} J^k_\mu &= \bar{\psi} \gamma_\mu \tau_k \psi \quad (k = 1, 2, 3) \quad \text{(isospin current)}, \\ j_\mu &= \bar{\psi} \gamma_\mu \psi \quad \text{(baryonic current)}, \\ J^k_{5\mu} &= \bar{\psi} \gamma_\mu \gamma_5 \tau_k \psi \quad (k = 1, 2, 3), \\ j_{5\mu} &= \bar{\psi} \gamma_\mu \gamma_5 \psi. \end{aligned} \quad (12.86)$$

The baryonic current is conserved even in the perturbed system defined by $\mathcal{L}$. The isospin current is conserved as long as $m_u = m_d$. The axial-vector currents are conserved only in the chiral-symmetric limit apart from possible anomalies discussed later.

How are these symmetries manifested in nature? First of all, $[SU(2)]_V$ and $[U(1)]_V$ are manifested directly as isospin conservation and baryon number conservation, respectively. In particular, hadrons fall into easily recognizable isospin multiplets.

On the other hand, a direct manifestation of $[SU(2)]_A$ would require that each isospin multiplet be accompanied by a mirror multiplet of the same mass, but with opposite parity. No hint of this can be found in the hadronic spectrum. For example, there is not even an approximate mirror image of the nucleon iso-doublet. Assuming that the real world is well-approximated by the chiral-symmetric limit, we must conclude that $[SU(2)]_A$ is spontaneously broken. This calls for an $I = 1$ pseudoscalar Goldstone boson, which we identify with the "unperturbed" pion. The physical pion then corresponds to the perturbed state of the Goldstone boson, whose mass comes from m_u and m_d.

The question of how $[U(1)]_A$ is manifested presents a puzzle, and will be discussed separately later.

What is the dynamical reason that $[SU(2)]_A$ manifests itself in the Goldstone mode? We do not have an answer. In the non-renormalizable pre-QCD model of Nambu and Jona-Lasinio[17], the cause of spontaneous symmetry breakdown is a direct nucleon-nucleon attraction built into the model, in analogy with the effective electron-electron attraction responsible for the formation of Cooper pairs in the theory of superconductivity[g]. A direct quark-quark interaction is ruled out by renormalizability in QCD; but an effective interaction could arise, just as the effective electron-electron attraction in superconductivity arises from the more fundamental electron-phonon interaction: It has been suggested that instantons play a role in such an effective quark-quark interaction.[18]

Whatever the dynamical mechanism, the spontaneous breakdown of chiral symmetry will lead to non-vanishing vacuum expectation values $\langle \bar{u}u \rangle$ and $\langle \bar{d}d \rangle$, which furnish the scale for hadronic masses in the chiral limit (dimensional transmutation). One could also attribute to u and d some sort of effective masses—what one would term "constituent" quark masses. Thus we see that the dynamical problem of chiral symmetry breaking cannot be dissociated from that of quark confinement.

3 PCAC

Let the pion state be denoted by $|\pi^j\rangle$, where $j = 1, 2, 3$ is the isospin index. The operator $J_{5\mu}^k$ can annihilate states of the same quantum numbers as the pion, and hence connects the pion state to the vacuum. By Lorentz invariance and isospin conservation, we can write

$$\langle 0|J_{5\mu}^k(x)|\pi^j\rangle = i\,\delta_{jk}f_\pi p_\mu\,e^{-ip\cdot x}, \qquad (12.87)$$

where p_μ is the pion 4-momentum, and f_π is a constant. We immediately obtain

$$\langle 0|\partial^\mu J_{5\mu}^k(x)|\pi^j\rangle = \delta_{jk}f_\pi m_\pi^2\,e^{-ip\cdot x}. \qquad (12.88)$$

This is consistent with the view that the pion is a Goldstone boson in the chiral limit, for $\partial^\mu J_{5\mu}^k = 0$ implies $m_\pi = 0$.

Define

$$\phi_\pi^k(x) \equiv \frac{1}{m_\pi^2 f_\pi}\,\partial^\mu J_{5\mu}^k(x). \qquad (12.89)$$

Then

$$\langle 0|\phi_\pi^k(x)|\pi^j\rangle = \delta_{jk}\,e^{-ip\cdot x}. \qquad (12.90)$$

Thus, $\phi_\pi^k(x)$ can be used as a pion field operator. It is composed of quark operators, reflecting the bound-state nature of the pion. The content of "PCAC" is a rule for using $\phi_\pi^k(x)$ in the chiral limit ($m_\pi \to 0$).

---

[g] In the theory of superconductivity, the Goldstone boson is "eaten" by the electromagnetic field through the Higgs mechanism, and the photon becomes massive in a superconductor (Meissner effect).

[18] R. D. Carlitz, *Phys. Rev.*, D17, 3225 (1978).

Consider the reaction $a \to b + \pi^k$. Using the reduction formula[19]

$$\langle \pi^k, b^{\text{out}} | a^{\text{in}} \rangle = i \int d^4x \, e^{ip \cdot x} (\Box^2 + m_\pi^2) \langle b | \phi_\pi^k(x) | a \rangle, \quad (12.91)$$

where p is the pion 4-momentum, we can obtain the transition amplitude by extracting the coefficient of $i(2\pi)^4 \delta^4(p + p_b - p_a)$:

$$\text{Amp}\,(a \to b + \pi^k) = (m_\pi^2 - p^2) \langle a | \phi_\pi^k(0) | b \rangle. \quad (12.92)$$

The implementation of "PCAC" consists of defining the off-mass-shell amplitude to be

$$T_{ab}^k(p^2) \equiv \frac{m_\pi^2 - p^2}{f_\pi p^2} \langle a | \partial^\mu J_{5\mu}^k(0) | b \rangle. \quad (12.93)$$

The chiral limit is approached by first letting $m_\pi \to 0$ to obtain the off-mass-shell amplitude in the chiral limit, and then taking the mass-shell limit $p^2 \to 0$ eventually. Thus,

$$T_{ab}^k(p^2) \xrightarrow[\text{limit}]{\text{chiral}} -\frac{1}{f_\pi} \langle a | \partial^\mu J_{5\mu}^k(0) | b \rangle. \quad (12.94)$$

We now determine the constant f_π. In QCD, none of the currents in (12.86) are coupled to dynamical fields. When we enlarge the system to include the electroweak interactions, J_μ^k and $J_{5\mu}^k$ ($k = 1, 2$) become part of the charge-changing weak currents coupled to W_μ^k ($k = 1, 2$). Thus, the components $k = 1, 2$ in (12.87) are involved in the matrix element for charged pion decay ($\pi \to \mu + \nu'$). Consequently, f_π can be determined from the charged pion lifetime. Without going through the derivation, we merely quote the result:[14]

$$\text{Rate}\,(\pi \to \mu + \nu') = \frac{1}{4\pi} f_\pi^2 (G \cos \theta)^2 m_\pi m_\mu^2 \left(1 - \frac{m_\mu^2}{m_\pi^2}\right)^2, \quad (12.95)$$

where G is the Fermi constant, θ the Cabibbo angle, and m_π and m_μ are respectively the physical mass of the pion and the muon. Using the value 2.6×10^{-8} s for the lifetime of the charged pion, one obtains

$$f_\pi = 93 \text{ MeV}. \quad (12.96)$$

This is called the "pion decay constant".

4 The Decay $\pi^0 \to 2\gamma$

We apply (12.94) to the electromagnetic decay of π^0 into two photons. The π^0 operator is proportional to

$$J_{5\mu}^3 = \bar{u} \gamma_\mu \gamma_5 u - \bar{d} \gamma_\mu \gamma_5 d, \quad (12.97)$$

---

[19] K. Huang and H. A. Weldon, *Phys. Rev.*, D11, 257 (1975) show that the reduction formula is valid for bound-state operators.

which has an axial anomaly. From (11.36) we obtain

$$\partial^\mu J^3_{5\mu} = 2m_u(\bar{u}\gamma_5 u) - 2m_d(\bar{d}\gamma_5 d) + \frac{\xi\alpha}{2\pi}\widetilde{F}\cdot F, \qquad (12.98)$$

where

$$\xi = 3\left[\left(\frac{2}{3}\right)^2 - \left(\frac{1}{3}\right)^2\right] = 1. \qquad (12.99)$$

The factor 3 accounts for the quark colors. In the chiral limit we put $m_u = m_d = 0$, so that only the anomaly survives in (12.98).

Let the 4-momenta and polarization vectors of the two final photons be denoted respectively by k_1, ε_1 and k_2, ε_2. The off-mass-shell decay amplitude in the chiral limit is, by (12.94) and (12.98),

$$\begin{aligned}T(\omega) &= -\frac{1}{f_\pi}\frac{\alpha}{2\pi}\langle k_1\varepsilon_1, k_2\varepsilon_2|\widetilde{F}\cdot F|0\rangle \\ &= -\frac{2\alpha}{\pi f_\pi}\varepsilon_{\mu\nu\alpha\beta}\varepsilon_1^\mu\varepsilon_2^\nu k_1^\alpha k_2^\beta \qquad (12.100) \\ &= -\frac{4\alpha}{\pi f_\pi}\omega^2|\boldsymbol{\varepsilon}_1\times\boldsymbol{\varepsilon}_2|,\end{aligned}$$

where the 3-vectors are those in the c.m. frame, with

$$\begin{aligned}k_1 &= (\omega, \mathbf{k}), \quad k_2 = (\omega, -\mathbf{k}), \quad |\mathbf{k}| = \omega, \\ \varepsilon_1 &= (0, \boldsymbol{\varepsilon}_1), \quad \varepsilon_2 = (0, \boldsymbol{\varepsilon}_2).\end{aligned} \qquad (12.101)$$

Using (12.100) we obtain, after a standard calculation,

$$\text{Rate}\,(\pi^0 \to 2\gamma) = \frac{\alpha^2}{64\pi^3}\left(\frac{m_\pi}{f_\pi}\right)^2 m_\pi, \qquad (12.102)$$

which gives for the π^0 lifetime

$$\tau = (8.5\text{ eV})^{-1}, \qquad (12.103)$$

in agreement with the experimental value[20]

$$\begin{aligned}\tau_{\text{expt}} &= [(7.95 \pm 0.55)\text{ eV}]^{-1} \\ &= 0.828\times 10^{-16} \pm 0.057\text{ s}.\end{aligned} \qquad (12.104)$$

If we had not known about the anomaly, we would have obtained zero for the decay rate in the chiral limit, and would have been puzzled by the failure of "PCAC" here, when it had been successful in other applications. The puzzle did exist historically, and was the main motivation behind the discovery of the axial anomaly.

---

[20] Particle Data Group, *Rev. Mod. Phys.*, **52**, S1 (1980).

If we had forgotten about color, the decay rate we calculated would have been too small by a factor of 3. However, this is not evidence for color by itself, for we could have gotten the right answer by making the current out of nucleons instead of quarks.

We can view the amplitude (12.100) from another point of view, namely it is the 4-divergence of the triangle graph with respect to the axial-vector vertex. In the chiral limit $m_u = m_d = 0$, the anomaly due to the u and d quark loops must be reproduced by massless bound states if quarks are confined, according to 't Hooft's principle (Sec. 11.7). We have assumed that the limiting chiral symmetry is spontaneously broken. Hence the massless bound state must be the Goldstone boson π^0, whose coupling to two photons is entirely determined by the anomaly. The triangle graph also receives contributions from quarks of higher flavors, but they do not contribute to the anomaly that the π^0 has to reproduce, because they remain massive in the chiral limit we are considering. This is just a formal way of saying that the pion is composed mostly of $\bar{u}u$ and $\bar{d}d$, but very little $\bar{s}s$, $\bar{c}c$, etc.

5 Extension to Pion Octet

Since there is strong evidence for approximate unitary symmetry in hadronic physics, we might entertain the idea of an extended chiral-symmetric limit corresponding to

$$m_u = m_d = m_s = 0. \tag{12.105}$$

By the same reasoning as before, we would conclude that this extended chiral symmetry is spontaneously broken, and is manifested through the existence of a Goldstone boson identifiable with the pion octet. To consider this limit, all we have to do is to enlarge (12.82) and (12.83) by the re-definition

$$\Psi = \begin{pmatrix} u \\ d \\ s \end{pmatrix}, \tag{12.106}$$

and replace the Pauli matrices τ_k in (12.85) and (12.86) by the Gell-Mann matrices λ_a ($a = 1, \ldots, 8$). We then have an "unperturbed" system with global symmetry group

$$[SU(3)]_V \times [SU(3)]_A \times [U(1)]_V \times [U(1)]_A, \tag{12.107}$$

with $[SU(3)]_V$ realized directly in the "eight-fold way", and $[SU(3)]_A$ realized in the Goldstone mode.

The mass splittings in the pion octet, as well as those in all other flavor $SU(3)$ multiplets, arise from the perturbation Lagrangian density

$$-(m_u \bar{u}u + m_d \bar{d}d + m_s \bar{s}s), \tag{12.108}$$

whose effect is usually treated by "chiral perturbation theory"[21], which is a combination of "current algebra" methods and "extended PCAC". Using such

---

[21] H. Pagels, *Phys. Reports*, **16C**, 219 (1975).

Quantum Chromodynamics

techniques, one can relate quark mass ratios to those involving pions and kaons. We quote the results of Weinberg:[22]

$$\frac{m_d}{m_u} \cong \frac{m^2(K^0) - m^2(K^+) + m^2(\pi^+)}{2m^2(\pi^0) + m^2(K^+) - m^2(K^0) - m^2(\pi^+)} = 1.80,$$

$$\frac{m_s}{m_d} \cong \frac{m^2(K^0) + m^2(K^+) - m^2(\pi^+)}{m^2(K^0) - m^2(K^+) + m^2(\pi^+)} = 20.1.$$

(12.109)

Using the observed mass splittings in flavor $SU(3)$ multiplets as a guide for an estimate of m_s, Weinberg suggests[23]

$$m_s = 150 \text{ MeV},$$
$$m_d = 7.5 \text{ MeV},$$
$$m_u = 4.2 \text{ MeV}.$$

(12.110)

12.5 The $U(1)$ Puzzle

If the symmetry $[U(1)]_A$ were manifested directly, then in the chiral limit all massless hadrons would have a massless partner of opposite parity. In the real world we would expect a scalar counterpart of the pion, of roughly the same mass. Since there does not appear to be such a particle, we assume that the symmetry is spontaneously broken. But then there should be an $I = 0$ pseudoscalar Goldstone boson, whose perturbed state should have about the same mass as the pion. Using chiral perturbation theory, Weinberg[24] has estimate the mass to be less than $\sqrt{3}m_\pi$. Among the known hadrons, the only candidates with the right quantum numbers are $\eta(549)$ and $\eta'(985)$. Both violate the Weinberg bound. Besides, $\eta(549)$ has already been claimed by the pion octet. The $U(1)$ puzzle is: *where is the extra Goldstone boson?*

't Hooft removed the puzzle by showing that the expected Goldstone boson is not a physical particle, due to instanton effects. We shall not go into the detailed analysis, but merely mention some relevant points.

The current j_5^μ is not conserved, due to a QCD axial anomaly:[h]

$$\partial_\mu j_5^\mu = \frac{N_f g_0^2}{16\pi^2} \mathcal{F} \cdot \mathcal{F},$$

(12.111)

where N_f is the number of quark flavors taken into account in the chiral limit ($N_f = 2$ or 3), and we use the abbreviation $\mathcal{F} \cdot \mathcal{F} = \mathcal{F}_a^{\mu\nu} \mathcal{F}_{a\mu\nu}$. Now, we know

---

[h] See the remarks at the end of Sec. 11.4. The form of the anomaly is an obvious generalization from the Abelian case. The coefficient is $N_f/2$ times that in the Abelian case—the same factor that appears in the GL function, as discussed in footnote c.

[22] S. Weinberg, in *A Festschrift for I. I. Rabi*, Ed. L. Motz (New York Academy of Science, New York, 1977).
[23] For another suggestion see T. D. Lee, *Particle Physics and Introduction to Field Theory* (Harwood Publishers, Chur, Switzerland, 1981), p. 584.
[24] S. Weinberg, *Phys. Rev.*, D11, 3583 (1975).

that $\tilde{\mathcal{F}} \cdot \mathcal{F}$ is a total 4-divergence [see (5.6)]:

$$\tilde{\mathcal{F}} \cdot \mathcal{F} = \partial_\mu \bar{X}^\mu, \tag{12.112}$$

whose integral over all space-time is proportional to the topological charge. We can define a conserved but non-gauge-invariant current

$$J_5^\mu \equiv j_5^\mu - \frac{N_f g_0^2}{16\pi^2} X^\mu, \tag{12.113}$$

which is the non-Abelian generalization of (11.39). The generator of the $[U(1)]_A$ symmetry may be taken to be

$$Q_5 \equiv \int d^3x \, J_5^0 = \int d^3x \left[\psi^\dagger \gamma_5 \psi - \frac{N_f g_0^2}{16\pi^2} X^0 \right]. \tag{12.114}$$

In the Abelian case, Q_5 is gauge invariant, because of the absence of topological charge. This is no longer true here, and Q_5 is not a physical quantity. In fact, Q_5 is not even conserved, because of the existence of instantons. To see this, integrate the equation $\partial_\mu J_5^\mu$ over Euclidean 4-space. The result can be presented in the form

$$\int_{-\infty}^{\infty} dt \, \frac{dQ}{dt} = 2N_f q[G], \tag{12.115}$$

where

$$q[G] = \frac{g_0^2}{32\pi^2} \int d^4x \, \tilde{\mathcal{F}} \cdot \mathcal{F} \tag{12.116}$$

is the topological charge, a functional of the gauge field $G_a{}^\mu$. For $G_a{}^\mu$ corresponding to one instanton, $q[G] = 1$. Therefore, in this case, the boundary values of Q_5 in Euclidean time differ by

$$\Delta Q_5 = 2N_f q[G]. \tag{12.117}$$

This can be attributed to the fact that an instanton interpolates (in Euclidean time) between two gauge-field configurations differing by one unit of topological charge. (See Sec. 8.6). Thus, there is no reason to expect $[U(1)]_A$ to have physical manifestations. 't Hooft's detailed analysis explains what becomes of the would-be Goldstone boson, which we shall not go into.

The $U(1)$ puzzle is not a mathematical paradox, but rather a frustration of cherished beliefs. The work of 't Hooft did remove the puzzle in its original form, but generates new questions. For example, does "instanton physics" alter conventional thinking in low-energy hadronic physics, in particular the spontaneous breaking of $[SU(2)]_A$? To go into these questions would embroil us in unsettled and controversial arguments. In the next section, we discuss only one

12.6 θ-Worlds in QCD

1 Euclidean Action

We recall that there are "large" and "small" gauge transformations. (See Sec. 8.6). We can make the vacuum state invariant under all gauge transformations, including the "large" ones, by adding to the action a term proportional to the topological charge. [See (8.140)]. Ignoring the heavy quarks c, b, t, we consider the Minkowski action

$$S_\theta = \int d^4x \left[-\tfrac{1}{4}\mathcal{G}^2 + \bar{\psi}(i\slashed{D} - M)\psi\right] - \theta q[G], \qquad (12.118)$$

where ψ is given by (12.106), and M is a mass matrix to be discussed in detail later. The parameter θ multiplying the topological charge $q[G]$ is an unknown constant. In quarkless QCD, it labels different θ-worlds that are physically distinct. We shall see that introducing quarks into the theory changes the meaning of θ, both mathematically and physically.

We shall work in Euclidean 4-space. To continue from Minkowski space to the latter, we replace the quantities in the left column of Table 12.2 by the

Table 12.2
CONTINUATION FROM
MINKOWSKI TO EUCLIDEAN 4-SPACE

	Minkowski	Euclidean
Coordinate	x^0	$-ix_E^4$
	x^k	x_E^k ($k = 1, 2, 3$)
Momentum	p^0	ip_E^4
	p^k	p_E^k
Gauge fields	$G_a^{\,0}$	$i(G_E)_a^{\,0}$
	$G_a^{\,k}$	$(G_E)_a^{\,k}$
Invariants	$\mathcal{G}^2$	$\mathcal{G}_E^{\,2}$
	$\mathcal{G} \cdot \tilde{\mathcal{G}}$	$i\mathcal{G}_E \cdot \tilde{\mathcal{G}}_E$
Topological charge	$q[G]$	$q[G_E]$
Dirac matrices	γ^0	$\gamma_E^{\,4}$
	γ^k	$-i\gamma_E^{\,k}$
	$\gamma^\mu\gamma^\nu + \gamma^\nu\gamma^\mu = 2g^{\mu\nu}$	$\gamma_E^{\,\mu}\gamma_E^{\,\nu} + \gamma_E^{\,\nu}\gamma_E^{\,\mu} = 2\delta_{\mu\nu}$
	$\gamma_5 \equiv -i\gamma^0\gamma^1\gamma^2\gamma^3$	$(\gamma_E)_5 \equiv \gamma_E^{\,1}\gamma_E^{\,2}\gamma_E^{\,3}\gamma_E^{\,4}$
	$\slashed{p} \equiv \gamma^\mu p_\mu$	$\slashed{p}_E \equiv i\gamma_E^{\,\mu} p_E^{\,\mu}$
θ-Action	S_θ	$i(S_E)_\theta$

corresponding ones in the right column. Note that there is no distinction between Euclidean upper and lower indices. The Euclidean Dirac matrices are all hermitian, and γ_5 is represented by the same matrix in both spaces. From now on we drop the subscript E denoting Euclidean quantities. The Euclidean action is written as

$$S_\theta = \int d^4x \left[\tfrac{1}{4}\mathcal{G}^2 + \bar{\psi}(\slashed{D} + M)\psi\right] + i\theta q[G]. \tag{12.119}$$

The "partition function", from which Euclidean quark Green's functions can be obtained, is given by

$$Z_\theta[\eta, \eta'] = \mathcal{N} \int (DG)(D\bar{\psi})(D\psi) \exp[-S_\theta - (\bar{\psi}, \eta) - (\bar{\eta}, \psi)], \tag{12.120}$$

where η and η' are anti-commuting c-number sources, and we have left understood gauge-fixing and ghost terms in the exponent. We use the abbreviation $(f, g) = \int d^4x\, f \cdot g$, where $f \cdot g$ is a product contracted in all indices, if any.

2 The Axial Anomaly and the Index Theorem

In the massless limit, S_θ is invariant under chiral transformations. Since all quantities in (12.120) are classical, one might expect to find the classical result $\partial_\mu j_5^\mu = 0$, and wonder where the axial anomaly comes from. Fujikawa[25] showed that it comes from the non-chiral-invariance of the fermionic measure

$$d\mu \equiv (D\bar{\psi})(D\psi). \tag{12.121}$$

To see this, define a complete set of spinor eigenfunctions of $\slashed{D}$:[i]

$$\slashed{D}\phi_n(x) = E_n\phi_n(x),$$
$$\int d^4x\, \phi_n^\dagger(x)\phi_m(x) = \delta_{nm}, \tag{12.122}$$
$$\sum_n \phi_n(x)\phi_n^\dagger(y) = \delta^4(x - y),$$

where $\phi_n(x)$ is a functional of G_a^μ, by virtue of the dependence of $\slashed{D}$ on the latter. We expand the quark fields ψ and $\bar{\psi}$ of each flavor as follows (with the flavor index omitted for brevity):

$$\psi(x) = \sum_n a_n\phi_n(x),$$
$$\bar{\psi}(x) = \sum_n b_n\phi_n^\dagger(x), \tag{12.123}$$

where $\{a_n\}$ and $\{b_n\}$ are sets of independent anti-commuting c-numbers. We take

$$d\mu = \prod_f \prod_n db_n\, da_n, \tag{12.124}$$

---

[i] We normalize ϕ_n to unity in a 4-dimensional Euclidean volume $\Omega = 1$. The infinite-volume limit is taken by letting the unit of volume approach zero.

[25] K. Fujikawa, *Phys. Rev. Lett.*, **42**, 1195 (1979).

where f denotes flavor. Under a local chiral transformation,

$$\psi(x) \to \psi'(x) = e^{-i\gamma_5\alpha(x)} \psi(x),$$
$$\bar{\psi}(x) \to \bar{\psi}'(x) = \bar{\psi}(x) e^{-i\gamma_5\alpha(x)}, \quad (12.125)$$

and the coefficients a_n and b_n transform linearly:

$$a_n \to a'_n = \sum_n C_{nm} a_m,$$
$$b_n \to b'_n = \sum_n C_{nm} b_m, \quad (12.126)$$

where C_{nm} is an ordinary number:

$$C_{nm} = \int d^4x \, \phi_n^\dagger(x) \, e^{-i\gamma_5\alpha(x)} \, \phi_m(x). \quad (12.127)$$

Since γ_5 anticommutes with $\slashed{D}$, we can choose ϕ_n to be states of definite chirality. For each $E_n \neq 0$, there are two degenerate states of opposite chirality. In such a "chiral representation" C_{nm} is diagonal, and we can easily see that

$$\prod_n da'_n = (\det C)^{-1} \prod_n da_n. \quad (12.128)$$

For infinitesimal $\alpha(x)$ we have

$$(\det C)^{-1} = \prod_n \left[1 - i \int d^4x \, \alpha(x) \phi_n^\dagger(x) \gamma_5 \phi_n(x) \right]$$
$$= \exp \cdot i \int d^4x \, \alpha(x) \sum_n \phi_n^\dagger(x) \gamma_5 \phi_n(x). \quad (12.129)$$

This result is independent of the representation. Thus, under an *infinitesimal local* chiral transformation,

$$d\mu \to d\mu' = e^{i\Delta} \, d\mu, \quad (12.130)$$

where

$$\Delta = 2N_f \int d^4x \, \alpha(x) \sum_n \phi_n^\dagger(x) \gamma_5 \phi_n(x), \quad (12.131)$$

where N_f is the number of quark flavors under consideration. This shows the non-invariance of the fermionic measure.

The sum in the integral of (12.131) is ambiguous, and we define it as

$$\sum_n \phi_n^\dagger(x) \gamma_5 \phi_n(x) \equiv \lim_{\Lambda \to \infty} \sum_n e^{-E_n^2/\Lambda^2} \phi_n^\dagger(x) \gamma_5 \phi_n(x), \quad (12.132)$$

where Λ is an energy cutoff, which respects chiral symmetry. The calculation of the right-hand side is straightforward but somewhat tedious. We refer the reader to Ref. 26, and just quote the result:

$$\sum_n \phi^\dagger(x)\gamma_5\phi_n(x) = \frac{g_0^2}{32\pi^2} \mathfrak{F} \cdot \mathfrak{F}. \tag{12.133}$$

Substituting this into (12.131), we obtain

$$\Delta = \frac{N_f g_0^2}{16\pi^2} \int d^4x \, \alpha(x) \mathfrak{F} \cdot \mathfrak{F}. \tag{12.134}$$

In the absence of sources, Z_θ should be invariant under a local chiral transformation, since all the spinor components of ψ and $\bar\psi$ are independently integrated over. Using (12.134), we find that, under an infinitesimal local chiral transformation, Z_θ changes by

$$\delta Z_\theta = i \int (DG) \, d\mu \, e^{-S_\theta}.$$

$$\int d^4x \, \alpha(x) \left[-\partial_\mu j_5^\mu + 2\bar\psi M \gamma_5 \psi + \frac{N_f g_0^2}{16\pi^2} \mathfrak{F} \cdot \mathfrak{F} \right]. \tag{12.135}$$

Hence the quantity in the brackets must vanish, and we have the axial anomaly as expressed in (12.111).

The above derivation of the axial anomaly is no more rigorous than that of Chapter XI, for we have not shown that the result is independent of the cutoff procedure. Thus, we have not improved on the argument (given in Sec. 11.3) that there are no radiative corrections to the anomaly.

Let us go back to (12.133) and integrate both sides over Euclidean 4-space. The left-hand side, being made well-defined by the cutoff procedure (12.132), receives contributions only from states with $E_n = 0$, which we call "zero-modes". For $E_n \neq 0$, the contributions of the two degenerate chiral states cancel each other. The right-hand side gives the topological charge. Hence

$$n_+ - n_- = q, \tag{12.136}$$

where $n_\pm$ are respectively the number of zero-modes of chirality ± 1 (of a given flavor), in the presence of a background gauge field G_a^μ, and q is the topological charge of the background field. This is the *Atiyah-Singer index theorem*[26], for which our derivation is of the nature of a "poor man's proof". The theorem is rigorous.

For a background field consisting of one instanton, we have $q = 1$, and the index theorem tells us $n_+ - n_- = 1$. This shows that there is at least one zero-mode for each quark flavor in a one-instanton field. 't Hooft[27] has shown that there is exactly one zero-mode with positive chirality in the instanton field, with normalizable wave function

$$\phi_{0\text{-mode}}(x) = (1 + x^2)^{-3/2} u, \tag{12.137}$$

where $x^2 = x^\mu x^\mu$, and u is a constant Dirac spinor.

---

[26] M. Atiyah and I. Singer, *Ann. Math.*, **87**, 484 (1968).
[27] G. 't Hooft, *Phys. Rev. Lett.*, **37**, 8 (1976); *Phys. Rev.*, **D14**, 3432 (1976). See also S. Coleman in *The Ways of Subnuclear Physics*, Ed. A. Zichichi (Plenum, New York, 1980).

Quantum Chromodynamics

3 Chiral Limit: Collapse of the θ-Worlds

The chiral limit is defined by setting $M = 0$ in the action:

$$S_\theta^{(0)} = \tfrac{1}{4}(\mathcal{F}, \mathcal{F}) + (\bar{\psi}, \slashed{D}\psi) + i\theta q[G]. \tag{12.138}$$

The partition function is then

$$Z_\theta^{(0)}[\eta, \bar{\eta}] = \mathcal{N} \int (DG) \exp\{-\tfrac{1}{4}(\mathcal{F}, \mathcal{F}) - i\theta q[G]\} f_0[G, \eta, \bar{\eta}],$$

$$f_0[G, \eta, \bar{\eta}] = \int d\mu \, \exp[-(\bar{\psi}, \slashed{D}\psi) - (\eta, \bar{\psi}) - (\bar{\eta}, \psi)]. \tag{12.139}$$

Changing the variables of integration by performing an *infinitesimal global* chiral transformation, we have

$$\int d\mu \, \exp[-(\bar{\psi}, \slashed{D}\psi) - (\eta, \bar{\psi}) - (\bar{\eta}, \psi)]$$

$$= \int d\mu' \, \exp[-(\bar{\psi}', \slashed{D}\psi') - (\eta, \bar{\psi}') - (\bar{\eta}, \psi')] \tag{12.140}$$

$$= e^{2i\Delta} \int d\mu \, \exp[-(\bar{\psi}, \slashed{D}\psi) - (\eta', \bar{\psi}) - (\bar{\eta}', \psi)],$$

where

$$\eta' = e^{-i\alpha\gamma_5} \eta,$$
$$\bar{\eta}' = \bar{\eta} \, e^{-i\alpha\gamma_5}. \tag{12.141}$$

Therefore

$$f_0[G, \eta, \bar{\eta}] = e^{2i\Delta} f_0[G, \eta', \bar{\eta}'], \tag{12.142}$$

where, by (12.134),

$$\Delta = 2\alpha N_f q[G]. \tag{12.143}$$

With this, we have

$$Z_\theta^{(0)}[\eta, \bar{\eta}] = \mathcal{N} \int (DG) \exp\{-\tfrac{1}{4}(\mathcal{F}, \mathcal{F}) - i(\theta - 2\alpha N_f) q[G]\} \cdot f_0[G, \eta', \bar{\eta}']. \tag{12.144}$$

Hence

$$Z_\theta^{(0)}[\eta, \bar{\eta}] = Z_{\theta - 2\alpha N_f}^{(0)}[\eta', \bar{\eta}']. \tag{12.145}$$

This is also valid for a finite global chiral transformation, because the α's corresponding to successive infinitesimal transformations are additive, owing to the group property of chiral transformations. Therefore, we can always reduce θ to zero by making a global chiral transformation on the fermion sources, with $\alpha = \theta/2N_f$. Such a chiral transformation does not change the physics, because all Green's functions are evaluated in the sourceless limit. Thus, all θ-worlds are physically equivalent to one another, when the theory contains massless quarks. In particular, the theory is invariant under CP, in contradistinction to quarkless QCD with $\theta \neq 0$.

Any quark field left out in the above discussion has no effect on the conclusion, for they are spectators that do not participate in the chiral transformation. Thus, for the θ-worlds to "collapse", thereby restoring CP, it is sufficient to have one massless quark.

4 Quark Mass Matrix

We now take into account the mass term

$$\mathcal{L}_{\text{mass}} = -\bar{\psi} M \psi. \tag{12.146}$$

The partition function becomes

$$Z_\theta[\eta, \bar{\eta}; M] = \mathcal{N} \int (DG) \exp\{-\tfrac{1}{4}(\mathcal{F}, \mathcal{F}) - i\theta q[G]\} f_M[G, \eta, \bar{\eta}],$$
$$f_M[G, \eta, \bar{\eta}] = \int d\mu \exp[-(\bar{\psi}, (\slashed{D} + M)\psi) - (\eta, \bar{\psi}) - (\bar{\eta}, \psi)]. \tag{12.147}$$

The angle θ can again be absorbed into the fermion sector, through the chiral transformation (12.141); but now it has a physical presence in the mass matrix. The relation (12.142) is now replaced by

$$f_M[G, \eta, \bar{\eta}] = e^{2i\Delta} f_{M'}[G, \eta', \bar{\eta}'], \tag{12.148}$$

where M' is the transformed mass matrix defined by

$$M' = M \, e^{-2i\alpha\gamma_5}. \tag{12.149}$$

Thus,

$$Z_\theta[\eta, \bar{\eta}; M] = Z_{\theta - 2\alpha N_f}[\eta', \bar{\eta}'; M']. \tag{12.150}$$

By choosing $\alpha = \theta/2N_f$, we have

$$Z_\theta[\eta, \bar{\eta}; M] = Z_0[\eta', \bar{\eta}'; M']. \tag{12.151}$$

Since η', $\bar{\eta}'$ are arbitrary sources, the physical meaning of θ resides solely in the θ-dependence of M'.

To parametrize the mass matrix, it is convenient to decompose ψ into its right and left-handed components R and L. The most general mass term can be written in the form

$$\mathcal{L}_{\text{mass}} = -\bar{L}\mathcal{M}R - \bar{R}\mathcal{M}^\dagger L, \tag{12.152}$$

where $\mathcal{M}$ is an arbitrary complex 3×3 matrix. The mass matrix then takes the form

$$M = A + i\gamma_5 B, \tag{12.153}$$

where A and B are hermitian matrices given by

$$A = \frac{1}{2}(\mathcal{M} + \mathcal{M}^\dagger),$$
$$B = \frac{i}{2}(\mathcal{M} - \mathcal{M}^\dagger).$$
(12.154)

It is clear that the transformed mass matrix M' is also of the form (12.153):

$$M' = A' + i\gamma_5 B',\qquad(12.155)$$

where A' and B' are 3×3 hermitian matrices.

The symmetry group $[SU(3) \times U(1)]_V \times [SU(3) \times U(1)]_A$ of the unperturbed problem is equivalent to $[SU(3) \times U(1)]_L \times [SU(3) \times U(1)]_R$, which acts in the following manner:

$$[SU(3) \times U(1)]_L: L \to e^{-i\omega_a \lambda_a/2 - i\delta} L,$$
$$[SU(3) \times U(1)]_R: R \to e^{-i\rho_a \lambda_a/2 - i\varepsilon} R.$$
(12.156)

If there were no restrictions on these transformations, then through them we can make $\mathcal{M}$ diagonal, with non-negative diagonal elements. In particular, the term $i\gamma_5 B$ in the mass matrix, which violates CP invariance, can be transformed away. This was what we did in Chapter VI in the context of the Weinberg-Salam model. We now recognize that there are constraints previously ignored, arising from

(a) the existence of the topological charge,
(b) the assumption that $[SU(3)]_A$ is spontaneously broken.

The existence of the topological charge leads to the term $i\theta q[G]$ in the gauge-field sector, and the transformation (12.148) in the fermion sector. Thus, the theory can be CP-conserving only if the value of θ is chosen in a special way in conjunction with parameters in the original mass matrix M. In general, therefore, the theory violates CP.

The constraint imposed by the spontaneous breaking of $[SU(3)]_A$ was pointed out by Dashen[28] and Nuyts[29]. In the chiral limit, the vacuum is infinitely degenerate, labelled by a phase angle ξ. An $[SU(3)]_A$ transformation changes ξ, and takes one vacuum state into another. The perturbation $\mathcal{L}_{\text{mass}}$ explicitly violates $[SU(3)]_A$, and singles out a particular value of ξ. When we turn off the perturbation, the vacuum is left in the state labelled by that ξ. If we choose a vacuum with the wrong value of ξ, then $\mathcal{L}_{\text{mass}}$ cannot be considered a small perturbation, for its iterative effect will tend to rotate the vacuum to the correct one. Thus, to be able to use chiral perturbation theory, we must start with the correct vacuum, out of the infinitely many degenerate ones.

An analogy may be made with a ferromagnet. In the absence of an external field, the magnetization can point along any direction in space. The effect of an external magnetic field, however small, can be treated as a small perturbation only if its direction coincides with that of the unperturbed magnetization.

---

[28] R. Dashen, *Phys. Rev.*, D3, 1879 (1971).
[29] J. Nuyts, *Phys. Rev. Lett.*, 26, 1604 (1971).

In the present context, the condition that $\mathcal{L}_{\text{mass}}$ can be treated as a small perturbation is that it must not create Goldstone bosons from the unperturbed vacuum:

$$\langle 0|\mathcal{L}_{\text{mass}}|\pi^k\rangle = 0, \qquad (12.157)$$

where π^k ($k = 1,\ldots,8$) refers to any member of the unperturbed pion octet π, η, κ. Since π^k is a pseudoscalar particle, this imposes a constraint only on the term $i\gamma_5 B'$ in (12.155):

$$\langle 0|i\bar{\psi}\gamma_5 B'\psi|\pi^k\rangle = 0 \quad (k = 1,\ldots,8). \qquad (12.158)$$

This requires $\bar{\psi}i\gamma_5 B'\psi$ to be a singlet with respect to $[SU(3)]_A$. Hence B' must be proportional to the unit 3×3 matrix:

$$B' = \omega \begin{pmatrix} 1 & 0 & 0 \\ 0 & 1 & 0 \\ 0 & 0 & 1 \end{pmatrix}. \qquad (12.159)$$

Thus

$$\begin{aligned}\bar{\psi}M'\psi &= \bar{\psi}(A' + i\gamma_5\omega)\psi \\ &= \bar{L}(A' + i\omega)R + \bar{R}(A' - i\omega)L.\end{aligned} \qquad (12.160)$$

We are no longer free to make $[SU(3)]_A$ transformations, but we can still make $[SU(3)]_V$ transformations, through which A' can be brought to diagonal form, with eigenvalues a_f ($f = 1, 2, 3$). The current quark masses are then $(a_f^2 + \omega^2)^{1/2}$ ($f = 1, 2, 3$).

To determine ω, we note that the transformation (12.149), which takes θ out of the gauge-field sector into the fermion sector, gives rise to a phase factor in the determinant of the left-handed part of M', given by

$$e^{2i\alpha N_f} = e^{i\theta}. \qquad (12.161)$$

There may be other phase factors coming from the parameters in the original mass matrix M; but we shall ignore them, because they merely cause a shift of θ, which is arbitrary anyway. Diagonalizing A', and treating ω as small, we have

$$\begin{aligned}\det(A' - i\omega) &= (m_u - i\omega)(m_d - i\omega)(m_s - i\omega) + O(\omega^2) \\ &= m_u m_d m_s - i\omega(m_u m_d + m_u m_s + m_d m_s) + O(\omega^2).\end{aligned} \qquad (12.162)$$

Equating the phase factor of the above to $e^{i\theta}$, we obtain

$$\omega = -\frac{\theta}{\dfrac{1}{m_u} + \dfrac{1}{m_d} + \dfrac{1}{m_s}} + O(\theta^2). \qquad (12.163)$$

Hence

$$\mathcal{L}_{\text{mass}} = -(m_u \bar{u}u + m_d \bar{d}d + m_s \bar{s}s) + \mathcal{L}_{\text{CP}}, \qquad (12.164)$$

Quantum Chromodynamics

with the CP violating term $\mathcal{L}_{CP}$ given to lowest order in θ by

$$\mathcal{L}_{CP} = i\zeta\theta(\bar{u}\gamma_5 u + \bar{d}\gamma_5 d + \bar{s}\gamma_5 s),$$

$$\zeta = \frac{m_u m_d m_s}{m_u m_d + m_u m_s + m_d m_s}. \quad (12.165)$$

If one of the quark masses vanishes, then $\mathcal{L}_{CP} = 0$, as expected. The coefficient ζ was first derived in pre-QCD language by Bég[30], and in the present form by Baluni[31].

5 Strong CP Violation

The term $\mathcal{L}_{CP}$ in (12.164) gives rise to CP violation by the strong interactions, which may be avoided in two ways:

(a) We may set one of the quark masses equal to zero, and it would be most natural to take $m_u = 0$. While not ruled out by experimental facts, such a choice is deemed extremely unpalatable from a phenomenological point of view.[32]

(b) We have remarked that θ may be shifted by parameters in the original mass matrix M. By constructing M from a Higgs field, θ can be banished from QCD into the Higgs sector, with attendant delights for theorists[33]. However, there is as yet no indication that such a mechanism operates in nature.

Taking $\mathcal{L}_{CP}$ as given by (12.165), one can calculate its observable effects through the use of standard chiral perturbation theory. We merely quote certain results exact in the chiral limit.[34]

One of the effects of $\mathcal{L}_{CP}$ is to induce a CP violating term in the effective pion-nucleon coupling:

$$\mathcal{L}_{\pi NN} = g_{\pi NN}(\bar{N}\gamma_5 \tau N) \cdot \pi + g'_{\pi NN}(\bar{N}\tau N) \cdot \pi,$$
$$g'_{\pi NN} = -\theta\zeta f_\pi^{-1}(m_\Xi - m_N), \quad (12.166)$$

where m_Ξ and m_N are respectively the mass of Ξ and the nucleon. Numerically,

$$|g_{\pi NN}| \cong 13.4,$$
$$|g'_{\pi NN}| \cong 0.038|\theta|, \quad (12.167)$$

where (12.109) has been used to calculate ζ.

Using the effective pion-nucleon coupling, one can calculate the neutron electric dipole moment D_n exactly in the chiral limit. The physical reason is as follows: the neutron can dissociate virtually into a proton and a pion (among other possibilities). The charge separation in such a virtual state contributes to the neutron electric dipole moment, because there is a CP-violating pion-nucleon vertex. In the chiral limit $m_\pi \to 0$, this virtual state dominates over all others, because the virtual pion travels far from the virtual proton before

---
[30] M. A. B. Bég, *Phys. Rev.*, **D4**, 3810 (1971).
[31] V. Baluni, *Phys. Rev.*, **D19**, 2227 (1979).
[32] P. Langacker and H. Pagels, *Phys. Rev.*, **D19**, 2070 (1979).
[33] R. D. Peccei and H. R. Quinn, *Phys. Rev. Lett.*, **38**, 1440 (1977); S. Weinberg, *Phys. Rev. Lett.*, **40**, 223 (1978); F. Wilczek, *Phys. Rev. Lett.*, **40**, 279 (1978).
[34] R. J. Crewther, P. DiVecchia, G. Veneziano, and E. Witten, *Phys. Lett.*, **88B**, 123 (1979).

recombining, thus maximizing the contribution to the electric dipole moment. The result of a calculation based on such a picture gives

$$\frac{D_n}{m_N} = \frac{1}{4\pi^2} g_{\pi NN} g'_{\pi NN} \ln \frac{m_N}{m_\pi}, \qquad (12.168)$$

with numerical value

$$D_n = 5.2 \times 10^{-16} \, \theta \text{ cm}. \qquad (12.169)$$

Comparison with the experimental bound

$$|D_n| < 10^{-24} \text{ cm} \qquad (12.170)$$

yields

$$|\theta| < 10^{-9}. \qquad (12.171)$$

CODA

There was a lady from Squantum
Who had incredible momentum.
      She shook off the glue
      That colored her blue
And got asymptotic freedom.

There's a boson they call Goldstone
That's very hard to disown.
      You try electrocution,
      You get a transmutation:
The photon comes off the light-cone.

A theorist shows me how to mix
Technicolor $SU(6)$.
      Should you find in the stew
      A funny quark or two.
Just sweep them under the Higgs.

A mathematician named Anatole
Stepped on a magnetic monopole.
      He struggled with the string
      Dirac had tied to the thing,
And so became one fiber bundle.

And now my dear for change of pace
Take imaginary holidays
      Where the metric is good,
      No $i\varepsilon$'s intrude,
In the realm of Euclidean 4-space.

INDEX

adjoint representation 13, 63
Adler-Bell-Jackiw anomaly 220
anomalous dimension 195
anomalous magnetic moment 247
anomaly cancellation 234
anticommutation relations 144
anti-screening 176
asymptotic freedom 9, 176, 251
axial anomaly 219, 265, 268
axial gauge 152, 161, 164

baryons 3
ß-function, see Gell-Mann-Low function
Bogolubov-Parasiuk-Hepp renormalization scheme 179
bosons 3

Cabibbo angle 41, 117
Callan-Symanzik equation 194
canonical quantization 148
charge renormalization 173
charm 43f
charmonium 45
chiral current 219
chirality 6, 105
chiral symmetry 258
chiral transformation 219
color 2, 7, 28f, 38
color gyromagnetic ratio 249
color singlet 10, 31
color $SU(3)$ 7, 30f
conserved current 6, 47
contravariant vector 11
Coulomb gauge 101, 153, 159
coupling constant 6
covariant derivative 48
covariant vector 11
CPT theorem 117
CP violation 167, 275
CVC 42

Dirac matrices 11
Dirac monopole 103
dimensional transmutation 211
   in non-relativistic quantum mechanics 214

effective action 200
effective potential 202
electric quadrupole moment 113

electromagnetic field 6, 47f,
electromagnetic interaction 2, 12, 33f, 106
electromagnetic tensor 250
electron propagator 180
Euclidean space 133

Fadeev-Popov ghosts 164
Fadeev-Popov method 152
Fermi constant 111
fermions 3
Feynman gauge
Feynman graphs 42, 138
Feynman path integral 122
   in non-relativistic quantum mechanics 122
   in quantum field theory 127
Feynman propagator 136
Feynman rules 139
field strength tensor 48
fine structure constant 174
finite energy solutions, classical 55
fixed points 192
flavor mixing 8
flavors 7
flux quantization 55v, 96
fundamental representation 25, 63

gauge bosons 10
gauge fields 6, 7, 48
gauge-fixing 129
gauge invariance
   global 47, 67
   local 47, 69
   spontaneous breaking of global 50-52
   spontaneous breaking of local 53-55
gauge principle 6
gauge transformations 6, 47, 67
Gell-Mann-low function 190, 252
Gell-Mann matrices 17, 241
Gell-Mann-Okubo mass formula 15, 30
generating functional for Green's functions 131, 242
ghost field 144, 242
GIM mechanism 43
gluon fields 241
gluon propagator 251
gluons 9, 10
gluon self-interactions 245
Goldstone boson 53
Goldstone's theorem 53

grand unified theories 2, 10
gravitational interaction 2
Green's function 131

hadrons 3, 28
Higgs boson 10
Higgs fields 7, 8, 107
Higgs mechanism 53, 83
homotopy classes 91, 95, 166
hypercharge 12

index theorem 268, 270
infinite momentum frame 36
instanton 88, 165, 266
internal symmetries 8, 12
irreducible representation 2, 12, 20f
isospin 13, 23

J/ψ 45
Jacobi identity 62

$K \to \mu\mu$ decay 43f
kink 104
Kobayashi-Maskawa matrix 117

Landau gauge 158
leading logarithms 191
leptons 3
Lie algebra 61, 62
Lie groups 61
little group 80
loop expansion 204
Lorentz gauge 153, 157
Lorentz transformations 2

magnetic dipole moment 113
mass 2
mass matrix 114
mesons 3
metric tensor 10
minimal coupling 2
monopole 94f
multiplets 8, 12
muon 3

neutral currents 8
neutrinos 7
neutron electric dipole moment 275
Nielsen-Huges formula 257
Nielsen-Olesen unstable modes 256
Noether's theorem 47, 219
non-Abelian gauge theories 8, 61f

one-particle irreducible graph 177
one-particle irreducible Green's function 202

parity violation 6
parton model 35f
path integrals 122
  in Hamiltonian form 151
path representation 74
Pauli exclusion principle 30
Pauli matrices 11, 13
PCAC 261
photon, massless 50
  propagator 173, 180
pion decay constant f_π 262
Poincaré group 3, 29
Poincaré transformations 2
point-splitting method 187
polarization vector 249
preons 236
primitively divergent graph 177
pure gauge 72

quantum chromodynamics QCD 241
  chiral limit 271
  Feynman rules 242
  Lagrangian density 242
  θ-worlds 267
  three-gluon vertex 244
  vacuum, as a magnetic medium 254
quark confinement 10
quark mass matrix 272
quarks 1, 7, 12f

radiative corrections 225
renormalization group 187
renormalization point 174

scaling 35
screening 176
self-duality 93
skeleton graph 177
solitons 58, 86f, 119
space-time translations 2
spin 2, 3
spin-statistic theorem 3
spontaneous symmetry breaking 8, 50, 80
static gauge 87
strangeness 22, 42
strong interaction 2
structure constants 13, 61
$SU(n)$ 14
$SU(2)$ 12
$SU(3)$ 5, 17, 18, 30
$SU(6)$ 29
superficial degree of divergence 177

τ lepton 3
θ action 170
θ vacuum 167

Index

temporal gauge 76, 153, 161
topological charge 88, 266
tree-graphs 138
triangle graph 220
't Hooft's principle 236

$U(1)$ 6
unitary gauge 75, 101, 107
unitary symmetry 15
U-spin 18
Υ 45

V-A Coupling 6
vacuum polarization tensor 173
vacuum-vacuum amplitude 130
V-spin 18
vorticons 120

W-bosons 40
W-propagator 43
Ward-Takahashi identity 179
weak hypercharge 7, 12, 108
weak interactions 2, 12, 40
weak isospin 7
weight diagrams 25f
Weinberg angle 109, 111
Weinberg-Salam model 7, 105f
winding number 91, 166

Yang-Mills fields 6, 61f
   quantization of, 162
Yang-Mills theory 6
Young's tableaux 18f

Z°-boson 109